Sexual Differentiation of the Brain

Sexual Differentiation of the Brain

edited by
Akira Matsumoto, Ph.D.

CRC Press
Boca Raton London New York Washington, D.C.

Library of Congress Cataloging-in-Publication Data

Sexual differentiation of the brain / edited by Akira Matsumoto.
    p. cm.
    Includes bibliographical references and index.
    ISBN 0-8493-1165-9 (alk. paper)
    1.Neuroendocrinology. 2. Brain—Differentiation. 3. Sex.
differentiation. I. Matsumoto, Akira, 1943–
QP356.4.S49   1999
612.8'2—dc21
        99-37571
        CIP

No claim to original U.S. Government works
International Standard Book Number 0-8493-1165-9
Library of Congress Card Number 99-37571
Printed in the United States of America 1 2 3 4 5 6 7 8 9 0
Printed on acid-free paper

Preface

This monograph is dedicated to Dr. Yasumasa Arai who retired from the professorship in anatomy at the Juntendo University School of Medicine in March 1998. He is an internationally established scientist, contributing to the advance of neuroendocrinology and neuroscience for more than 30 years. He is a member of the Editorial Boards of *Hormones and Behavior* and the *Journal of Neuroendocrinology*. He has conducted important research in the understanding of sexual differentiation of the brain and sex steroid influences on neural and behavioral functions.

Sex steroids exert influences on modulating neural development and neural circuit formation in developing the sex steroid–sensitive neuroendocrine brain. Estrogen or aromatizable androgen can act as a neurotrophic factor on neural tissues, stimulating axonal and dendritic growth and synapse formation. The development of sexual dimorphism in neural structures such as nuclear volume and synaptic organization may reflect sex steroid–modulating neural development and synapse formation during the perinatal period. Recent neurobiological and molecular biological studies in this field have provided new information about mechanisms underlying sexual differentiation of the brain. Moreover, attempts have been made to clarify sex differences in the human brain by using anatomical and noninvasive techniques such as MRI or PET. Psychological studies also have contributed in revealing the difference. All of the contributors, leading scientists in this field, have reviewed the most recent advances in studies on sexual differentiation of the brain. The monograph will provide a valuable insight into the current state of knowledge about this field.

Akira Matsumoto

Editor

Akira Matsumoto, Ph.D., is an associate professor in the Department of Anatomy, Juntendo University School of Medicine, Tokyo (Japan). Dr. Matsumoto received a B.S. degree (biology, 1968) and a Ph.D. degree (biology, 1974) from the University of Tokyo, Tokyo (Japan). He has served on the faculty of the Department of Anatomy, Juntendo University School of Medicine from 1974 to present. He received an International Society of Andrology Award (1993) and a Zoological Science Award (1997). He served as coeditor of the *Atlas of Endocrine Organs* (Springer–Verlag, 1992). He is a member of the Society of Neuroscience, the International Brain Research Organization, the International Society of Neuroendocrinology, the Society of Behavioral Neuroendocrinology, the International Society of Andrology, the Japanese Association of Anatomists, the Japanese Association of Endocrinology, the Japan Neuroscience Society, the Japanese Society for Comparative Endocrinology, and the Zoological Society of Japan. His research interests concern the organizational and activational effects of sex steroids on neural function and structures in the mammalian neuroendocrine brain. He has a number of publications on topics in this field in prominent journals.

Contributors

Laurie A. Abler, B.S., Wisconsin Regional Primate Research Center, University of Wisconsin, Madison, Wisconsin

Paola Agrati, Center Milano Molecular Pharmacology Laboratory, Institute of Pharmacological Sciences, University of Milan, Milan, Italy

Yasumasa Arai, Ph.D., Laboratory of Medical Education, Juntendo University School of Medicine, Tokyo, Japan

Arthur P. Arnold, Ph.D., Departments of Physiological Science and Neurobiology, Laboratory of Neuroendocrinology of the Brain Research Institute, Mental Retardation Research Center, University of California, Los Angeles, California

Laura Bolzoni, Center Milano Molecular Pharmacology Laboratory, Institute of Pharmacological Sciences, University of Milan, Milan, Italy

S. Marc Breedlove, Ph.D., Department of Psychology, University of California, Berkeley, California

Alessia Brusadelli, Center Milano Molecular Pharmacology Laboratory, Institute of Pharmacological Sciences, University of Milan, Milan, Italy

Gloria Patricia Cardona-Gomez, Instituto Cajal, Consejo Superior de Investigaciones Cientificas, Madrid, Spain

Julie Ann Chowen, Ph.D., Instituto Cajal, Consejo Superior de Investigaciones Cientificas, Madrid, Spain

Scott E. Christensen, M.A., Department of Psychology, University of California, Berkeley, California

Paolo Ciana, Ph.D., Center Milano Molecular Pharmacology Laboratory, Institute of Pharmacological Sciences, University of Milan, Milan, Italy

David Crews, Ph.D., Institute of Reproductive Biology, Section of Integrative Biology, University of Texas, Austin, Texas

Maria Carmen Fernandez-Galaz, Ph.D., M.D., Instituto Cajal, Consejo Superior de Investigaciones Cientificas, Madrid, Spain

Luis Miguel Garcia-Segura, Ph.D., Instituto Cajal, Consejo Superior de Investigaciones Cientificas, Madrid, Spain

Roger A. Gorski, Ph.D., Department of Neurobiology, University of California Los Angeles School of Medicine, Los Angeles, California

Elizabeth Hampson, Ph.D., Department of Psychology and Neurosciences Program, University of Western Ontario, London, Ontario, Canada

Melissa Hines, Ph.D., Department of Psychology, City University, London, United Kingdom

John B. Hutchison, Ph.D., St. John's College, University of Cambridge, Cambridge, United Kingdom CB21TP

Cynthia L. Jordan, Ph.D., Department of Psychology, University of California, Berkeley, California

Adriana Maggi, Ph.D., Center Milano Molecular Pharmacology Laboratory, Institute of Pharmacological Sciences, University of Milan, Milan, Italy

Elena Marini, Center Milano Molecular Pharmacology Laboratory, Institute of Pharmacological Sciences, University of Milan, Milan, Italy

Luciano Martini, M.D., Department of Endocrinology, University of Milan, Milan, Italy

Akira Matsumoto, Ph.D., Department of Anatomy, Juntendo University School of Medicine, Tokyo, Japan

Clara Meda, Center Milano Molecular Pharmacology Laboratory, Institute of Pharmacological Sciences, University of Milan, Milan, Italy

Shizuko Murakami, Ph.D. Department of Anatomy, Juntendo University School of Medicine, Tokyo, Japan

Paola Negri-Cesi, Ph.D., Department of Endocrinology, University of Milan, Milan, Italy

Sonoko Ogawa, Ph.D., Laboratory of Neurobiology and Behavior, The Rockefeller University, New York, New York

Cesare Patrone, Ph.D., Center Milano Molecular Pharmacology Laboratory, Institute of Pharmacological Sciences, University of Milan, Milan, Italy

Michael C. Penlington, Center Milano Molecular Pharmacology Laboratory, Institute of Pharmacological Sciences, University of Milan, Milan, Italy

Donald W. Pfaff, Ph.D., Laboratory of Neurobiology and Behavior, The Rockefeller University, New York, New York

Flavio Piva, Ph.D., Department of Endocrinology, University of Milan, Milan, Italy

Angelo Poletti, Ph.D., Department of Endocrinology, University of Milan, Milan, Italy

Giuseppe Pollio, Ph.D., Center Milano Molecular Pharmacology Laboratory, Institute of Pharmacological Sciences, University of Milan, Milan, Italy

Monica Rebecchi, Center Milano Molecular Pharmacology Laboratory, Institute of Pharmacological Sciences, University of Milan, Milan, Italy

Jon Sakata, Institute for Neuroscience, University of Texas, Austin, Texas

Yoshie Sekine, B. Human S., Department of Anatomy, Juntendo University School of Medicine, Tokyo, Japan

Nancy M. Sherwood, Ph.D., Department of Biology, University of Victoria, Victoria, British Columbia, Canada

Rodolfo H. Sialino, Center Milano Molecular Pharmacology Laboratory, Institute of Pharmacological Sciences, University of Milan, Milan, Italy

Richard B. Simerly, Ph.D., Division of Neuroscience, Oregon Regional Primate Research Center, Beaverton, Oregon and Program in Neuroscience and Department of Cell and Developmental Biology, Oregon Health Sciences University, Portland, Oregon

Ei Terasawa, Ph.D., Wisconsin Regional Primate Research Center and Department of Pediatrics, University of Wisconsin, Madison, Wisconsin

Jose Luis Trejo, Instituto Cajal, Consejo Superior de Investigaciones Cientificas, Madrid, Spain

Elisabetta Vegeto, Ph.D., Center Milano Molecular Pharmacology Laboratory, Institute of Pharmacological Sciences, University of Milan, Milan, Italy

James C. Woodson, M.A., Department of Psychology, University of California Los Angeles, Los Angeles, California

Harold H. Zakon, Ph.D., Section of Neurobiology and Institute for Neuroscience, University of Texas, Austin, Texas

Contents

1

Sexual Differentiation of the Brain: A Historical Review

Yasumasa Arai

CONTENTS

I. Introduction

Sexually dimorphic regulation of pituitary gonadotropin secretion was postulated in the early 1900s. Marshall and Jolly[1] reported that an ovary transplanted into a gonadectomized adult female was able to ovulate, whereas an ovary transplanted into a gonadectomized adult male did not show ovulation, containing only follicles without corpora lutea. The mechanism for release of gonadotropins in mammals was originally conceived as a simple feedback of gonadal steroids to the pituitary.[2] However, if this were the only determining factor, external and internal conditions such as photoperiods, availability of food, social contacts, etc. could not affect the reproductive activity. Channels must exist, therefore, through which these modifying influences enter the control system and become integrated with or supersede the ovarian feedback mechanism. This can only take place in the brain.

Hohlweg and Junkmann[3] demonstrated that the pituitary tissue transplanted into the renal capsule failed to show castration cells after gonadectomy, whereas the occurrence of castration cells in the pituitary *in situ* was clearly seen following castration. Since then, evidence has gradually accumulated to indicate the existence of nerve centers which are sensitive to gonadal steroids and, in turn, control the release of gonadotropins. However, a prevailing concept of gonadal feedback did not include the central nervous system in the 1930s.

In 1932, Pfeiffer[4] showed that neonatal ovariectomy did not prevent the female rat from developing into an adult capable of maintaining cyclic ovulation in a transplanted ovary. On the other hand, the implantation of neonatal testes into female rats at birth caused a permanent loss of the ability to ovulate. Thus, in the genetic female, the development of the ability to ovulate does not require the presence of an ovary, but that of a testis can prevent it.

Pfeiffer also showed that genetic males castrated at birth developed into adults which showed the cyclic release of ovulatory gonadotropins when an ovarian graft was given in adulthood. He concluded that both females and males have the potential to develop a female-type cyclic ovulatory mechanism. If orchidectomy was performed at birth, the resultant adult developed a female-type ovulatory mechanism. Testes in rats of either genetic sex during the neonatal period induce a male-type, noncyclic mechanism. He thought that the anterior pituitary was the target for sexual differentiation according to Moor and Price.[2]

However, Harris and Jacobsohn[5] found that male pituitaries could support estrous cycle, mating, pregnancy, and milk production when transplanted into the median eminence region in hypophysectomized adult female rats. They established that the pituitary was not sexually differentiated, but sex differences existed in the brain.

II. Sex Steroids and Anovulatory Persistent Vaginal Estrus Syndrome

Takasugi[6] reported that neonatal treatment of female rats with estrogen induced persistent vaginal estrus. He was of the opinion that the neural substrates for the estrogen feedback system were highly sensitive to estrogen in neonatal females, and the development of the estrogen feedback system was irreversibly disturbed by estrogen given neonatally. Since these were reported long before the introduction of "aromatization theory,"[7] Takasugi's findings were interpreted as a pharmacological phenomenon, and were not estimated correctly at that time.

By 1960, it was known that the sterilizing effects of a testis transplanted into neonatal females could be replaced more conveniently by a single injection of androgen and that the critical period for inducing persistent anovulatory syndrome in the female rat was limited to the first week after birth.[8] Based on these findings, the attention of endocrinologists was focused on sexually differentiated mechanisms in the brain.

Accumulation of data from stimulation and lesion experiments had shown that cyclic ovulatory mechanisms in the rat depended on the integrity of the preoptic area (POA). Barraclough and Gorski[9] found that electrical stimulation of the POA failed to induce ovulation in androgenized female rats. They suggested that androgen given neonatally irreversibly inhibited the development of the ovulatory trigger mechanism located in the POA. Figure 1.1 summarizes a concept proposed by Gorski.[10] This concept is still alive, although partly modified.

The tonic center in Figure 1.1 may correspond to the GnRH pulse generator. Since pulsatile luteinizing hormone (LH) secretion in male and female rats could be stimulated by naloxone, the pulse generator contains at least gonadotrophin-releasing hormone (GnRH) neurons and opioid neurons located in the medial basal hypothalamus, where GnRH neurons expressing c Fos in response to naloxone are abundant.[11] However, the cyclic center may be identical to the GnRH surge generator in which GABA neurons play a role in restraining the activity of GnRH neurons, because the GABA receptor antagonist or agonist can modify LH surge. GnRH neurons that express c Fos in response to bicuculline administration show a sexually dimorphic distribution. Most of these GnRH neurons are concentrated in the POA in female rats, but not in males.[12]

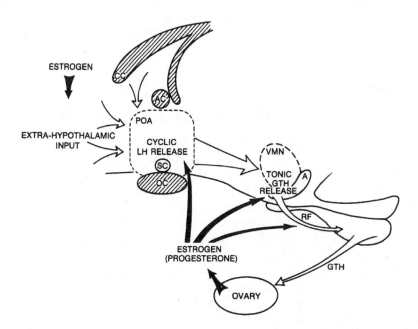

FIGURE 1.1
A model of the localization of the hypothalamic control of the secretion of LH in the normal female rat. (From *Frontiers in Neuroendocrinology*, Martini, L. and Ganong, W., Eds., Oxford University Press, New York, 1971. With permission.)

III. Behavioral Sexual Differentiation

The concept of sexual differentiation can be applied as well to the organization of the neural substrates for sexual behavior. The presence or absence of androgen during a critical period induces behavioral sexual differentiation.[13]

In the case of the behavioral defeminization in males and androgenized females, one possible interpretation is that testicular or exogenous androgen abrogated a mechanism for displaying lordosis. However, an attempt to recover the lost ability to show lordosis in defeminized animals has been successful. The neural transection between the POA and the septum (roof deafferentation, RD) enabled males to show lordosis.[14] Similar recovery of lordotic response also occurred following RD in neonatally androgenized females.[15]

As shown in Figure 1.2, the lordotic activity in androgenized females with RD is negatively correlated with the dose of testosterone propionate (TP). Since males and these androgenized females can display lordosis after the surgical removal of the forebrain influence, the minimal neural structure needed for the performance of lordosis exists in the male brain. One of the organizing actions of testicular androgen in behavioral defeminization is to establish an inhibitory neural circuit from the lordosis-mediating system.[16]

IV. Morphological Effects on Neuroendocrine Brain

By the late 1960s, there was evidence for permanent effects of neonatal castration or neonatal estrogen treatment on neuronal nuclear size in specific hypothalamic nuclei.[17,18] In the 1970s, two interesting and important findings indicating the effect of gonadal steroids on brain morphology were reported. Raisman and Field[19] provided the first electron microscopic evidence for a sex difference in the type of dendritic synapses found in the rat dorsomedial POA. The number of dendritic spine synapses of nonamygdaloid origin in females is greater than in males. This difference can be reversed appropriately by neonatal manipulation of androgen level — removal of neonatal testes or injection of androgen. The second is the discovery of the sexually dimorphic nucleus of the POA (SDN-POA). Gorski and his co-workers[20] have described a region with a striking sex difference in the rat POA. The SDN-POA occupies a greater volume and contains more neurons in males than in females. These two findings have stimulated the interest of neuroscientists. Evidence for anatomical sexual dimorphism in the brain was rapidly accumulated in 1980s. Wiring programs in the neural circuit network of the

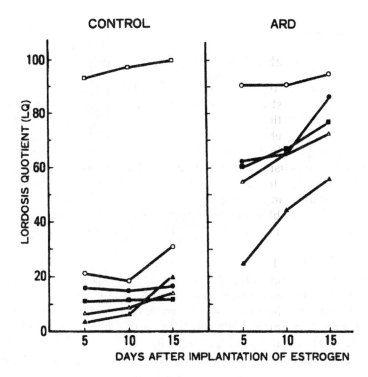

FIGURE 1.2

Mean lordosis quotients (LQs) in three successive tests after estrogen implantation on rats with or without anterior roof deafferentation (ARD). Females with ARD (right in figure) or without ARD (control, left in figure) which had been injected with various doses (0 to 1000 µg) of TP neonatally were subjected to behavioral tests after estrogen implantation at 11 weeks of age. (□) 0 µg TP, (○) 100 µg TP, (●) 250 µg TP, (△) 500 µg TP, (▲) 1000 µg TP, (■) male. (From Kondo, Y., Shinoda, A., Yamanouchi, K., and Arai, Y., *Physiol. Behav.*, 37, 495, 1986. With permission.)

neuroendocrine brain were found to be determined by the organizational action of gonadal steroids.[21,22]

Regarding regulation of neuron number in the sexually dimorphic neuron groups, there is recent evidence suggesting that gonadal steroids regulate apoptotic cell death in these groups. Androgen or estrogen given perinatally has a facilitatory effect on the incidence of apoptotic cell death in the anteroventral periventricular nucleus of the POA (AVPvN-POA),[23,24] whereas these steroids exert an inhibitory effect in the SDN-POA.[24,25] As shown in Figure 1.3, a single injection of estradiol benzoate (EB) given to Day 5 female rats is capable of facilitating apoptosis in the AVPvN-POA within 24 h, while EB effectively inhibits apoptotic cell death of the SDN-POA neurons.[24] Exposure to EB for 5 h is enough to get a significant inhibition of apoptosis in the SDN-POA. This suggests the possible involvement of a rapid molecular process in this phenomenon.

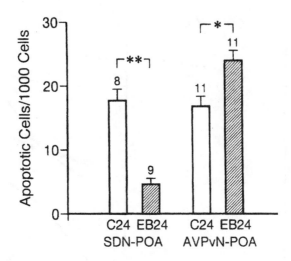

FIGURE 1.3

Effect of a single injection of 25 µg of EB on apoptotic cell death in the developing SDN-POA and AVPvN-POA of Day 5 female pups. The hatched bars represent the mean (+ SEM) of TUNEL-positive cells / 1000 cells in the pups sacrificed 24 h after EB injection (EB24). The open bars represent that of control pups for EB24 (C24). Numbers on the bars indicate the number of animals used in each group. *$P < 0.005$, **$P < 0.0002$. (From Arai, Y., Sekine, Y., and Murakami, S., *Neurosci. Res.*, 25, 403, 1996. With permission.)

It is not known how sex steroids act on the developing neuroendocrine brain to produce sex differences in neuronal circuitry. There is evidence indicating that estrogen stimulates axonal and dendritic growth and synapse formation.[26] Estrogen markedly enhances synaptogenesis in the arcuate nucleus[27] and the medial amygdala.[28] This estrogen action on neuronal differentiation has also been confirmed by using cultured cells into which estrogen receptor gene are transfected.[29,30] Furthermore, the importance of developmental estrogen–neurotrophin interactions has been suggested in the study of these mechanisms underlying estrogen actions during neuronal differentiation. Recent achievements will be described and discussed in the following chapters.

Finally I would like to present a photograph (Figure 1.4) taken at a conference in Tokyo to dedicate to the memory of Dr. R. W. Goy who passed away 14 January 1999. He was a great scientist, a distinguished pioneer of behavioral neuroendocrinology, and a great human being. We all will miss him so much. His untimely death is an inestimable loss to his many friends, to all neuroendocrinologists, and to all neurobiologists.

FIGURE 1.4
Participants of a conference on long-term effects of perinatal sex steroid administration organized by K. Takewaki and H.A. Bern in Tokyo, 1972. Dr. Goy is the 5th from the left, front row.

References

1. Marshall, F.H.L. and Jolly, W.A., Results of removal and transplantation of varies, *Trans. R. Soc. Edinburgh*, 45, 589, 1907.
2. Moor, C.R. and Price, D., Gonad hormone functions, and the reciprocal influence between gonads and hypophysis with its bearing on the problem of sex hormone antagonism, *Am. J. Anat.*, 50, 13, 1932.
3. Hohlweg, W. and Junkmann, K., Die hormonal-nervase Regulierung der Funktion des hypophysenvorderlappens, *Klin. Wochenschr.*, 11, 321, 1932.
4. Pfeiffer, C.A., Sexual differences of the hypophyses and their determination by the gonads, *Am. J. Ant.*, 58, 195, 1936.
5. Harris, G.W. and Jacobsohn, D., Functional grafts of the anterior pituitary gland, *Proc. Roy. Soc. London*, 139, 263, 1952.
6. Takasugi, N., Veranderungen der hypophsearen, Gonadotropen von Gebrut an Zwei gemische Steroiden injiziert wurden, *J. Fac. Sci. Univ. Tokyo Sect. IV*, 7, 299, 1954.

7. Naftolin, F., Ryan, K.J., Davies, I.J., Reddy, V.V., Flores, F., Petro, Z., Kuhn, M., White, R.J., Takaoka, Y., and Wolin, L., The formation of estrogens by central neuroendocrine tissues, *Recent Prog. Horm. Res.*, 31, 259, 1975.

8. Barraclough, C.A., Production of anovulatory, sterile rats by single injections of testosterone propionate, *Endocrinology*, 68, 62, 1961.

9. Barraclough, C.A. and Gorski, R.A., Evidence that the hypothalamus is responsible for androgen-induced sterility in the female rat, *Endocrinology*, 68, 68, 1961.

10. Gorski, R.A., Gonadal hormones and the perinatal development of neuroendocrine function, in *Frontiers in Neuroendocrinology*, Martini, L. and Ganong, W.F., Eds., Oxford University Press, New York, 1971, 237.

11. Funabashi, T., Jinnai, K., and Kimura, F., Fos expression by naloxone in LHRH neurons of the mediobasal hypothalamus and effects of pentobarbital sodium in the proestrous rat, *J. Neuroendocrinol.*, 9, 87, 1997.

12. Funabashi, T., Jinnai, K., and Kimura, F., Bicuculine infusion advances the timing of Fos expression in LHRH neurons in the preoptic area of proestrous rats, *NeuroReport*, 8, 1997.

13. Goy, R.W. and McEwen, B.S., Sexual differentiation of the brain, MIT Press, Cambridge, MA, 1980.

14. Yamanouchi, K. and Arai, Y., Female lordosis pattern in the male rat induced by estrogen and progesterone: effect of interruption of the dorsal inputs to the preoptic area and hypothalamus, *Endocrinol. Jpn.*, 22, 243, 1975.

15. Kondo, Y., Shinoda, A., Yamanouchi, K., and Arai, Y., Recovery of lordotic activity by dorsal deafferentation of the preoptic area in male and androgenized female rats, *Physiol. Behav.*, 37, 495, 1986.

16. Yamanouchi, K. and Arai, Y., Presence of a neural mechanism for expression of female sexual behavior in the male rat brain, *Neuroendocrinology*, 40, 393, 1985.

17. Pfaff, D.W., Morphological changes in the brain of adult male rats after neonatal castration, *J. Endocrinol.*, 36, 415, 1966.

18. Arai, Y. and Kusama, T., Effect of neonatal treatment with estrone on hypothalamic neurons and regulation of gonadotrophin secretion, *Neuroendocrinology*, 3, 107, 1968.

19. Raisman, G. and Field, P.M., Sexual dimorphism in the neuropil of the preoptic area of the rat and its dependence on neonatal androgen, *Brain Res.*, 54, 1, 1973.

20. Gorski, R.A., Gordon, J.H., Shryne, J.E., and Southam, A.M., Evidence for a morphological sex difference within the medial preoptic area of the rat brain, *Brain Res.*, 148, 333, 1978.

21. Arai, Y., Synaptic correlates of sexual differentiation, *Trends Neurosci.*, 4, 291, 1981.

22. Arai, Y., Matsumoto, A., and Nishizuka, M., Synaptogenesis and neuronal plasticity to gonadal steroids: implication for the development of sexual dimorphism in the neuroendocrine brain, *Curr. Top. Neuroendocrinol.*, 7, 291, 1986.

23. Arai,Y., Murakami, S., and Nishizuka, M., Androgen enhances degeneration in the developing preoptic area: apoptosis in the anteroventral periventricular nucleus (AVPvN-POA), *Horm. Behav.*, 28, 313, 1994.

24. Arai, Y., Sekine, Y., and Murakami, S., Estrogen and apoptosis in the developing sexually dimorphic preoptic area in female rats, *Neurosci. Res.*, 25, 403, 1996.

25. Davis, E.C., Popper, P., and Gorski, R.A., The role of apoptosis in sexual differentiation of the rat sexually dimorphic nucleus of the preoptic area, *Brain Res.*, 734, 10, 1996.

26. Toran-Allerand, C.D., On the genesis of sexual differentiation of the central nervous system: morphogenic consequences of steroid exposure and possible role of a-fetoprotein, *Prog. Brain Res.*, 61, 63, 1984.

27. Arai, Y. and Matsumoto, A., Synapse formation of the hypothalamic arcuate nucleus during post-natal development in female rat and its modification by neonatal estrogen treatment, *Psychoneuroendocrinology*, 3, 31, 1978.

28. Nishizuka, M. and Arai, Y., Synase formation in response to estrogen in the medial amygdala developing in the eye, *Proc. Natl. Acad. Sci. U.S.A.*, 79, 7024, 1982.

29. Ma, Z.Q., Spreafico, E., Pollio, G. Santagati, S., Conti, E., Cattanco, E., and Maggi, A., Activated estrogen receptor mediates growth arrest and differentiation of a neuroblastoma cell line, *Proc. Natl. Acad. Sci. U.S.A.*, 90, 3740, 1993.

30. Ustig, R.H., Sex hormone modulation of neural development in vitro, *Horm. Behav.*, 28, 383, 1994.

2

Genetic Contributions to the Sexual Differentiation of Behavior

Sonoko Ogawa and Donald W. Pfaff

CONTENTS

I. Gene/Behavior Relations

Two major forces in the applications of modern molecular genetics to the understanding of brain function and behavior have led to unfortunate misunderstandings of how explorations of gene/behavior relations would evolve. First, geneticists who work with the fruit fly, *Drosophila*, have a penchant for making simple lists of genes and behaviors and writing as though these would lead the way toward the delineation of one-to-one gene-to-behavior causalities. Second, the legitimate search for "disease genes" — genes whose mutations lead to serious neurological or psychiatric disorders — has led many readers to believe that following explication of gene structure and promoter analysis, *pari passu,* behavior would be understood. Nothing could be farther from the truth. The highly evolved, complex integrative possibilities offered by the mammalian central nervous system require a significantly different approach for the *systematic* understanding of the *lawful* relations between particular genes and specific behaviors. Quite aside from the obvious subtleties of highly interconnected and modulated neuronal circuits in the mammalian brain, two fundamental views of the difficulties of genetic applications, outlined below, render the above warning unavoidable. Moreover, recent results in the genetic analysis of mammalian reproductive behaviors (Sections II and III) illustrate, even at this early date, the difficulties of simpleminded thinking as could result from an uncritical application of genetic knowledge.

Some of the reasons that tracing causal chains from genes to mammalian behaviors will be extremely difficult lie in the nature of the genes themselves. The pleiotropy of genetic actions (cartooned in Figure 2.1) tells us that any single gene may have many different effects, especially if its actions during different portions of the life cycle are considered. Beyond that, it is simply not true that any individual physiological function, even in its subcomponents, depends exclusively on one gene. That is, redundancy among the functions of different genes will often frustrate the investigator seeking to demonstrate the role of any particular gene for any particular behavior (see Figure 2.1). A third difficulty follows from the unfortunate truth that we understand little of how the penetrance of a dominant allele is controlled among heterozygotes. This can be different for different genes and, for a single gene, different for different circumstances. Therefore, the task of constructing gene dose–response relationships using heterozygotes will be daunting, indeed.

Another approach to describing the difficulties of unraveling genetic influences on behavior is to chart all possible causal routes with broad categorization (Figure 2.2). While most behavioral biologists wish, at least for strategic purposes, to analyze direct actions of genes on integrative mechanisms in the adult brain, a large number of other routes must also be admitted. Not only must we consider all the indirect causal routes through development, but also the important role of gene expression outside the brain as it influences behavior in the adult. Gene expression for the product from fat tissue, leptin, provides an excellent example.

In the face of these difficulties, what can be stated about genetic influences on reproductive behaviors (Section II) and what lessons can we derive (Section III)?

II. Findings for Reproductive Behaviors

Despite the difficulties summarized above, there has been some success with genetic analyses of the simplest forms of mammalian reproductive and aggressive behaviors. In particular, these studies built upon substantial structures of knowledge about the mechanisms of lordosis behavior, studied at the neuroanatomical, neurophysiological, and neurochemical levels, accompanied by many studies of hormone-dependent messenger RNA fluctuations.[1,2] Consistent with neuroendocrine findings, the gene for the classical estrogen

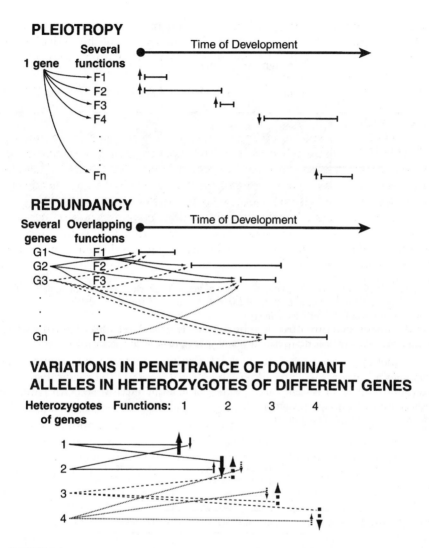

FIGURE 2.1
Features of genetic function which add complexity to the discernment of gene/behavior causal relationships. (From Pfaff, D.W., *Genetic Influences on Neural and Behavioral Functions*, CRC Press, Boca Raton, FL, 1999.)

receptor, ERα, is absolutely required for lordosis behavior (Figure 2.3).[3] Instead of performing normal reproductive behaviors, estrogen receptor knockout (ERKO) female mice refuse to allow stud males to mount and intromit properly and instead show high frequencies of aggressive behaviors.

The opposite is true for ERKO male mice.[4] ERKO males show reduced frequencies of intromissions and rarely are able to ejaculate. Their aggressive behaviors are reduced and their emotional responses to standardized tests that involve anxiety are somewhat feminized.

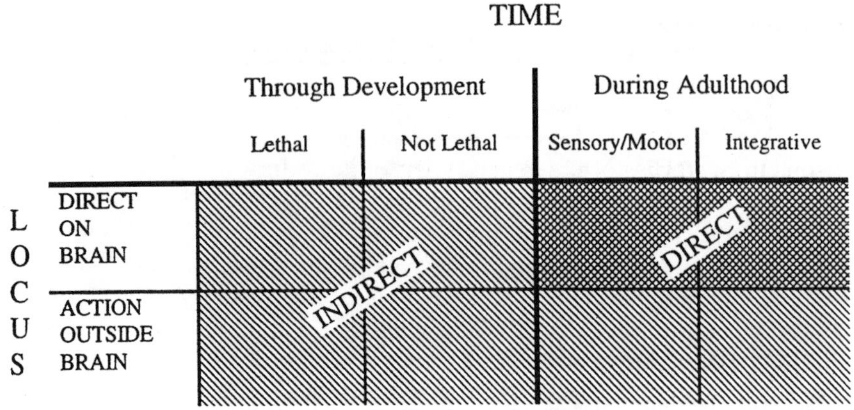

TIME

	Through Development		During Adulthood	
	Lethal	Not Lethal	Sensory/Motor	Integrative

L O C U S

DIRECT ON BRAIN

ACTION OUTSIDE BRAIN

Note:　Considering adult behavior, all actions of genes on development, including mutations of hormone receptors and alterations of enzymes for steroid hormone synthesis and metabolism would be considered indirect.

　　　　Direct effects would include genes expressed in the brain in adulthood necessary for a particular behavior, including genes for hormone receptors and hormone metabolism.

FIGURE 2.2
Schematic chart of all possible causal routes, not mutually exclusive, between genes and behaviors. (From Ogawa, S., Gordon, J.D., Taylor, J., Lubahn, D., Korach, K., and Pfaff, D.W., *Horm. Behav.*, 1996. With permission.)

	Incidence of female reproductive behavior	Incidence of female - female aggression
Wild type	normal	2/21 mice
Estrogen Receptor KnockOut	none	10/25 mice*

*Aggression exhibited by ERKO females mainly offensive attacks typical of intermale aggression.

FIGURE 2.3
Genetic knockout evidence that the classical ER gene is absolutely required for normal lordosis behavior. Interestingly, it also affects aggressive behavior in female mice. (From Ogawa, S., Taylor, J., Lubahn, D., Korach, K., and Pfaff, D.W., *Neuroendocrinology*, 64, 467, 1996. With permission.)

Thus, we have the unusual situation that a single gene, that for the classical estrogen receptor, is required for the normal performance of both female-typical and male-typical reproductive behaviors.

III. Lessons from Studies of Reproductive Behaviors

In concert with the general warnings in Section I above, certain complex comparisons from studies of the genetic basis of reproductive behaviors lead us to lessons for the systematic understanding of the genetic basis of mammalian behaviors. Consider the findings in Section II. After all, in the continuum of neuroendocrine mechanisms in which extreme feminine-typical functions are at one end and extreme masculine-typical functions are at the other, we ordinarily would expect a genetic or physiological manipulation to push the result in one direction or the other, but not both. Counterintuitively, the loss of the estrogen receptor gene pushes the results from the genetic females toward masculinization, yet pushes the results for the genetic males in the feminine-typical direction.[3,4] *Thus, the actions of a gene can depend on the gender in which the gene is expressed.*

A more complexly derived lesson comes from the comparison of genetic knockout results to antisense DNA results. The partial masculinization of the behaviors of ERKO female mice emanates from a genetic manipulation which has its effect throughout the body and throughout the entire lifetime of the female mouse. Yet, interruption of the estrogen receptor gene product by antisense DNA microinjection into the hypothalamus of neonatal females had the opposite result: reduction of the masculinization of females subsequent to a neonatal testosterone injection.[5] This antisense DNA manipulation obviously is not applied throughout the body and is effective for only a brief portion of the life cycle — here, a neonatal period. *Therefore, the effect of a gene on behavior depends upon exactly where and when it is expressed.*

We conclude, as a result of the considerations in Section I and these complex comparisons just summarized, that it is virtually impossible to reason simply from the structure and function of a gene what its behavioral effects will be. Instead, we must possess detailed neural and physiological mechanisms of adaptive behavioral functions and then, by genetic analysis, see how an individual gene participates, either during development or in adulthood. This path of discovery is much more complex than one would anticipate from studies of *Drosophila*[6] and has a substantially different logic from the search for disease genes.

IV. Endocrine Analyses of Gene Knockout Phenotypes

Could the phenotypes of ERKO females and ERKO males summarized in Section II simply result from altered circulating hormone levels? To answer this question, we gonadectomized ERKO females and males, administered the appropriate hormones experimentally, and measured their behaviors.

Despite the administration of estrogen or estrogen plus progesterone, gonadectomized female mice did not show any lordosis behavior.[7] Detailed behavioral analyses revealed that even when exogenous gonadal hormones were supplied, ERKO females were deficient in sexual behavior interactions preceding lordosis and were extremely rejective toward attempted mounts by stud male mice which, in turn, could show no intromissions (Figure 2.4).[7] As well, aggressive behavior differences among genotypes persisted after gonadectomy.

Male ERKO mice and their wild-type controls were castrated and the virtually complete suppression, in the ERKO, of male-typical offensive attacks was further confirmed, despite the daily injection of testicular hormones.[8] Regarding reproductive behaviors, as expected, gonadectomy reduced sexual behaviors in all genotypes, but the serious deficiencies in both intromissions and ejaculations for ERKO males compared to wild-type controls persisted (Figure 2.5).[8]

We conclude that the effects of classical estrogen receptor gene disruption on reproductive behaviors in females and males are not simply due to alterations in circulating gonadal hormone levels, but instead must be registered in the central nervous system.

V. How Specific Genes Influence Particular Reproductive Behaviors

Even under circumstances where gene/behavior relations have been documented for mammals, the causal routes might remain obscure. Here we summarize briefly the causal routes delineated for classical ER (ERα) and thyroid hormone receptor (TRβ).

We know that estrogens act through both the classical estrogen receptor (ERα) and ERβ to stimulate the synthesis of at least six gene products which, in turn, foster reproductive behavior: the progesterone receptor, adrenergic α-1 receptors, muscarinic receptors, enkephalins (and their opioid receptors), oxytocin (and its receptor), and gonadotropin-releasing hormone (and its receptor) (summarized in Reference 11). Therefore, we can state with confidence that the role the classical estrogen receptor gene plays in the production of a specific behavior, lordosis behavior, is to act as a transcription factor in hypothalamic neurons facilitating — upon estrogen binding — the synthesis of gene products that will promote that behavior.

A second gene for which we can state a causal route is that for TRβ. Gel-shift evidence of DNA binding, transfection experiments in CV1 cells revealing transcriptional mechanisms, and studies of estrogen-stimulated messenger RNA elaboration in brain tissue all show that thyroid hormones, acting

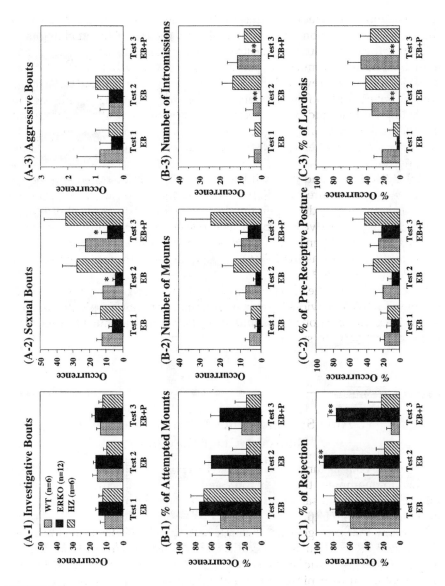

FIGURE 2.4

Social behavior phenotype of ovariectomized ERKO females primed with EB or estradiol plus progesterone (EB+P). Despite normal frequencies of social investigations, lordosis behavior was virtually abolished and the stud male could never achieve intromission. (From Ogawa, S., Eng, V., Taylor, J., Lubahn, D., Korach, K., and Pfaff, D.W., *Endocrinology*, 239, 5070, 1998. With permission.)

through thyroid hormone receptors, can interfere with estrogen effects on reproductive behavior,[9-14] New studies with TRα and TRβ knockout mice show that TRβ knockout mice actually have stronger responses to estrogenic stimulation of reproductive behavior than their wild-type controls.[15]

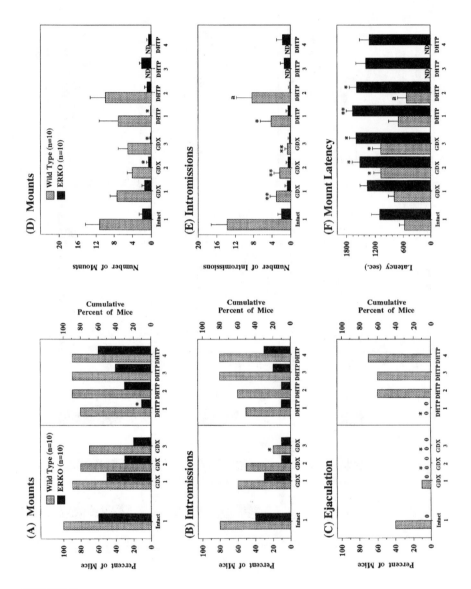

FIGURE 2.5
Phenotype of castrated ERKO males (GDX) primed with androgenic hormones (dihydrotest-osterone propionate, DHTP). Even with androgen hormone replacement, ERKO male mounting tended to be reduced, intromissions were much suppressed, and ejaculations were absent. (From Ogawa, S., Washburn, T., Taylor, J., Lubahn, D., Korach, K., and Pfaff, D.W., *Endocrinology*, 139, 5058, 1998. With permission.)

Therefore, the TRβ gene has as its role in reproductive behavior the reduction of estrogenic transcriptional potency with a mechanism including, but not limited to, competitive DNA binding as would disrupt estrogen receptor action. Future studies will determine whether or not, in playing this molecular role, thyroid hormones and their receptors signal important environmental or metabolic events which would naturally be assumed to limit reproduction.

Dedication and Acknowledgments

This chapter is dedicated to honoring the scientific work and career of Professor Yasumasa Arai, of the Department of Anatomy at Juntendo University School of Medicine in Tokyo. Over the years, Professor Arai and his colleagues have made splendid contributions to our understanding of sex differences in brain and behavior. Moreover, the wisdom, civility, and leadership demonstrated by Yasumasa Arai as an internationally recognized participant in the neuroscientific endeavor provide us with a model that we all could be proud to emulate.

The original experiments reviewed here were supported by NIH Grants HD-05751 and MH-38273 to Donald W. Pfaff and by an H.F. Guggenheim Foundation Grant and NSF Grant IBN-9728579 to Sonoko Ogawa. We thank Lucy Frank of the Rockefeller University for editorial assistance.

References

1. Pfaff, D.W., McCarthy, M., Schwartz-Giblin, S., and Kow, L.M., Female reproductive behavior, in *The Physiology of Reproduction*, Vol. 2, E. Knobil and J. Neill, Eds., Raven, New York, 1994, 107–220.
2. Pfaff, D.W., *Drive: Neural and Molecular Mechanisms for Sexual Motivation*, MIT Press, Cambridge, MA, 1999.
3. Ogawa, S., Taylor, J., Lubahn, D., Korach, K., and Pfaff, D., Reversal of sex roles in genetic female mice by disruption of estrogen receptor gene, *Neuroendocrinology*, 64, 467–470, 1996.
4. Ogawa, S., Lubahn, D.B., Korach, K.S., and Pfaff, D.W., Behavioral effects of estrogen receptor gene disruption in male mice, *Proc., Nat. Acad. Sci.*, 94, 1476–1481, 1997.
5. McCarthy, M., Schlenker, E., and Pfaff, D., Enduring consequences of neonatal treatment with antisense oligonucleotides to estrogen receptor mRNA on sexual differentiation of rat brain, *Endocrinology*, 133, 433–443, 1993.
6. Pfaff, D., Hormones, genes, and behavior, *Proc. Natl. Acad. Sci. U.S.A.*, 94, 14213–14216, 1997.
7. Ogawa, S., Eng, V., Taylor, J., Lubahn, D., Korach, K., and Pfaff, D., Roles of estrogen receptor-alpha gene expression in reproduction-related behaviors in female mice, *Endocrinology*, 139, 5070–5081, 1998.

8. Ogawa, S., Washburn, T., Taylor, J., Lubahn, D., Korach, K., and Pfaff, D., Modifications of testosterone-dependent behaviors by estrogen receptor-alpha gene disruption in male mice, *Endocrinology*, 139, 5058–5069, 1998.

9. Dellovade, T., Kia, H., Zhu, Y.-S., and Pfaff, D., Thyroid hormone coadministration inhibits the estrogen-stimulated elevation of preproenkephalin mRNA in female rat hypothalamic neurons, *Neuroendocrinology*, in press.

10. Dellovade, T., Zhu, Y., Krey, L., and Pfaff, D., Thyroid hormone and estrogen interact to regulate behavior, *Proc. Natl. Acad. Sci. U.S.A.*, 93, 12581–12586, 1996.

11. Zhu, Y., Yen, P., Chin, W., and Pfaff, D., Estrogen and thyroid hormone interaction on regulation of gene expression, *Proc. Natl. Acad. Sci. U.S.A.*, 93, 12587–12592, 1996.

12. Morgan, M., Dellovade, T., Ogawa, S., and Pfaff, D., Female mouse sexual behavior is regulated by thyroid hormones and estrogen, *Soc. Neurosci. Abstr.*, 1997.

13. Zhu, Y., Ling, Q., Cai, L., Imperato-McGinley, J., and Pfaff, D., Regulation of preproenkephalin (PPE) gene expression by estrogen and its interaction with thyroid hormone, *Soc. Neurosci. Abstr.*, 23, 798, 1997.

14. Dellovade, T., Zhu, Y., and Pfaff, D., Thyroid hormone and estrogen affect oxytocin gene expression in hypothalamic neurons, *J. Neurodendocrinol.*, in press, 1998.

15. Dellovade, T. et al., *Nature*, Submitted.

3

In Vitro Study for Effects of Estrogen on Estrogen Receptor-Transfected Neuroblastoma Cells

Paola Agrati, Laura Bolzoni, Alessia Brusadelli, Paolo Ciana, Elena Marini, Clara Meda, Cesare Patrone, Michael C. Penlington, Giuseppe Pollio, Monica Rebecchi, Rodolfo H. Sialino, Elisabetta Vegeto, and Adriana Maggi

CONTENTS

I. Introduction

A. The Complexity of Estradiol Activities in Neural Cells

17β-Estradiol (E_2) is a well-known hormone and differentiating agent covering a fundamental role in the functioning and development of reproductive and nonreproductive tissues. Similarly to what reported in other, better-studied target organs, the effects of E_2 in neural cells are pleiotropic. During brain development, E_2 has a pivotal role in the differentiation of specific sexually dimorphic areas.[1-4] In the mature brain, besides the well-established role in the control of endocrine functions and reproductive behavior,[5-8] E_2 affects cognitive and affective functions.[9-11] Finally, in the aging brain, the presence of this sex hormone has been associated with protection against the onset of neurodegenerative diseases such as Alzheimer's.[12]

The increasing evidence of a beneficial effect of estrogens in the aging brain in both maintaining mnemonic and cognitive functions as well as in neurodegenerative diseases will drive to the use of this hormone in replacement therapies for postmenopausal and aging women. However, the simple use of estradiol in the therapy has undesirable side effects. It is well known that estradiol induces several cell types to hyperproliferate (e.g., mammary gland cells and uterine cells), and additional unwanted effects of the hormone could be unleashed in some of its other numerous targets. The rapidly evolving pharmacology of intracellular receptors, especially with regard to estrogen receptor (ER) agonists and antagonists, points to the possibility of generating synthetic ER ligands endowed with relevant specificity of action; unraveling the biochemical events induced by E_2 in specific target cells would certainly help in the research and development of ligands acting on a very restricted milieu.

To this aim, the acquisition of the comprehension of the molecular details of estrogen activity in specific brain areas and cells is mandatory. In addition, the consequences of the molecular events triggered by the hormone should be clarified at the physiological level. This task is particularly arduous due to the nature of the tissue taken in consideration and even more difficult because of the panoply of effects of the hormone on neural cells. The two ER subtypes (ERα and ERβ) appear to be expressed in areas of relevance for the control of sexual activities as well as in limbic and motor areas as shown in Figure 3.1.[13-18] The existence and differential distribution of two receptors binding E_2 is part of the diverse effects of the hormone in brain. In addition, in the various brain areas, E_2 can target different cell types as suggested by the observation of ER expresssion in both neurons and glial cells (oligodendrocytes, and astrocytes type I and type II).[19]

Along this line of thought, several years ago we proposed that the way to investigate the physiological significance of estrogen action on neural cells

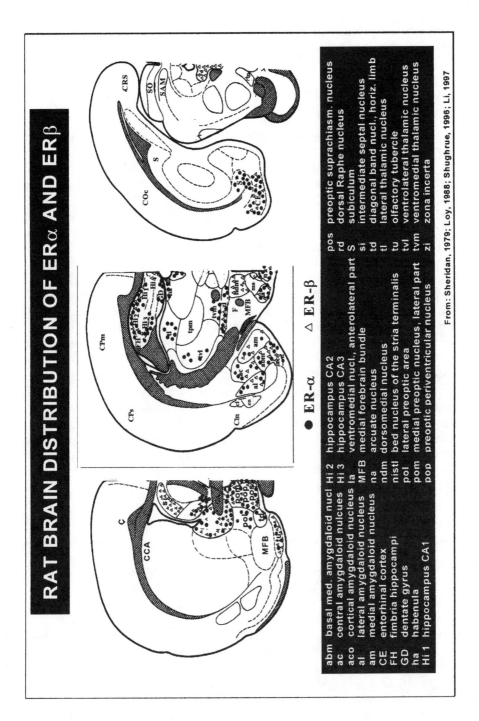

FIGURE 3.1
Distribution of ERα and ERβ in adult rat brain by immunocytochemical and *in situ* studies.

was to use molecular genetic tools; the results here presented will review some of the efforts made in our laboratory in this area.

B. Estrogens May Modulate Neural Cell Functions by Different Mechanism of Action

From the biochemical point of view, estrogens are believed to signal to the target cells by two mechanisms. The first is the "genomic" mechanism of action which involves the diffusion of the hormone across the neural cell membrane, binding of one or both of the two well-characterized intracellular receptors (ERα and ERβ) and activation of transcriptional regulation of target genes that ultimately results in protein synthesis.[20] In this way, the effects of the hormone on cellular activities can take several hours to develop. In contrast, estrogen has also been found to produce very short term effects ("nongenomic" mechanism) involving the electrical properties of neurons. The rapid onset of these effects implies that they arise from a direct interaction of estradiol on membrane receptors or other cellular components whose identification remains elusive.[21] The understanding of the molecular basis of E_2 nongenomic activities must await a better definition of the molecules leading the specific binding of the hormone to the cell membrane and the consequent unraveling of the transduction pathway responsive to the hormone. On the other hand, the study of the genomic activities of E_2 in neural cells is facilitated by the recent progress in the comprehension of the mode of action of intracellular receptors and by the availability of technologies which allow the generation of specific model systems. Recent development in the field provided the key to the understanding of some of the aspects of cells specificity of E_2 activity showing that the E_2-induced modulation of gene transcription is the result of a series of protein–protein interactions occurring at the promoter of each single target gene.

II. SK-ER3 Neuroblastoma Cells as a Model System for the Understanding of Estrogen Effects in Cells of Neural Origin

The generation of specific model systems in which the activity of the hormone can be evaluated in a typical setting will provide a restricted, but biochemically well-defined, view of the intracellular signaling evoked by E_2 treatment. Once obtained, this information will be put together as tiles of a mosaic for the generation of a more comprehensive view of hormonal activities in brain. Following this strategy, we devised a model system aimed at the analysis of the biochemical events resulting from the interaction of estradiol with ERα, in neuronlike cells specifically. We chose to transfect the hERα

cDNA in a neuroblastoma cell line. Neuroblastoma cells might be the ideal system to study the activity of ER and its interactions with other neural-specific, accessory proteins because it is known that neural crest derivatives express ER.[22] In addition, we decided to select a neuroblastoma cell line that conceivably retains considerable developmental potential, allowing ER to display fully its differentiation activity (if any). The SK-N-BE cells selected for our study undergo phenotypic interconversion between two distinct cell types, the neuroblast and the melanocyte, which are known to separate from each other at a very early stage in the differentiation pathway described for neural crest cell derivatives.[23-24] The hERα cDNA was therefore stably transfected in the SK-N-BE human neuroblastoma cell line not expressing the intracellular ERs.[25]

We first ensured that the hERα was expressed in SK-ER3 at a concentration compatible with an activity of the hormone by establishing that the number of ERα receptors in SK-ER3 and other known target for the hormone (e.g., MCF-7 cell line) was similar. Next, we demonstrated that the receptor, once activated by the hormone, could induce the transcription of specific target genes. This was done by assessing the effect of E_2 on the transcriptional activity of an exogenous, transiently transfected reporter gene (chloramphenicol-acetyltransferase under the control of a promoter carrying the *estrogen responsive element*) or of endogenous genes like *c-fos*[26] and opioid receptors.[27] The results obtained led us to conclude that the transcriptional activity of the transfected receptor in SK-ER3 cells and of the endogenous receptor in other cell lines was indistinguishable.[25,28]

A. SK-ER3 Neuroblastoma Cells: An Insight into the Molecular Events Elicited by Estradiol in Cells of Neural Origin

In SK-ER3 cells no interconversion between the melanocyte and neural phenotype can be observed; the actual morphology of these cells is slightly different with respect to the parental cell line (Figure 3.2). All the cells in the culture appear to have a neural phenotype with fuse-shaped somata and neurites much longer than in the parental SK-N-BE cells. The explanation for this phenomenon was provided by a series of studies in which we showed that treatment of SK-ER3 cells with estradiol (or growth factors known to activate the ER in the absence of the ligand, such as IGF-I or insulin) causes differentiation of these cells along the neuronal pathway.[29-31] Upon treatment with a low concentration of E_2 (10 nM) SK-ER3 cells rapidly undergo a series of morphological changes: the pericaryon acquires a fuselike shape and begins to grow numerous neurites which keep elongating and branching with time in culture. After 6 to 7 days of the hormonal treatment, immunocytochemical examination of SK-ER3 cells shows the newly acquired expression of proteins typical of the activity of mature neurons, (obviously unexpressed in SK-N-BE or E_2-untreated SK-ER3). E_2-treated neuroblastoma cells express TAU, a neurofilament of mature neurons or sinaptophysin, a

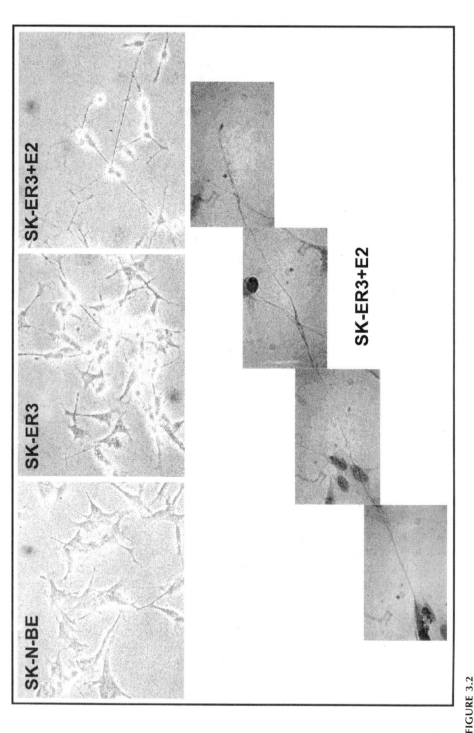

FIGURE 3.2
The parental human neuroblastoma cell line SK-N-BE was stably transfected with hERα cDNA to generate the ER-positive neuroblastoma SK-ER3. SK-ER3 cells following treatment with 1 nM 17-β-estradiol change their morphology and assume a dopaminergic phenotype as shown by immunostaining with antityrosine hydroxylase antibodies (lower panel).

protein present in the secretion granules.[25] Most interesting, however, is the observation that the E_2-induced SK-ER3 cells acquire all the enzymatic apparatus typical of a dopaminergic cell: the key synthetic enzyme tyrosine hydroxylase, the DM-transporter for the neurotransmitter uptake, and the capability to release dopamine following treatment with KCl.[28] The body of observations carried out in SK-ER3 compared with the data so far available on estradiol involvement in the development of specific brain areas led us to conclude that our model system might recapitulate at least some of the events actually occurring in immature neural cells stimulated by E_2. Our model was therefore of interest for the continuation of our studies aimed at the characterization of target genes for estrogen activities in neural cells.

B. Identification of Estrogen-Responsive Genes in Neuroblastoma SK-ER3 Cells

Using the differential display PCR method (ddPCR) we started the identification of estrogen-regulated genes in SK-ER3 cells.[32] This method proved quite reproducible and allowed us to identify a group of early genes in the cascade of events triggered by the hormone. Among the DNA sequences we found, six corresponded to sequences identified for the first time, whereas five showed 80 to 100% homology with previously reported genes. In particular, we found that two of the sequences identified code for the known proteins Nip2 and prothymosine (PTMA), previously associated with cell death protection and cell growth and differentiation. The pattern of estrogen regulation of the two mRNAs is opposite: PTMA is rapidly induced by the hormone, whereas Nip2 is significantly reduced. Both effects are transient and reverse by the fourth day after treatment. In spite of the fact that both genes had been described for quite some time, very little is known with regard to their physiological role. We became quite interested in the effect of the hormone on Nip2 which was recently described as one of the proteins interacting with the apoptosis inhibitors Bcl2 and E1B 19-kDa proteins.[33] Mutational analysis indicated that the protein no longer associates with Bcl2 mutants defective in suppression of cell death, suggesting a role of Nip2 in apoptotic cell death. With regard to the other mRNAs found as modulated by estrogens, we found of interest that two mitochondrial DNAs were increased by E_2. This observation was consistent with (1) previous studies in neural tissues[34] and in nonneural cells[35] and (2) the well-described increase in neuronal activity (i.e., increased firing) induced by estrogens.

C. From Genes to Function

To establish whether Nip2 could indeed influence cell survival or proliferation, we performed a series of studies in transient transfection in which Nip2 full-length cDNA in SK-N-BE cells was cotransfected with a tracer gene (beta-galactosidase) to allow the count of cells indeed expressing the two cDNAs. These initial experiments clearly showed that cells expressing Nip2 undergo death with some of the features typical of apoptosis. At few hours following Nip2 cDNA transfection cells started to die and at the third day following the transfection, all the cells expressing Nip2 had disappeared from the culture. The effect is dose- dependent: the more Nip2 cDNA was transfected, the higher the cell death observed. Untransfected cells survive and proliferate indicating that the expression of Nip2 does not imply the release agents toxic for the surrounding cells.

These data pointed to a proaoptotic activity for the gene found to be target for estrogen activity. The question we asked next was whether E_2, by virtue of its down-modulatory effect on Nip2 gene expression, might have played a role in cell death induced by proapoptotic agents. This hypothesis was tested in SK-ER3 cells treated with several proapototic agent (Ca^{++} ionophore, beta-peptide, etc.) in the presence or absence of E_2 at a concentration compatible with an receptor-mediated activity (1 nM). In a time course experiment we could show that E_2 significantly delays the onset of apoptotic death and that the treatment is maximally effective when cells are treated with the hormone for 6 to 2 h prior to exposure to the apoptotic agent (Figure 3.3).[36] E_2 antiapoptotic activity is ER-mediated because can be blocked by the presence of specific ER antagonists (ICI 182,780, tamoxifen) and is not observed in the parental cell line, SK-N-BE, in which the ER is not expressed.

III. Conclusions

The biochemical analysis of the effects triggered by E_2 in the SK-ER3 model system led us to the identification of function of estrogens in apoptosis which had been unexpected. Of course, the relevance of the study carried out in SK-ER3 cells had to be tested outside the cell system in other neural cells and, eventually, *in vivo* in the brain. Interestingly, a number of studies of recent publication came in support of our findings by demonstrating that E_2 indeed has a defined role in the modulation of neural cell apoptotic death induced by several agents.[37,40] The impetus to these studies was clinical reports pointing to a beneficial effect of E_2 replacement therapy in Alzheimer's or other mental deficits typical of aging which led us to hypothesize an effect of E_2 in the program of death of neural cells.

A distinct possibility is that the antiapoptotic activity exerted by E_2 in mature neural cells uses the same biochemical steps necessary for E_2 during the differentiation of sexually dimorphic brain areas which occurs during development. Our study, besides proving an effect of the hormone on SK-ER3 cell apoptosis, provides an insight on the mechanisms possibly involved in such an effect placing the basis for a better understanding of the overall significance of estrogen action in neural cells and of potential interventions

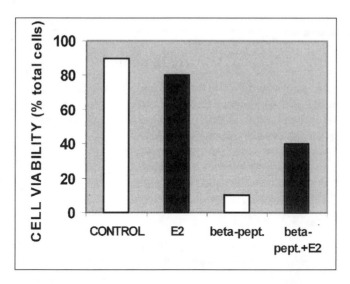

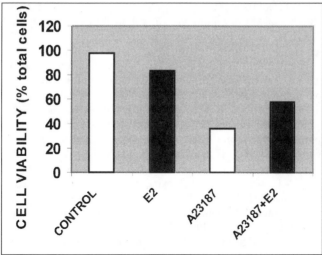

FIGURE 3.3

17-β-Estradiol exerts antiapoptotic activity in SK-ER3 cells. SK-ER3 cells were treated overnight with the calcium ionophore A23187 or amyloid β-peptide. Pretreatment with 1 nM 17-β-estradiol significantly decreased the percentage of cells undergoing apoptosis.

aimed at facilitating or increasing the beneficial effects of E_2 in neural cells. We believe that Nip2 is part of a chain of events triggered by the hormone, and further studies should be devoted to the exact understanding of the sequela of events sensitive to the presence of the hormone and relevant for neural cell life.

Acknowledgments

This work was supported by the European Economic Community (BIOMED Programme P1962286), the Italian Association for Cancer Research (AIRC), Telethon Italy (E600), Italian CNR Target Project Biotechnologies, Murst 40%, and Istituto Superiore di Sanita' (93/J/T61).

References

1. Gorski, R.A., Gordon, J.H., Shryne, J.F., and Southam, A.M., Evidence for a morphological sex difference within the medial preoptic area of the rat brain, *Brain Res.*, 148, 333, 1978.
2. De Vries, G.J., DeBruin, J.P.C., Uylings, H.B.M., and Corner, M.A., *Sex Differences in the Brain. Progress in Brain Research*, Vol. 61, Elsevier, Amsterdam, 1984.
3. McEwen, B.S., Davis, P.G., Parsons, B., and Pfaff, D.W., The brain as a target for steroid hormone action, *Annu. Rev. Neurosci.*, 2, 65, 1979.
4. Toran-Allerand, C.D., On the genesis of the sexual differentiation of the central nervous system: morphogenetic consequences of steroidal exposure and the possible role of α-fetoprotein, *Prog. Brain Res.*, 61, 63, 1984.
5. Maggi, A. and Perez, J., Role of female gonadal hormones in the CNS: clinical and experimental aspects, *Life Sci.*, 37, 893, 1985.
6. Lehrman, D.S., The reproductive behavior of ring doves, *Sci. Am.*, 211, 48, 1964.
7. McEwen, B.S., Davis, P.G., Parsons, B., and Pfaff, D.W., The brain as a target for steroid hormone action, *Annu. Rev. Neurosci.*, 2, 65, 1979.
8. McLusky, N.J. and Naftolin, F., Sexual differentiation of the central nervous system, *Science*, 211, 1294, 1981.
9. Tallat, P., Hormonal influences in developmental learning disabilities, *Psychoneuroendocrinology*, 16, 203, 1991.
10. Luine, V.N., Steroid hormone influences on the spatial memory, *Ann. N.Y. Acad. Sci.*, 14, 201, 1994.
11. Sherwin, R.B., Estrogenic effects on memory in women, in *Hormonal Restructuring of the Adult Brain*, Luine, V.N. and Harding, C.F., Eds., New York Academy of Sciences, New York, 1994, 213.
12. Tang, M.X., Jacobs, D., Stern, Y., Marder, K., Schofield, P., Gurland, B., Andrews, H., and Mayeux, R., Effect of estrogens during menopause on risk and age at onset of Alzheimer's disease, *Lancet*, 348, 429, 1996.
13. Sheridan, P.J., Estrogen binding in the neonatal neocortex, *Brain Res.*, 178, 201, 1979.
14. Maggi, A., Susanna, L., Bettini, E., Mantero, G., and Zucchi, I., Hippocampus: a target for estrogen action in mammalian brain, *Mol. Endocrinol.*, 3, 1165, 1989.

15. Loy, R., Gerlach, J.L., and McEwen, B.S., Autoradiographic localization of estradiol-binding neurons in the rat hippocampal formation and entorhinal cortex, *Brain Res.*, 467, 245, 1988.
16. Don Carlos, L.L., Developmental profile and regulation of estrogen receptor mRNA expression in the preoptic area of neonatal rats, *Dev. Brain Res.*, 94, 224, 1996.
17. Shughrue, P.J., Komm, B., and Merchenthaler, I., The distribution of estrogen receptor-beta mRNA in the rat hypothalamus, *Steroids*, 61, 678, 1996.
18. Li, X., Schwartz, P.E., and Rissman, E.F., Distribution of estrogen receptor-beta-like immunoreactivity in rat forebrain, *Neuroendocrinology*, 66, 63, 1997.
19. Santagati, S., Melcangi, R.C., Celotti, F., Martini, L., and Maggi, A., Estrogen receptor is expressed in different types of glial cells in culture, *J. Neurochem.*, 63, 2058, 1994.
20. Evans, R.M., The steroid and thyroid hormone receptor superfamily, *Science*, 240, 889, 1988.
21. Gu, Q. and Moss R.L., 17β-Estradiol potentiates kainate-induced currents via activation of the cAMP cascade, *J. Neurosci.*, 16, 3620, 1996.
22. Sohrabji, F., Miranda, R.S., and Toran-Allerand, D., Estrogen differentially regulates estrogen and nerve growth factor receptor mRNA in adult sensory neurons, *J. Neurosci.*, 14, 459, 1994.
23. Anderson, D.J., Molecular control of cell fate in the neural crest: the sympathoadrenal lineage, *Annu. Rev. Neurosci.*, 16, 129, 1993.
24. LeDouarin, N.M., Ziller, C., and Couly, G.F., Patterning of neural crest derivatives in the avian embryo: *in vivo* and *in vitro* studies, *Dev. Biol.*, 159, 24, 1993.
25. Ma, Z.Q., Spreafico, E., Pollio, G., Santagati, S., Cattaneo, E., and Maggi, A., Activated estrogen receptor mediates growth arrest and differentiation of a neuroblastoma cell line, *Proc. Natl. Acad. Sci. U.S.A.*, 90, 3740, 1993.
26. Santagati, S., Ma, Z.Q., Ferrarini, C., Pollio, G., and Maggi, A., Expression of early genes in estrogen induced phenotypic conversion of neuroblastoma cells, *J. Neuroendocrinol.*, 7, 875, 1995.
27. Maggi, R., Ma, Z.Q., Pimpinelli, F., Maggi, A., and Martini, L., Decrease of the number of opioid receptors and the responsiveness to morphine during neuronal differentiation induced by 17beta-estradiol in estrogen receptor-transfected neuroblastoma cells (SK-ER3), *Neuroendocrinology*, 69, 54, 1999.
28. Agrati, P., Ma, Z.Q., Patrone, C., Picotti, G.B., Pellicciari, C., Bondiolotti, G., Bottone, M.G., and Maggi, A., Dopaminergic phenotype induced by oestrogens in a human neuroblastoma cell line, *Eur. J. Neurosci.*, 9, 1008, 1997.
29. Ma, Z.Q., Santagati, S., Patrone, C., Pollio, G., Vegeto, E., and Maggi, A., Insulin-like growth factors activate estrogen receptor to control the growth and differentiation of the human neuroblastoma cell line SK-ER3, *Mol. Endocrinol.*, 8, 910, 1994.
30. Patrone, C., Ma, Z.Q., Pollio, G., Agrati, P., Parker, M.G., and Maggi, A., Cross-coupling between insulin and estrogen receptor in human neuroblastoma cells, *Mol. Endocrinol.*, 10, 499, 1996.
31. Patrone, C., Gianazza, E., Santagati, S., Agrati, P., and Maggi, A., Divergent pathways regulate ligand-independent activation of ER alpha in SK-N-BE neuroblastoma and COS-1 renal carcinoma cells, *Mol. Endocrinol.*, 12, 835, 1998.

32. Garnier, M., Di Lorenzo, D., Albertini, A., and Maggi, A., Identification of estrogen-responsive genes in neuroblastoma SK-ER3 cells, *J. Neurosci.*, 17, 4591, 1997.

33. Boyd, J.M., Malstrom, S., Subramanian, T., Venkatesch, L.K., Schaeoer, U., Elangovan, B., D'Sa-Eipper, C., and Chinnadurai, G., Adenovirus E1B 19 kDa and Bcl-2 proteins interact with a common set of cellular proteins, *Cell*, 79, 341, 1994.

34. Bettini, E. and Maggi, A., Estrogen induction of cytochrome *c* oxydase subunit III in rat hippocampus, *J. Neurochem.*, 58, 1923, 1992.

35. Van Itallie, T.J. and Dannies, P., Estrogen induces accumulation of the mitochondrial ribonucleic acid for subunit II of cytochrome oxidase in pituitary tumor cells, *Mol. Endocrinol.*, 2, 332, 1988.

36. Pollio, G., Agrati, P., Meda, C., Piepoli, T., and Maggi, A., *in preparation*.

37. Behl, C., Widmann, M., Trapp, T., and Holsboer F., 17- estradiol protects neurons from oxidative stress-induced cell death in vitro, *Biochem. Biophys. Res. Commun.*, 216, 473, 1995.

38. Green, P.S., Gridley K.E., and Simpkins, J. W., Nuclear estrogen receptor-independent neuroprotetion by estratrienes: a novel interaction with glutathione, *Neuroscience*, 84, 7, 1995

39. Goodman, Y., Bruce, A.J., Cheng, B., and Mattson M.P., Estrogens attenuate and corticosterone exacerbates excitoxicity, oxidative injury, and amyloid β-peptide toxicity in hippocampal cells, *J. Neurochem.*, 66, 1836, 1996.

40. Garcia-Segura, L.M., Cardona-Gomez, P., Naftolin, F., and Chowen, J.A., Estradiol upregulates Bcl-2 expression in adult brain neurons, *NeuroReport*, 9, 593, 1998.

4

Steroid Metabolism in the Brain: Role in Sexual Differentiation

Paola Negri-Cesi, Angelo Poletti, Luciano Martini, and Flavio Piva

CONTENTS

I. Introduction

The development and the differentiation of the brain involve an extremely large complex of events, which begins during gestation and continues, at least in some species, in the early postnatal period. One of the major aspects of these processes is represented by the imprinting mechanisms which induce permanent structural/morphological and functional sexual differences in some specialized structures of the central nervous system (CNS). A large group of data, mainly obtained in rodents, indicates that a particular nucleus localized in the preoptic area (the sexually dimorphic nucleus of the preoptic area, SDN-POA) shows a high degree of sexual dimorphism, being several times greater in males than in females [see Reference 1 for references]. A structural dimorphism is present also in a series of other brain nuclei; among these, one may quote the accessory olfactory bulb, some subnuclei of the bed nucleus of the stria terminalis, and the anteroventral periventricular nucleus. The volumes of these nuclei differ between males and females.[1] There are also examples of sexual dimorphism in the motoneuron system such as, for instance, that occurring in the spinal nucleus of the bulbocavern- osus, a group of motoneurons in the lumbar spinal cord of mammals inner- vating the bulbocavernosus and the levator ani muscles, which obviously develop only in male animals.[1] Some sexual-related dimorphic morphologi- cal structures have been found also in humans. It is known, for example, that at least two of the interstitial nuclei of the anterior hypothalamus (INAH-2 and -3, which are located in the SDN-POA area) are greater in men than in women,[1] and that INAH-3 is smaller in homosexual than in heterosexual men.[2] Dimorphism has been shown also for other human brain nuclei; how- ever, the data have not been fully validated, since some of the studies have been performed on autoptic material obtained from AIDS patients; it is known that HIV is a virus, which may directly induce morphological alter- ations of brain structures.[3] It is also interesting to note that some pathologies seem to occur more frequently in one of the two sexes. For instance, Parkin- son's disease is a characteristic of men, while depression, insomnia and other psychological disturbances (e.g., anorexia nervosa) are the prerogative of women.

In addition to these and possibly other morphological differences, several gender-specific physiological responses have been reported. These include endocrine as well as behavioral functions, such as the dimorphic regulation of gonadotropin,[4] of somatostatin,[5] and of GH[6] (growth hormone) secretion, the dimorphic response of the adrenal axis to stressors,[7] the dimorphism of sexual and aggressive behavior.[8] All of these may depend not only on a dif- ferential morphological development of the CNS, but also on sex-related dif- ferences of one or more neurotransmitter systems. A sexual dimorphism has been indeed reported for several central neurotransmitters, for example, adrenaline, noradrenaline, dopamine, and serotonin;[9-13] more recently, data have also been reported showing that the pattern of organization of a variety of brain neuropeptide systems is sex related. This has been demonstrated for substance P,[14] CCK (cholecystokinin),[15] vasopressin,[16] and VIP (vasoactive intestinal polypeptide).[17]

On the basis of the observation that testicular hormones, like testosterone and dihydrotestosterone (DHT), are fundamental for the development of the mammalian internal and external genitalia towards a male pattern, about 40 years ago it was proposed that androgens, acting in the very early periods of the fetal or neonatal life, could be responsible also for the permanent organi- zation of the developing CNS toward masculine patterns.[18] The concept that the prenatal exposure of the brain to endogenous testosterone might direct the organization of the CNS toward a male direction is supported by the old observation that the implantation of testicular tissue, or the administration of exogenous testosterone, to neonatal female rats may masculinize their brain both anatomically and functionally. For instance, the very early administration of testosterone leads to a larger, more masculine-like organization of the SDN-POA;[19] moreover, the same steroid induces, when given very precociously, a male pattern of gonadotropin and GH secretion, and of sexual behavior.[19,20] Conversely, orchidectomy, performed in the neonatal male rat, results in a smaller, feminine-like appearance of the SDN-POA,[19,20] and induces permanently a female type of gonadotropin release and of sex behavior.[21] Similar data have been reported for many nonprimate species [see Reference 22 for references]. It is still controversial whether the same mechanism also operates in primates[23] and in humans[1,24] (see below).

An important aspect, which has emerged in the last 30 years, is that the brain in general, and some specialized CNS structures in particular, may metabolize hormonal steroids. Several enzymatic systems have been described, and some of them have been fully characterized. Among these, some have the peculiarity of totally changing the molecular behavior of the substrate (e.g., the aromatization of testosterone to estradiol), while others have the property of enhancing its activity (e.g., the 5α-reductase, 5α-R, which converts testosterone to the more androgenic compound 5α-DHT); finally, some enzymes (e.g., the 3α-hydroxy steroid dehydrogenase, 3α-HSD) may modify steroids so that the resulting metabolites may eventually interact with receptors other than the classical intracellular steroid receptors (e.g., the 3α-hydroxylated derivatives of 5α-dihydroprogesterone, 5α-DHP, and 5α-deoxycorticosterone, 5α-DOC, which bind and activate the GABA$_A$ receptor,[25,26] see below).

A. Mechanism of Action of Hormonal Steroid Metabolites in the Developing Brain

Steroids may affect brain functions, both during the fetal–perinatal period and in adulthood, because several types of steroid-binding sites are present in the CNS. The best studied among these are the classical intracellular receptors which are ligand-activated transcription factors; however, more recently, specific membrane binding sites for some families of steroid hormones have been described.[27] These may be of importance for explaining some rapid (seconds or fraction of seconds) and transient effects (e.g., electric phenomena) induced by steroids, which cannot be explained by the slower genomic action; they may also justify some estrogen-mediated expression of proteins whose genes do not contain estrogen-responsive elements (ERE). Finally, hormonal steroids and their metabolites may interact with receptors proper of some neurotransmitters. The example of the 5α-, 3α-reduced derivatives of progesterone and of DOC acting on the GABA$_A$ receptor has already been mentioned, but there are also interactions with the sigma receptor[28] and with the receptors of endogenous opioids.[29] It should also be recalled that steroid effects may also be obtained on classical receptors in the absence of the specific ligands. The cross-talk of steroid receptors with growth factors,[30,31] neurotransmitters,[32] as well as with cAMP,[33] and other second messengers are of particular interest at the moment. Another interesting interaction occurs between steroids and neurotrophic peptides through the cross-coupling of their signaling pathways.[34] For instance, there is an increasing body of evidence suggesting that estrogens are able to activate the MAP (mitogen activating protein) kinase cascade (one of the major signal transduction cascades activated by growth factors and involved in differentiative processes) by activating ERK-1 (extracellular-signal regulated kinase) and ERK-2,[35] with a mechanism that does not seem to involve the binding of estradiol with the estrogen receptor (ER),[36] at least the ERα, which is the only one studied so far in this respect (see below). Moreover, it has been demonstrated that in those CNS structures in which ERs and the membrane-bound receptors for neurotrophic peptides are coexpressed,[37,38] there is the possibility of a reciprocal regulation (see Reference 39 for references).

Classical intracellular receptors for each family of hormonal steroids have been shown to be present not only in the adult brain, but also in many structures of the developing rodent brain. For instance, the mRNA for the glucocorticoid receptor (GR) is expressed in the fetal striatum;[40] progesterone receptors (PR) are present in a typical dimorphic structure of the brain, like the medial preoptic nucleus;[41] the receptors for androgens (AR) are widely distributed in the fetal/neonatal brain even if they show peak concentrations in the areas controlling the reproductive system.[42,43] There is increasing information on the presence and on the distribution in the brain of the classical ER (now called ERα) as well as of the recently discovered ER subtype β.[39,44] The ERβ, which is particularly expressed in the periphery at the level of the prostate, the ovary, the bladder, and the lungs,[45] is more largely represented in the brain than ERα. Because of this, the possibility exists of synthesizing estrogenic compounds that, by interacting with this receptor subtype, might act specifically on the CNS, possibly without affecting other estrogen-dependent structures.[46] The presence of the ERβ in the brain and particularly in the hypothalamic-preoptic area[44] suggests that this receptor might play a role in the control of the sexual differentiation of the central structures, even if the expression of the two ER isoforms seems to overlap in many regions of the developing rat brain.[39] Since ERα and ERβ can act on the ERE of the responding genes not only as homodimers, but also as heterodimers,[46] it is indeed possible that some aspects of brain differentiation and/or function could be triggered only if the ERβ is present. Interestingly, it seems that the β form is the only ER expressed in the embryonal mouse CNS [Gufstafsson, J.-A., personal communication, 1999]. However, the precise role of ERβ in the mechanisms leading to the sexual differentiation of the brain is still unclear, since the first generation of ERβ-knockout male mice appear to exhibit normal male reproductive functions and male sexual behavior.[47] On the contrary,

ERα-knockout (ERKO) male mice display severe reproductive and behavioral abnormalities, including an alteration of aggressive behavior, and this despite the presence of a normally functional ERβ.[48] It is interesting to note that while reproductive behaviors (as, for example, mounting, intromission, etc.) are restored in adult castrated ERKO male mice by the administration of testosterone, the same treatment is ineffective in inducing aggressive behavior.[48] These findings indicate that the interaction of testosterone-derived estrogens with ERα is essential to control "motivational," but not "consummatory" aspects of male sexual behavior, which, instead, seem to be regulated either by the interaction of estradiol with the ERβ or by the 5α-reduced metabolites of testosterone.[48] Finally, it is interesting to note that some xeno- and phytoestrogens can act as specific agonists or antagonists of the ERβ, depending on their chemical structure.[49] This could account for the putative effects of estrogen-like environmental chemicals on the organization of some specific central mechanisms involved in controlling reproduction.[50] However, the possible influence of xenoestrogens on the sexual development of the brain is far from clear, as it will also appear from the data quoted in another section of this chapter. This discussion on the two types of ERs is particularly relevant since, as it will appear from the data quoted below, the majority of the masculinizing effects of androgens are due to their local conversion into estrogens, via the process of aromatization catalyzed by the enzyme aromatase (Aro).

It is interesting to note that the presence of receptors for hormonal steroids is not limited to neurons, but that these are also found, in similar concentrations, in glial elements (e.g., astrocytes, oligodendrocytes, Schwann cells).[51-54] This observation may have important physiological implications because of the close cross-talk existing between neurons and glial elements.

The present chapter will describe in some detail the most important metabolic pathways involved in mediating the effects of steroid hormones in the developing CNS. The chapter will deal in particular with the Aro, with the two isoforms of the 5α-R and with the 3α-HSD, taking into account particularly the data obtained in the authors' laboratory. Some data obtained in the authors' laboratory, and others recently appeared in the literature regarding the interaction between estrogens, androgens, neurotransmitters, and external influences on brain sexual differentiation will also be briefly reviewed.

II. Androgen Metabolizing Enzymes

A. General Overview

The importance of steroid metabolism in the brain is supported by the fact that both the aromatizing and the 5α-reductase pathways are phylogenetically well conserved and can be traced back millions of years. Aro has been detected in brain tissue from representatives of each major vertebrate group back to cyclostomes, and its activity appears to be restricted to the phylogenetically older basal forebrain regions. The available phylogenetic evidence indicates that 5α-R is even older than Aro, being present in the neural tissue of cyclostomes and invertebrates;[55] apparently, this enzyme is also more widely distributed in the brain than Aro.

The physiological functions of Aro are relatively well known, at least in rodents, since estrogens derived intracellularly from androgens represent the major effectors of brain masculinization, acting as the permanent organizers of the different brain systems involved. The data supporting the "aromatization hypothesis" are as follows: (1) aromatizable androgens mimic the masculinizing effect of testosterone in the nervous centers of neonatally treated female rats,[56] and are also able to prevent the demasculinizing effect of neonatal orchidectomy in the male rat;[57] (2) 19-hydroxytestosterone (an intermediate in the aromatization process) is a more potent masculinizing agent than testosterone when administered to neonatal female rats;[58] (3) estradiol and other steroidal estrogens,[59] given to neonatal females, have a masculinizing effect at doses much lower than those of testosterone, especially if administered directly into the hypothalamus;[60] in particular, estrogens appear to masculinize the SDN-POA, an effect which may be antagonized by antiestrogens;[61] (4) pretreatment of newborn females with antiestrogens,[62] with antisense oligonucletides to the ER mRNA,[63] or with inhibitors of the aromatization process[64] counteract many aspects of testosterone-induced masculinization; (5) estrogens and aromatizable androgens stimulate neurite outgrowth and differentiation in culture slices of the hypothalamic/preoptic area of the newborn mouse and rat (see Reference 39 for references), whereas nonaromatizable androgens are ineffective.[65]

B. The Aromatase

The transformation of androgens into estrogens is catalyzed by Aro, an enzymatic complex present in several regions of the developing brain, especially in the preoptic/hypothalamic area and in the limbic system.[66] Aro is a membrane-bound enzymatic complex, located in the endoplasmic reticulum, which binds with high affinity (in the nanomolar range) and converts delta4, 3oxo-androgens into estrogens. Testosterone and androstenedione are the major physiological substrates which are converted into estradiol and estrone, respectively.

Aro is the product of the CYP 19 gene, and is a member of the P-450 cytochrome superfamily. The CYP 19 gene (isolated from a human genomic DNA library) is located on the long arm of chromosome 15, spans at least 70 Kbp, is composed of 10 exons, the first of which (exon I) is untranslated and presents a high degree of heterogeneity.[67] As a matter of fact, exon I presents several subtypes,[68,69] that give origin by alternative splicing to different mRNAs.

The process is triggered by different tissue-specific promoters, which are located upstream of each exon I subtype. Many studies on the human, monkey, and rat brain have revealed that the transcript of a unique exon I (exon I-f) of Aro is found in the brain.[69,70] The specific promoter region of this exon contains a potential androgen/glucocorticoid binding site,[68] as well as a consensus sequence for SF-1, a factor involved in the regulation of the genes of all steroidogenic enzymes, and which during differentiation acts directly not only on steroid-producing glands, but also on the pituitary and on some CNS nuclei.[71] The possible interactions of SF-1 with estrogens and other steroids in the control of the sexual differentiation of the brain deserve further attention.

Until recently Aro — at variance with the 5α-R — seemed to be expressed in physiological conditions only in neurons.[72-74] However, it has been shown that Aro is present also in glial cells of the developing zebra finch telencephalon;[75] moreover, following a neurotoxic lesion of the rat brain, the aromatization of androgens may take place also in reactive glial elements, mainly composed of astrocytes.[76] Since ERs are localized not only in neurons, but also in astrocytes, and in ependymal and in endothelial cells,[53,77] estrogens may participate in the control of the sexual differentiation of the brain by acting in a very complex manner; in other words, they may act simultaneously as intracrine or paracrine modulators of neuronal and nonneuronal functions.

The colocalization of Aro and estrogen receptors has been studied in different animal species;[78,79] unfortunately, at the time of those studies the existence of the ERβ was unknown. The two proteins appear to be colocalized in the majority of fetal and neonatal neurons in some species, but not in others; therefore, during embryogenesis, an intracrine rather than a paracrine role of estrogens in neurons deserves further investigation. Moreover, the colocalization of Aro with ERβ should be carefully evaluated.

Many studies have been devoted to quantify the possible sex-specific differences in the expression and activity of Aro during development. It is generally agreed that the enzyme levels are higher in the male than in the female brain,[80-82] possibly not because of a higher expression of Aro per neuron,[83] but because males possess a higher number of neurons expressing the enzyme.[84] As it will be reported also for the 5α-R type 2 (see below), the expression of Aro is generally higher in the perinatal than in the adult brain of experimental animals.[66,84,85] This appears to be true also in humans, as shown by Honda and co-workers[68] using a quantitative RT-PCR analysis.

During development, Aro mRNA expression and activity may show some variations in different regions of the brain. It has been reported that in the fetal rat preoptic/hypothalamic area, both Aro mRNA and activity, which are low on gestational days (GD)15-16, increase to peak on GD19-20 and then gradually decrease during the perinatal period, reaching the low levels observed in adulthood.[66,74,85] These studies, however, have been conducted without grouping the data for the two sexes. To the authors' knowledge, there is only one paper in the literature in which Aro expression has been evaluated separately by *in situ* hybridization in male and in female embryos.[86] With this technique, no differences have been observed in the two sexes at GD18 and GD20. In our laboratory, the semiquantitative expression of the enzyme was evaluated by RT-PCR, coamplifying mRNAs of Aro and of a typical neuronal marker (the isoform 2c of the microtubular-associated protein, MAP2c), in the hypothalamic area of rat embryos, whose sex had been identified by the genetic screening of the marker gene (SRY) of the Y-chromosome.[87] The results obtained on different days of gestation indicate that Aro shows a dimorphic pattern of expression during embryogenesis. In the female hypothalamus, Aro expression is steadily low throughout the whole gestational period considered; on the contrary, in males, the enzyme is low at GD16, presents a peak of expression on GD18, drops to very low levels at GD19, and recovers its high levels on GD21.[88] This result, if confirmed, would be of extreme importance because it would clearly indicate that the masculinizing enzyme is specifically present in males at the time which is critical for the sexual differentiation of the brain.[20,21] The dimorphic expression of hypothalamic Aro finds support also in the observation that in the mouse the expression of the enzyme is higher in males than in females in the postnatal period.[89] Unfortunately, this paper did not take into consideration the gestational period between GD17 and the day of birth when, in our experiments, the typical male peak of expression of Aro takes place. As opposed to the hypothalamus, Aro expression in other brain regions, like, for example, the bed nucleus of the stria terminalis and the amygdala does not show any variation during the prenatal period but remains at the same levels throughout development and adulthood.[66,74] These results indicate that the expression of this key enzyme within the brain is controlled by regional-specific factors.

Despite the above-mentioned patterns of expression of Aro during the embryogenesis, the factors involved in regulating the enzyme are still poorly understood, at least in species other than the mouse. Because of the close relationship between testosterone secretion by the fetal testes and Aro expression, many studies have been performed to assess the possible dependence of the enzyme on androgens. However, the data present in the literature are conflicting regarding the effect exerted by androgens on Aro activity/expression (stimulatory vs. inhibitory).[82,90-93] In particular, Lephart and co-workers,[92] using rat fetal hypothalamic explants maintained *in vitro* for 48 h, showed a significant and dose-dependent decrease in Aro activity induced either by testosterone or DHT; this effect seems to be specific since, under identical incubation conditions and steroid concentrations, estradiol, progesterone, and corticosterone were ineffective.[92] On the contrary, in the mouse, a stimulatory effect of testosterone has been thoroughly demonstrated.[89,93] Finally, it has also been shown that Aro expression, evaluated in cultured mouse diencephalic neurons, is insensitive to testosterone.[94] These opposite results are probably due to differences either in the species used or in the experimental approaches adopted. To assess whether the testosterone surge (which in male rat embryos begins on GD14[95]) might modulate the expression of hypothalamic Aro, we have performed a study in which the antiandrogen flutamide was given daily to pregnant rats beginning on GD13 at a dose known to cross the placental barrier and to block androgen binding to AR[87,96] (see below). Aro expression was then evaluated in the hypothalamus of male and female embryos on GD18, i.e., when in males the expression of Aro is maximal. The results obtained indicate that this *in utero* treatment is unable to affect the Aro surge in male fetuses.[88] This finding suggests that factors different from androgens, produced by neurons themselves or by other brain cell populations, might modulate the expression of Aro. In this context, it is interesting to recall that the brain Aro can be modulated by factors causing an increase of cAMP[92] or by β-receptor stimulation.[97] The developmental increase of Aro mRNA in the brain might also be a genetically determined endogenous characteristic. Indeed, it has been demonstrated that diencephalic neurons taken from 13-day-old mouse embryos, and maintained *in vitro* for a week, show a profile of Aro expression which goes in parallel with that found *in vivo*.[98] Finally, the activation of some transcription factors, for example SF-1, in the developing hypothalamus, might participate in the modulation of Aro expression. As already mentioned, a consensus sequence for SF-1 has been shown to exist on the brain-specific Aro promoter,[70] and the appearance of this transcription factor precedes that of Aro (around GD11, in the mouse) in those cells that will give origin to the hypothalamus.[99]

C. The 5α-Reductase, 3α-Hydroxysteroid Dehydrogenase Complex

The brain, like many other androgen-responsive tissues in the body (e.g., prostate, seminal vesicles, etc.) has been shown to possess the enzyme 5α-R. Actually, the activity of this enzyme is rather consistent not only in the CNS of mammals, but also in several other species.[22,100,101] This enzymatic activity is usually more pronounced in brain regions which, like the hypothalamus and the midbrain, are particularly rich in white matter; consequently, this preferential localization of 5α-R might be due to the peculiar presence of the enzyme in the myelin.[100] A possible role for 5α-R and its derivatives in the sexual differentiation of the brain has been suggested by the observation that, in the prenatal period, the enzyme is present in high concentrations not only in the structures just mentioned, but also in the cerebral cortex.[102,103] The enzyme activity appears to diminish in the weeks following birth.[102,103] However, as previously mentioned, the role of nonaromatizable androgenic steroids in the control of the sexual differentiation of the brain is still obscure, although DHT, which is a nonaromatizable androgen formed in the brain via the 5α-reduction of testosterone, has been shown to exert specific organizational roles on selected neuronal populations[104,105,106] (see below). Therefore, this steroid might be involved in the processes of sexual differentiation of some brain regions.

Two isoforms of the enzyme 5α-R have been cloned; they catalyze the same reaction, but possess different biochemical and pharmacological properties, distinct cell and tissue-specific patterns of expression,[107,108] and different subcellular distribution.[109,110] While the type 1 isoform is widely expressed in various tissues (with the highest levels in the liver), the type 2 isoform appears to be selectively concentrated in "classical" androgen-dependent structures, like the prostate, the seminal vesicles, etc.;[111] hence, the two isozymes may play distinct physiological functions.

Very few data were available until recently[87] regarding the presence and the distribution of the two 5α-R isoforms in the brain of prenatal, neonatal, and adult animals. Also the possible mechanisms of their control in different phases of life were unknown. The expression of the two isoforms of the 5α-R in the rat brain has been recently analyzed in this laboratory in studies performed *in vivo* at different stages of development.[87] The gene expression of the two 5α-R isozymes has been analyzed in fetal, postnatal, and adult rat brains by RT-PCR followed by Southern analysis. It has been found that the 5α-R type 1 mRNA is always detectable in the rat brain (from GD14 to adultood) and shows only a small increase at time of birth. On the contrary, the 5α-R type 2 mRNA expression is undetectable on GD14, increases after GD18, peaks on postnatal Day 2, then decreases gradually, becoming low in adulthood. Since this pattern of expression appears to be correlated with the rate of production of testosterone by the testis, the possible control by androgens of the gene expression of the two isozymes has been studied in brain tissues of animals exposed *in utero* to the androgen receptor antagonist flutamide; in these experiments, the sex of the animals was determined by the genetic screening of SRY gene on the Y chromosome. In the brain of male embryos, flutamide treatment inhibited the expression of 5α-R type 2, an effect which was much less pronounced in females; on the contrary, 5αR type 1 expression was not influenced by flutamide (and consequently by androgens) either in male or in female animals.[87] The latter finding is reminiscent of the lack of effect of endogenous androgens in the control of the expression of Aro (see above). Androgens may then represent the triggering element for the expression of the type 2 isoform in males, while other control mechanisms are probably involved in the control of 5α-R type 1 in males, and of both isoforms in females (growth factors? neurotransmitters? other steroids? etc.). The data obtained *in vivo* and showing that the 5α-R type 1 is constitutively expressed in the whole brain of the rat CNS at all stages of brain development agree with recent results obtained in mice.[89] To confirm whether androgens might be essential in controlling the expression of the 5α-R type 2, a more direct approach was chosen: we have treated cultured rat hypothalamic neurons, which in basal conditions are known to be devoid of 5α-R type 2 expression,[73] with testosterone or its active metabolites (DHT and estradiol); the results of the RT-PCR analysis have shown that 5α-R type 2 in hypothalamic neurons is highly induced by testosterone or DHT treatment, while estradiol was completely ineffective. The same treatments did not modify 5α-R type 1 expression in these cells.[87]

Taken as a whole, these results underline the existence of a selective ontogenetic pattern of regulation of the two 5α-R genes in the brain by androgens, and generally agree with the data obtained in classic androgen-dependent structures of the urogenital tract. In fact, in those structures androgens induce

5α-R type 2 mRNA, but not 5α-R type 1.[112] This seems to be peculiar of the perinatal period since it has been shown that in the prostate of adult rats, both isozymes are controlled by androgens.[111,112]

It is therefore possible that the 5α-R type 2 may be involved, along with the Aro, in the control of brain sexual differentiation during a very critical period, when the androgen-organizing effects are thought to take place in the CNS. Although it is generally accepted that the aromatization of testosterone into estrogens is responsible for the sexual differentiation of the brain toward the male pattern, DHT seems to be essential, alone or together with estrogens, for the development and the organization of selected neuronal populations, such as the SDN of the accessory olfactory tract[113] and the sexually dimorphic spinal nucleus of the bulbocavernosus.[106] Moreover, testosterone and DHT are morphogenetic signals for developing hypothalamic neurons containing Aro,[105,106] and may influence the plasticity and the synaptic connectivity of hypothalamic Aro positive neurons.[114] These effects are AR mediated, because they are suppressed by the antiandrogen flutamide, but not by the antiestrogen tamoxifen.[114] Moreover, the fact that AR concentration is particularly elevated in the rat brain around birth, and is higher in male than in female neonatal animals,[42] supports the concept of a direct effect of androgens on some developmental phenomena. It is the authors' opinion that the role of DHT in brain differentiation has not yet been fully elucidated, possibly because its formation in the CNS is transient, at least for the amounts contributed by the high-affinity type 2 isozyme, and probably limited to some specific CNS structures. It is conceivable that testosterone acts as differentiating agent via estrogens in some brain areas, nuclei, or even single neurons, and via the formation of DHT in others. Moreover, due to the close temporal relationship between the expression of 5α-R type 2 and of Aro (see above) a possible control of DHT on Aro could not be excluded. Very little is known on the participation of steroids other than androgens and estrogens in the differentiation of the brain. Very old studies have indicated that the administration of progesterone might counteract the masculinizing effect of testosterone in the neonatal or perinatal rat.[115] At that time it was thought that progesterone might be endowed with antiandrogenic properties. The data may now be interpreted in a different way, since it is known that progesterone is a preferential substrate for both 5α-R type 1 and 2.[111,116] The information on the possible roles of gluco- and minerocorticoids is even smaller. The study of this problem should be reinitiated, especially on the basis of the fact that progesterone and gluco- and minerocorticoids are substrates of the 5α-Rs and may, consequently, be transformed into their respective 5α-reduced derivatives. As previously mentioned, the conversion of testosterone into DHT, of progesterone into DHP, and of the gluco- and minerocorticoids into the respective 5α-reduced metabolites prepares the substrates for the second enzyme: the 3α-HSD.

The 3α-HSD catalyzes the 3-hydroxylation of 5α-reduced steroids such as DHT, DHP, and gluco- and minerocorticoids to the corresponding tetrahydroderivatives. This enzyme is widely distributed in the adult rat brain with the highest levels in the olfactory bulb, and its main localization is in the type 1 astrocytes (see Reference 117 for references). To the authors' knowledge, the ontogenetic profile of this enzyme has not been studied yet; however, indirect evidence obtained in neuronal and glial cell cultures obtained from fetal/neonatal brain indicates that it is already functional around the time of birth.[118] The presence of this enzyme might be of paramount importance for linking the steroidal effects occurring via the classical intracellular steroid receptors with other effects of steroids due to the binding of their metabolites to neurotransmitter receptors. It is indeed known that the 3α-HSD is responsible for the formation of 3α-hydroxy, 5α-pregnan-20-one (allopregnanolone or tetrahydroprogesterone, THP), and of 3α,21-dihydroxy-5α-pregnan-20-one (tetrahydrodeoxycorticosterone, THDOC), which are two potent neurally active steroids. These two 5α-reduced-3α-hydroxylated compounds are unable to interact with the classical intracellular receptors for progesterone and for the minerocorticoids, but their action may be mediated by the GABA$_A$ receptor.[25] The activation of this receptor may explain their potent anesthetic/anxiolytic activities. The peak of expression of 5α-R type 2 around the time of birth might therefore be crucial in providing the 3α-HSD with possible precursors for the final formation of active anxiolytic/anesthetic steroids involved in modulating the stress responses which occur at the time of parturition, i.e., a time at which also progesterone levels are particularly elevated.[25,119] It is interesting that during gestation the GABA$_A$ receptors are highly sensitive to modulators,[120] and that the hormonal manipulations which alter the sexual differentiation of the perinatal brain (castration in males, testosterone administration in females) modify the normal response of the animals to stressor stimuli.[121]

The finding that 5α-R type 1 is constantly present at all stages of development, combined with the observation that this isoform actively metabolizes androgens, progesterone, and the corticoids only when they reach high concentrations inside the cells (because of the low affinity for the substrates which is in the micromolar range)[111] leads to the hypothesis that this isoform might play essentially a catabolic function, protecting neurons from the excess of harmful levels of steroid hormones.[122,123] A protective role for 5α-R type 1 has been recently demonstrated in transgenic mice carrying a mutated inactive form of the 5α-R type 1.[124,125] These animals have higher mortality during gestation due to an excess of estradiol levels, derived from increased testosterone bioavailability for Aro, not removed by 5α-reduction.[125] The lethal effect of estrogens, which is probably linked to the negative morphological effects reported to be induced by excessive doses of these steroids in the brain by Naftolin,[126] is in essence contradictory to the universally accepted organizational effects of estradiol. Obviously a localized vs. a generalized effect, the doses, etc. might be crucial factors in directing these dual roles (organization vs. lethality) of estradiol.

III. Possible Influence of Xenoestrogens on Sex-Brain Differentiation

In addition to the "physiological" steroids, the increasing number of artificial hormonoids which is accumulating in the environment may contribute to modify or modulate the physiological development of the neuroendocrine brain. These may be represented by molecules that reach the body through the diet (mainly represented by herbal compounds, phytoestrogens), or by anthropogenic chemicals prepared for a variety of purposes. They may act on the neuroendocrine brain either as the original molecules, or through their metabolites that still retain a biological activity. Among phytoestrogens, particular attention has been devoted to the isoflavone genistein and its derivatives, and to the resorcilic acid lactone zearalenone. Both classes of these compounds are present in a variety of plants. They induce, in female rats, an enlargement of the SDN-POA which reaches a dimension similar to that observed in male animals.[127] This is reminiscent of the effect exerted by aromatizable androgens on the SDN-POA, when given perinatally to female rats. As already mentioned, this effect takes place after the conversion of androgens to estrogens. An estrogenic effect is displayed also by the insecticide chlordecone, which induces persistent vaginal estrus in female rats treated neonatally with this molecule.[128,129]

It is interesting to add that many synthetic compounds prepared for a variety of applications and possessing estrogenic properties have been shown to affect sexual differentiation in nonmammalian species also, such as fish, alligators, turtles, seagulls, etc.[130]

Many other anthropogenic compounds appear to act as antiandrogens.[129,131,132] For example, insecticides that are widely distributed in the environment such as DDT and its metabolite DDE, as well as methoxychlor, are probably responsible for the feminization of the western gulls of the Los Angeles area and of Florida alligators.[133-135] Also the fungicide vinchlozolin and its metabolites, which possess antiandrogenic properties, bring about male offspring with reduced mounting behavior, when given to pregnant rats.[136] These data may be taken as an indirect proof that, at least in these species, androgens might act as such in some processes linked to the sexual differentiation of the brain; the inactivation of ARs eliminates indeed the possibility of a real androgenization.

A demasculinizing, feminizing effect has been reported also for the herbicide dioxin. Male offspring from pregnant rats treated with this toxin present low levels of circulating lutenizing hormone (LH); furthermore, when males are castrated and treated with estrogen and progesterone, they exhibit a lordosis quotient similar to that recorded in females.[137-140] However, it has not been clarified yet whether dioxin acts on the neuroendocrine brain directly, or via a modification of androgen production from the fetal testes.

Finally, polychlorinated biphenyls (PCBs), synthesized mostly for use in electrical equipment, affect adult sex differentiation in turtles; eggs exposed to PCBs produce only females even if incubated at a temperature that would produce males.[141] The capacity of PCBs to bind estrogen receptors may be the basis of this sex shift.[142]

IV. Role of Testosterone Metabolites in the Sexual Dimorphism of Neurotransmitter Systems

In the 1980s, a series of papers was published suggesting that the hypothalamic organization of the opioid system might also be sexually dimorphic.[143,144] This issue has been elucidated in several experiments performed in the authors' laboratory. In a first series of studies, it was decided to analyze whether the response of prolactin (PRL) secretion to opioid modulation was similar in male and female rats, since there was evidence that the neural mechanisms controlling PRL secretion were different in the two sexes. It has been found that an acute administration of naloxone, an aspecific blocker of the different subclasses of opioid receptors (μ, κ, and δ), induces a significant decrease of serum PRL levels in adult and prepubertal male rats, but not in females of matching age. In order to analyze whether this dimorphic response to naloxone was also organized in the neonatal brain through the classical prenatal effect of estrogens, an experiment was performed on four groups of animals: (1) normal males, (2) normal females, (3) females treated during the neonatal critical period with testosterone, and finally (4) males castrated at birth. The results obtained have shown that the acute injection of naloxone significantly decreased serum PRL levels in normal males and in androgenized females at all ages considered before and after puberty (16, 26, and 60 days of age). On the contrary, the opioid antagonist was always ineffective in normal females and in neonatally castrated males. These data suggest that, in the rat, a sexual difference exists in the opiatergic control of PRL secretion; apparently, the central opioid systems which regulate PRL secretion develop toward a male pattern because of the presence of androgens in the neonatal period.[145]

The role of androgens in determining the sex-linked differentiation of the brain opioid systems was further substantiated by the following experiments performed in our laboratory.[146] It has been investigated whether the presence or the absence of testosterone at time of birth might induce changes of the binding characteristics of the hypothalamic μ-opioid receptors; the maximal binding capacity (B_{max}, an index of the number of receptors), and the affinity constant (K_a) of the specific μ-ligand dihydromorphine were evaluated in hypothalamic plasma membrane preparations derived from normal male rats, normal female rats, male rats orchidectomized 2 days after birth, and female rats treated neonatally with testosterone. The results obtained have shown that at 60 days of age, neonatally castrated male rats had a number of hypothalamic μ-receptors similar to that measured in normal females, and significantly higher than that recorded in the hypothalami of normal males and androgen treated females. The variations here reported took place without any change of the K_a of dihydromorphine for the μ-receptors. These data show a sexual dimorphism of hypothalamic μ-receptor distribution, and suggest that their ontogenetic development may be linked to the presence or the absence of androgens at the time of birth.

V. Are Testosterone Metabolism and the Presence of Aromatase Crucial Also for the Sexual Differentiation of the Human Brain?

Some abnormalities in humans, either pathological or iatrogenic have been studied to evaluate the possible role of fetal androgens on the sexual differentiation of the male brain. Some examples of these pathologies or treatments will be summarized here.

A rather common form of hyperandrogenism in women is represented by *congenital adrenal hyperplasia*; this may occur in two forms, one precocious and the other called late-onset; they are due to an inherited defect of the P-450$_{c21}$ hydroxylase, one of the key enzymes involved in the biosynthesis of cortisol. The deficiency of this enzyme results in a decreased capability of the adrenal gland to form cortisol, and in an accumulation of cortisol precursors which may be converted into androgens.[147] Because of the deficiency of cortisol, the feedback regulation of ACTH secretion is lost, and this increases adrenal size and leads to the overproduction of androgens. The excess of adrenal androgens in the mother during pregnancy induces the virilization of the external genitalia of the female fetuses. Females born to these mothers (and not treated *in utero*, see below) present signs of tomboyism, as well as other aspects of male-oriented behavior.[148] This fact would indicate that, as in rodents, in the human brain the imprinting due to the aromatization of testosterone overrides the normal differentiation of a female brain. However, since these young patients were often raised as males, the role of the sociocultural influence cannot be disregarded. Obviously, this should be a story of the past, since it is imperative now to treat these fetuses *in utero* with dexametasone to prevent the masculinization of the external genitalia and possibly of the brain. The same phenomenon used to occur in female fetuses born to mothers treated with progestagens having androgenic potencies;[149] this treatment has now been abandoned.

Another interesting disease is the *complete androgen insensitive syndrome*, or testicular feminization. This is an X-linked recessive disorder in which there are qualitative and/or quantitative defects in the AR status; genetically XY individuals possess a complete female phenotype, in spite of the presence of a normal testicular steroidogenesis.[150,151] Psychologically, they behave like females, and desire to be assigned to the feminine sex. This occurs even if their brains possess a full aromatization capability and are exposed to an excess of estrogen precursors; obviously, the only sex hormone receptor present is the ER. This type of disease underlines the importance of the socio-cultural experience in determining sex-related behavior in men.

The *genetic deficiency of the 5α-reductase type 2*, or Imperato–MacGinley syndrome, is another critical disease for understanding the mechanisms of the sexual organization of the human brain. The syndrome is due to a point mutation of 5α-R type 2 gene which results in the synthesis of a mutated enzyme unable to convert testosterone into DHT in the androgen-dependent tissues where this isozyme is the predominant form. Males affected by this pathology present ambigous external genitalia at birth and, at the beginning, are brought up as females; however, at the time of puberty, the increased secretion of testosterone from the testes, and the presence of a functional 5α-R type 1, overcomes the deficency of the type 2 isozyme, and leads to a normal male development of the secondary sex characteristics.[152] At the time of puberty, these subjects then claim to be reassigned to the male gender.[153] Since prenatally Aro was working in their brains, these subjects may be quoted as an example of the validity of the theory of aromatization also in humans; moreover, they also prove that the estrogenic imprinting prevails over the social experience. Finally, this pathology also seems to indicate that DHT does not play an important role in the overall sexual differentiation of the fetal human male brain. It emerges from all this that the subjects affected by this disease should be brought up as males, and that their genitalia should undergo plastic surgery in a male direction.

A final answer to this problem was expected from the few cases of men carrying a *genetic defect of Aro*. In this pathology, a mutation in the Aro gene results in an inactive enzyme which is unable to convert androgens into estrogens. Surprisingly, the male subjects carrying this mutation do not present any phenotypic abnormality, except an excessive tallness and an increased level of gonadotropins.[154-156] Even in the absence of Aro, men affected by this pathology display a gender identification and a psychosexual orientation consistent with their genotypic sex.[157] This disease therefore seems to provide evidence showing that the conversion of androgens into estrogens does not exert the same pivotal role in the developing human brain that it plays in rodents.

In summary, the human data favoring the "aromatization hypothesis" appear to have the same weight as those contradicting it. As already mentioned, the sociocultural environment, which is pertinent to humans, may obviously exert major effects. Moreover, it is conceivable that, in higher vertebrates and particularly in primates, the process of sexual differentiation of the brain, which plays such a crucial role for the propagation of the species, might be controlled by multiple mechanisms. The presence and the activation of redundant control systems might allow a differentiation of the brain consistent with the genetic sex, and hence right gender identification and behavior, even when one of these mechanisms is missing or is malfunctioning.

References

1. Cooke, B., Hegstrom, D., Villeneuve, L.S., and Breedlove, S.M., Sexual differentiation of the vertebrate brain: principles and mechanisms, *Front. Neuroendocrinol.*, 19, 323, 1998.
2. LeVay, S.A., A difference in hypothalamic structure between heterosexual and homosexual men, *Science*, 253, 1034, 1991.
3. Lipton, S.A., HIV-related neuronal injury. Potential therapeutic intervention with calcium channel antagonists and NMDA antagonists, *Mol. Neurobiol.*, 8, 181, 1994.
4. Harris, G.B., Sex hormones, brain development and brain function, *Endocrinology*, 75, 627, 1964.
5. Murray, H.E., Simonian, S.X., Herbison, A.E., and Gilliesn G.E., Correlation of hypothalamic somatostatin mRNA expression and peptide content with secretion: sexual dimorphism and differential regulation by gonadal factors, *J. Neuroendocrinol.*, 11, 27, 1999.
6. Jansson, J.-O., Eden, S., and Isaksson, O., Sexual dimorphism in the control of growth hormone secretion, *Endocrine Rev.*, 6, 128, 1985.
7. McCormick, C.M., Furey, B.F., Child, M., Sawyer, M.J., and Donahue, S.M., Neonatal sex hormones have "organizational" effects on the hypothalamic–pituitary–adrenal axis of male rats, *Dev. Brain Res.*, 105, 295, 1998.
8. Ehrhardt, A.A. and Meyer-Bahlburg, H.F.L., Effects of prenatal sex hormones on gender-behavior, *Science*, 211, 1312, 1981.
9. Gorski, R.A., Sexual differentiation of the endocrine brain and its control, in *Brain Endocrinology*, Motta, M., Ed., Raven Press, New York, 1991, 71.
10. Döhler, K.D., The pre- and postnatal influence of hormones and neurotransmitters on sexual differentiation of the mammalian hypothalamus, *Int. Rev. Cytol.*, 131, 1, 1991.
11. Kolbinger, W., Trepel, M., Beyer, C., Pilgrim, C., and Reisert, I., The influence of genetic sex on sexual differentiation of diencephalic dopaminergic neurons *in vitro* and *in vivo*, *Brain Res.*, 544, 349, 1991.
12. Sibug, R., Küppers, E., Beyer, C., Maxon, S.C., Pilgrim, C., and Reisert, I., Genotype-dependent sex differentiation of dopaminergic neurons in primary cultures of embryonic mouse brain, *Dev. Brain Res.*, 93, 136, 1996.
13. Simerly, R.B., Zee, M.C., Pendleton, J.W., and Korach, K.S., Estrogen-receptor-dependent sexual differentiation of dopaminergic neurons in the preoptic region of the mouse, *Proc. Natl. Acad. Sci. U.S.A.*, 94, 14077, 1997.
14. Malsbury, C.W. and McKay, K., Neurotropic effects of testosterone on the medial nucleus of the amygdala in adult male rats, *J. Neuroendocrinol.*, 6, 57, 1994.

15. Micevych, P.E., Park, S.S., Akesson, T.R., and Elde, R., Distribution of the cholecystokinin-immunoreactive cell bodies in male and female rats, *J. Comp. Neurol.*, 225, 124, 1987.

16. De Vries, G.J., Bujis, R.M., and Sluiter, A.A., Gonadal hormone actions on the morphology of the vasopressinergic innervation of the adult rat brain, *Brain Res.*, 298, 141, 1984.

17. Zhou, J.N., Hofman, M.A., Gooren, L.J.G., and Swaab, D.F., A sex difference in the human brain and its relation to transexuality, *Nature*, 378, 68, 1995.

18. Phoenix, C.H., Goy, R.W., Gerall, A.A., and Young, W.C., Organizing action of prenatally administered testosterone propionate on the tissues mediating mating behavior in the female guinea pig, *Endocrinology*, 65, 369, 1959.

19. Gorski, R.A., Harlan, R.E., Jacobson, C.D., Shryne, J.E., and Southam, A.M., Evidence for the existence of a sexually dimorphic nucleus in the preoptic area of the rat, *Comp. Neurol.*, 193, 529, 1980.

20. Barraclough, C.A., Modifications in reproductive function after exposure to hormones during the prenatal and early postnatal period, in *Neuroendocrinology*, Vol. 2, Martini, L. and Ganong, W.F., Eds., Academic Press, New York, 1967, 61.

21. Gorski, R.A., Gonadal hormones and the perinatal development of neuroendocrine function, in *Front. Neuroendocrinol.*, Martini, L. and Ganong, W.F., Eds., Oxford Press, New York, 1967, 273.

22. Martini, L., The 5α-reduction of testosterone in the neuroendocrine structures. Biochemical and physiological implications, *Endocrine Rev.*, 3, 1, 1982.

23. Goy, R.W. and Resko, J.A., Gonadal hormones and behavior of normal and pseudohermaphrodite nonhuman female primates, *Recent Prog. Horm. Res.*, 28, 707, 1972.

24. Collaer, M.L. and Hines, M., Human behavioral sex differences: a role for gonadal hormones during early development? *Psychol. Bull.*, 118, 55, 1995.

25. Paul, S.M. and Purdy, R.H., Neuroactive steroids, *FASEB J.*, 6, 2311, 1992.

26. Majewska, M.D., Harrison, N.L., Schwartz, R.D., Barker, J.L., and Paul, S.M., Steroid hormone metabolites are barbiturate-like modulators of the $GABA_A$ receptor, *Science*, 232, 1004, 1986.

27. Ramirez, V.D. and Zheng, J., Membrane sex-steroid receptors in the brain, *Front. Neuroendocrinol.*, 17, 402, 1996.

28. Su, T.P., London, E.D., and Jaffe, J.H., Steroid binding to σ-receptors suggests a link between endocrine, nervous and immune system, *Science*, 240, 219, 1988.

29. Maggi, R., Pimpinelli, F., Casulari, L.A., Piva, F., and Martini, L., Antiprogestins inhibit the binding of opioids to μ-opioid receptors in nervous membrane preparations, *Eur. J. Pharmacol.*, 301, 169, 1996.

30. Ma, Z.Q., Santagati, S., Patrone, C., Pollio, G., Vegeto, E., and Maggi, A., Insulin-like growth factors activate estrogen receptor to control the growth and differentiation of the human neuroblastoma cell line SK-ER3, *Mol. Endocrinol.*, 8, 910, 1994.

31. Bunone, G., Briand, P.A., Miksicek, R.J., and Picard, D., Activation of unliganded estrogen receptor by EGF involves the MAP kinase pathway and direct phosphorilation, *EMBO J.*, 15, 2174, 1996.

32. Mani, S.K., Allen, J.M., Clark, J.H., Blaunstein, J.D., and O'Malley, B.W., Convergent pathways for steroid hormone- and neurotransmitter-induced rat sexual behavior, *Science*, 265, 1246, 1994.

33. Murphy, D.D. and Segal, M., Morphological plasticity of dendritic spines in central neurons is mediated by activation of cAMP response element binding protein, *Proc. Natl. Acad. Sci. U.S.A.*, 94, 1482, 1997.
34. Schule, R. and Evans, R.M., Cross-coupling of signal transduction pathways: zinc finger meets leucine zipper, *Trends Genet.*, 7, 377, 1991.
35. Toran-Allerand, C.D., Singh, M., and Sétáló, G., Jr., Novel mechanisms of estrogen action in the brain: new players in an old story, *Front. Neuroendocrinol.*, 20, 97, 1999.
36. Singh, M., Sétáló, G., Jr., Frain, D., Warren, M., and Toran-Allerand, C.D., Estrogen-induced ERK phosphorylation in the cerebral cortex of estrogen receptor-α knockout (ERKO) mice, *Soc. Neurosci. Abstr.*, 24, 1295, 1998.
37. Toran-Allerand, C.D., Miranda, R.C., Bentham, W., Sohrabji, F., Brown, T., Hochberg, R., and MacLusky, N., Estrogen receptors colocalize with low-affinity nerve growth factor receptors in cholinergic neurons of the basal forebrain, *Proc. Natl. Acad. Sci. U.S.A.*, 89, 4668, 1992.
38. Miranda, R.C., Sohrabji, F., and Toran-Allerand, C.D., Neuronal colocalization of mRNA for neurotrophins and their receptors in the developing central nervous system suggests a potential for autocrine interactions, *Proc. Natl. Acad. Sci. U.S.A.*, 90, 6439, 1993.
39. Toran-Allerand, C.D., Novel mechanism of estrogen action in the developing brain: role of steroid/neurotrophin interaction, in *Contemporary Endocrinology, Neurosteroids: A New Regulatory Function in the Nervous System*, Baulieu, E.E., Robel, P., and Schumacher, M., Eds., The Humana Press, Totowa, N. J., 1999, in press.
40. Diaz, R., Sokoloff, P., and Fuxe, K., Codistribution of dopamine D_3 receptors and glucocorticoid receptor mRNAs during striatal prenatal development in the rat, *Neurosci. Lett.*, 227, 119, 1997.
41. Wagner, C.K., Nakayama, A.Y., and De Vries, G.J., Potential role of maternal progesterone in the sexual differentiation of the brain, *Endocrinology*, 139, 3658, 1998.
42. Meaney, M.J., Aitken, D.H., Jensen, L.K., McGinnis, M.Y., and McEwen, B.S., Nuclear and cytosolic androgen receptor levels in the limbic brain of neonatal male and female rats, *Dev. Brain Res.*, 23, 179, 1985.
43. McEwen, B.S., Steroid hormone action in the brain: when is the genome involved? *Horm. Behav.*, 28, 396, 1994.
44. Kuiper, G.G.J.M., Shughrue, P.J., Merchenthaler, I., and Gustafsson, J.-A., The estrogen receptor β subtype: a novel mediator of estrogen action in neuroendocrine system, *Front. Neuroendocrinol.*, 19, 253, 1998.
45. Kuiper, G.G.J.M., Carlsson, B., Grandien, K., Enmark, E., Haggblad, J., Nilsson, S., and Gustafsson, J.-A., Comparison of the ligand binding specificity and transcript tissue distribution of estrogen receptors alpha and beta, *Endocrinology*, 138, 863, 1997.
46. McInerney, E.M., Weiss, K.E., Sun, J., Mosselman, S., and Katzenellenbogen, B. S., Transcription activation by the human estrogen receptor subtype beta (ER beta) studied with ER beta and ER alpha receptor chimeras, *Endocrinology*, 139, 4513, 1998.
47. Kriege, J.H., Hodgin, J.B., Couse, J.F., Enmark, E., Warner, M., Mahler, J.F., Sar, M., Korach, K.S., and Gustafsson, J.-A., Generation and reproductive phenotypes of mice lacking estrogen receptor beta, *Proc. Natl. Acad. Sci. U.S.A.*, 95, 15677, 1998.

48. Ogava, S., Lubahn, D.B., Korach, K.S., and Pfaff, D.W., Behavioral effects of estrogen receptor gene disruption in male mice, *Proc. Natl. Acad. Sci. U.S.A.*, 94, 11476, 1997.

49. Kuiper, G.G.J.M., Lemmen, J.G., Carlson, B., Corton, J.C., Safe, S.H., van der Saag, P.T., van der Burg , B., and Gustafsson, J.-A., Interaction of estrogenic chemicals and phytoestrogens with estrogen receptor β, *Endocrinology,* 139, 4252, 1998.

50. Nimrod, A.C. and Benson, W.H., Environmental estrogenic effects of alkyl-phenol ethoxylates, *Crit. Rev. Toxicol.*, 26, 335, 1996.

51. Vielkind, U., Walencewicz, A., Levine, J.M., and Bohn, M.C., Type II glucocor-ticoid receptors are expressed in oligodendrocytes and astrocytes, *J. Neurosci. Res.*, 27, 360, 1990.

52. Jung-Testas, I., Renoir, M., Bugnard, H., Greene, G.L., and Baulieu, E.E., Dem-onstration of steroid hormone receptors and steroid action in primary cultures of rat glial cells, *J. Steroid Biochem. Mol. Biol.*, 41, 621, 1992.

53. Langub, M.C. and Watson, R.E., Jr., Estrogen receptor-immunoreactive glia, endothelia, and ependyma in guinea pig preoptic area and median eminence: electron microscopy, *Endocrinology,* 130, 364, 1992.

54. Magnaghi, V., Cavarretta, I., Zucchi, I., Susani, L., Rupprecht, R., Hermann, B., Martini, L., and Melcangi, R.C., Po gene expression is modulated by androgens in the sciatic nerve of adult male rats, *Mol. Brain Res.*, 70, 36, 1999.

55. Callard, G.V., Aromatization in brain and pituitary, in *Metabolism of Hormonal Steroids in the Neuroendocrine Structures*, Celotti, F., Naftolin, F., and Martini, L., Eds., Raven Press, New York, 1984, 79.

56. Luttge, W.G. and Whalen, R.E., Dihydrotestosterone, androstenedione, test-osterone: comparative effectiveness in masculinizing and defeminizing repro-ductive system in male and female rats, *Horm. Behav.*, 1, 265, 1970.

57. Arai, Y., Effect of 5α-dihydrotestosterone on differentiation of masculine pat-tern of the brain in the rat, *Endocrinol. Jpn.*, 19, 389, 1972.

58. McDonald, P.G. and Doughty, C., Effects of neonatal administration of different androgens in the female rat: correlation between aromatization and the induc-tion of sterilization, *J. Endocrinol.*, 61, 95, 1974.

59. Doughty, C., Booth, J.E., McDonald, P.G., and Parrott, R. F., Inhibition, by the anti-oestrogen MER-25, of defeminization induced by the synthetic oestrogen RU 2858, *J. Endocrinol.*, 67, 459, 1975.

60. Docke, F. and Dorner, G., Anovulation in adult female rats after neonatal intracerebral implantation of oestrogens, *Endokrinologie*, 65, 375, 1975.

61. Döhler, K.D., Srivastava, S.S., Shryne, J.E., Jarzab, B., Sipos, A., and Gorski, R.A., Differentiation of the sexually dimorphic nucleus in the preoptic area of the rat brain is inhibited by postnatal treatment with an estrogen antagonist, *Neuroendocrinology,* 38, 297, 1984.

62. McDonald, P.G. and Doughty, C., Androgen sterilization in the neonatal female rat and its inhibition by an estrogen antagonist, *Neuroendocrinology,* 13, 182 , 1973.

63. McCarthy, M.M., Schlenker, E.H., and Pfaff, D.W., Enduring consequences of neonatal treatment with antisense oligodeoxynucleotides to estrogen receptor messenger ribonucleic acid on sexual differentiation of rat brain, *Endocrinology,* 331, 433, 1993.

64. Morali, G., Larsson, K., and Beyer, C., Inhibition of testosterone-induced sexual behavior in the castrated male rat by aromatase blockers, *Horm. Behav.*, 9, 203, 1977.
65. Negri-Cesi, P., Celotti, F., and Martini, L., Androgen metabolism in the brain: role in sexual differentiation and in the control of gonadotropin secretion, in *Monographs in Neural Sciences*, Vol. 12, Cohen, M.M., Ed., Karger, Basel, 1986, 7.
66. Lauber, M. E. and Lichtensteiger, W., Pre- and postnatal ontogeny of aromatase cytochrome P450 messenger ribonucleic acid expression in the male rat brain studied by *in situ* hybridization, *Endocrinology*, 135, 1661, 1994.
67. Corbin, J.C., Graham-Lorence, S., McPhaul, M., Mason, J.I., Mendelson, C.R., and Simpson, E.R., Isolation of a full-length cDNA insert encoding human aromatase system cytochrome P450 and its expression in nonsteroidogenic cells, *Proc. Natl. Acad. Sci. U.S.A.*, 85, 8948, 1988.
68. Honda, S., Harada, N., and Takagi, Y., Novel exon 1 of the aromatase gene specific for aromatase transcripts in human brain, *Biochem. Biophys. Res. Commun.*, 198, 1153, 1994.
69. Yamada-Mouri, N., Hirata, S., Hayashi, M., and Kato, J., Analysis of the expression and the first exon of aromatase mRNA in monkey brain, *J. Steroid Biochem. Mol. Biol.*, 55, 17, 1995.
70. Honda, S., Morohashi, K., Nomura, K., Takeya, H., Kitajima, M., and Omura, T., Ad4BP regulating steroidogenic P-450 gene is a member of steroid hormone receptor superfamily, *J. Biol. Chem.*, 268, 7494, 1993.
71. Ikeda, Y., Luo, X., Abbud, R., Nilson, J.H., and Parker, K.L., The nuclear receptor steroidogenic factor 1 is essential for the formation of the ventromedial hypothalamic nucleus, *Mol. Endo.*, 9, 478, 1995.
72. Negri-Cesi, P., Melcangi, R.C., Celotti, F., and Martini, L., Aromatase activity in cultured brain cells: difference between neurons and glia, *Brain Res.*, 589, 327, 1992.
73. Poletti, A., Negri-Cesi, P., Melcangi, R.C., Colciago, A., Martini, L., and Celotti, F., Expression of androgen activating enzymes in cultured cells of developing rat brain, *J. Neurochem.*, 68, 1298, 1997.
74. Tsuruo, Y., Ishimura, K., Fujita, H., and Osawa, Y., Immunocytochemical localization of aromatase-containing neurons in the rat brain during pre- and postnatal development, *Cell Tissue Res.*, 278, 29, 1994.
75. Schlinger, B.A., Amur-Umarjee, S., Shen, P., Campagnoni, T., and Arnold, A.P., Neuronal and non-neuronal aromatase in primary cultures of developing zebra finch telencephalon, *J. Neurosci.*, 14, 7541, 1994.
76. Garcia-Segura, L.M., Wozniak, A., Azcoitia, I., Rodriguez, J.R., Hutchison, R.E., and Hutchison, J. B., Aromatase expression by astrocytes after brain injury: implications for local estrogen formation in brain repair, *Neuroscience*, 89, 567, 1999.
77. Santagati, S., Melcangi, R.C., Celotti, F., Martini, L., and Maggi, A., Estrogen receptor is expressed in different types of glial cells in culture, *J. Neurochem.*, 63, 2058, 1994.
78. Dellovade, T.L., Rissman, E.F., Thompson, N., Harada, N., and Ottinger, M. A., Co-localization of aromatase enzyme and estrogen receptor immunoreactivity in the preoptic area during reproductive aging, *Brain Res.*, 674, 181, 1995.
79. Tsuruo, Y., Ishimura, K, and Osawa, Y., Presence of estrogen receptors in aromatase-immunoreactive neurons in the mouse brain, *Neurosci. Lett.*, 195, 49, 1995.

80. Sholl, S.A., Goy, R.W., and Kim, K.L., 5alpha-Reductase, aromatase, and androgen receptor levels in the monkey brain during fetal development, *Endocrinology*, 124, 627, 1989.

81. Sholl, S.A. and Kim, K.L., Aromatase, 5alpha-reductase, and androgen receptor levels in the fetal monkey brain during early development, *Neuroendocrinology*, 52, 94, 1990.

82. Roselli, C.E. and Resko, J.A., Aromatase activity in the rat brain: hormonal regulation and sex differences, *J. Steroid Biochem. Mol. Biol.*, 44, 499, 1993.

83. Beyer, C., Green, S.J., and Hutchison, J.B., Androgens influence sexual differentiation of embrionic mouse hypothalamic aromatase neurons "in vitro," *Endocrinology*, 135, 1220, 1994.

84. George, F.W. and Ojeda, S.R., Changes in aromatase activity in the rat brain during embryonic, neonatal, and infantile development, *Endocrinology*, 111, 522, 1982.

85. Lephart, E.D., Simpson, E.R., McPhaul M.J., Kilgore, M.W., Wilson, J.D., and Ojeda, S.R., Brain aromatase cytochrome P450 messenger RNA levels and enzyme activity during prenatal and perinatal development in the rat, *Mol. Brain Res.*, 16, 187, 1992.

86. Lauber, M.E., Sarasin, A., and Lichtensteiger, W., Transient sex difference of aromatase (CYP19) mRNA expression in the developing rat brain, *Neuroendocrinology*, 66, 173, 1997.

87. Poletti, A., Negri-Cesi, P., Rabuffetti, M., Colciago, A., Celotti, F., and Martini, L., Transient expression of the 5alpha-reductase type 2 isozyme in late fetal and early postnatal life, *Endocrinology*, 139, 2171, 1998.

88. Negri-Cesi, P., Colciago, A., Celotti, F., and Martini, L., Dimorphic expression of aromatase in the hypothalamic area of fetal rats during late gestation, *Endocrin. Soc. Abstr.*, 81, 338, 1999.

89. Karolczak, M., Kuppers, E., and Beyer, C., Developmental expression and regulation of aromatese and 5α-reductase type 1 mRNA in the male and female mouse hypothalamus, *J. Neurobiol.*, 10, 267, 1998.

90. Callard, C.V., Petro, Z., and Ryan, K.J., Aromatization of androgen to estrogen by cultured turtle brain cells, *Brain Res.*, 202, 117, 1980.

91. Paden, C.M. and Roselli, C.E., Modulation of aromatase activity by testosterone in transplant of fetal rat hypothalamus-preoptic area, *Dev. Brain Res.*, 33, 127, 1987.

92. Lephart, E.D., Simpson, E.R., and Ojeda, S.R., Effect of cyclic AMP and androgens on "in vitro" brain aromatase enzyme activity during development in the rat, *J. Neuroendocrinol.*, 4, 29, 1992.

93. Beyer, C. and Hutchison, J.B., Developmental profile and regulation of brain estrogen synthesis by aromatase, *Biomed. Rev.*, 7, 41, 1997.

94. Abe-Dohmae, S., Tanaka, R., and Harada, N., Cell-type and region-specific expression of aromatase mRNA in cultured brain cell, *Mol. Brain Res.*, 24, 153, 1994.

95. Weisz, J. and Ward, I.L., Plasma testosterone and progesterone titers of pregnant rats, their male and female fetuses and neonatal offspring, *Endocrinology*, 106, 306, 1980.

96. Bentvelsen, F.M., McPhaul, M.J., Wilson, J.D., and George, F.W., The androgen receptor of the urogenital tract of the fetal rat is regulated by androgen, *Mol. Cell. Endocrinol.*, 105, 21, 1994.

97. Raum, W.J., Marcano, M., and Swerdloff, R.S., Nuclear accumulation of estradiol derived from the aromatization of testosterone is inhibited by hypothalamic beta-receptor stimulation in neonatal female rat, *Biol. Reprod.*, 30, 388, 1984.

98. Abe-Domahe, S.A., Tanaka, R., Takagi, Y., and Harada, N., In vitro increase of aromatase mRNA in diencephalic neurons, *Mol. Neuroendocrinol.*, 63, 46, 1996.

99. Simpson, E.R., Dodson, M.M., Agarwal, V.R., Hinshelwood, M.M., Bulun, S.E., and Zhao, Y., Expression of CYP19 (aromatase) gene: an unusual case of alternative promoter usage, *FASEB J.*, 11, 29, 1997.

100. Celotti, F., Melcangi, R.C., and Martini, L., The 5α-reductase in the brain, molecular aspects and relation to brain function, *Front. Neuroendocrinol.*, 13, 163, 1992.

101. Poletti, A., Rabuffetti, M., and Celotti, F., The 5α-reductase in the rat brain, in *The Brain: Source and Target for Sex Steroid Hormones*, Gennazzani, A.R., Petraglia, F., and Purdy, R.H., Eds., Parthenon Publishing, London, 1996, 123.

102. Massa, R., Justo, S., and Martini, L., Conversion of testosterone into 5α-reduced metabolites in the anterior pituitary and in the brain of maturing rats, *J. Steroid Biochem.*, 6, 567, 1975.

103. Melcangi, R.C., Celotti, F., Ballabio, M., Castano, P., Poletti, A., Milani, S., and Martini, L., Ontogenetic development of the 5α-reductase in the rat brain: cerebral cortex, hypothalamus, purified myelin and isolated oligodendrocytes, *Dev. Brain Res.*, 44, 181, 1988.

104. Arnold, A.P. and Gorski, R.A., Gonadal steroid induction of structural sex differences in the central nervous system, *Annu. Rev. Neurosci.*, 7, 413, 1984.

105. Takani, K. and Kawashima, S., Culture of rat brain preoptic area neurons: effects of sex steroids, *Int. J. Dev. Neurosci.*, 11, 63, 1993.

106. Goldstein, L.A. and Sengelaub, D.R., Differential effects of dihydrotestosterone and estrogens on the development of motoneuron morphology in a sexually dimorphic rat spinal cord, *J. Neurobiol.*, 25, 878, 1994.

107. Wilson, J.D., Griffin, J.E., and Russell, D.W., Steroid 5α-reductase 2 deficiency, *Endocrine Rev.*, 14, 577, 1993.

108. Russell, D.W. and Wilson, J.D., Steroid 5α-reductase: two genes/two enzymes, *Annu. Rev. Biochem.*, 63, 25, 1994.

109. Poletti, A., Celotti, F., Motta, M., and Martini, L., Characterisation of rat 5α-reductases type 1 and type 2 expressed in yeast Saccharomyces cerevisiae, *Biochem. J.*, 314, 1047, 1996.

110. Bonkhoff, H., Stein, U., Aumuller, G., and Remberger, K., Differential expression of 5α-R isoenzymes in the human prostate and prostatic carcinomas, *Prostate*, 29, 261, 1996.

111. Normington, K. and Russell, D.W., Tissue distribution and kinetic characteristics of rat steroid 5α-reductase isozymes. Evidence for distinct physiological functions, *J. Biol. Chem.*, 267, 19548, 1992.

112. Berman, D.M., Tian, H., and Russell, D.W., Expression and regulation of steroid 5α-reductase in the urogenital tract of the fetal rat, *Mol. Endo.*, 9, 1561, 1995.

113. Valencia, A., Collado, P., Cales, J.M., Segovia, S., and Perez Laso, C., Rodriguez Zafra, M., Guillamon, A., Postnatal administration of dihydrotestosterone to the male rat abolishes sexual dimorphism in the accessory olfactory bulb: a volumetric study, *Dev. Brain Res.*, 8, 132, 1992.

114. Beyer, C. and Hutchison, J. B., Androgens stimulate the morphological maturation of embryonic hypothalamic aromatase-immunoreactive neurons in the mouse, *Dev. Brain Res.*, 98, 74, 1997.

115. Cagnoni, M., Fantini, F., Morace, G., and Ghetti, A., Failure of testosterone propionate to induce the "early-androgen" syndrome in rats previously injected with progesterone, *J. Endocrinol.*, 33, 527, 1965.

116. Massa, R. and Martini, L., Interference with the 5α-reductase system, *Gynecol. Invest.*, 2, 253, 1971/72.

117. Poletti, A., Celotti, F., Maggi, R., Melcangi, R.C., Martini, L., and Negri-Cesi, P., Aspects of hormonal steroid metabolism in the nervous system, in *Contemporary Endocrinology, Neurosteroids: A New Regulatory Function in the Nervous System*, Baulieu, E.E., Robel, P., and Schumacher, M., Eds., The Humana Press, Totowa, N.J., 1999, in press.

118. Celotti, F., Melcangi, R.C., Negri-Cesi, P., and Poletti, A., Testosterone metabolism in brain cells and membranes, *J. Steroid Biochem. Mol. Biol.*, 40, 673, 1991.

119. Poletti, A., Coscarella, A., Negri-Cesi, P., Colciago, A., Celotti, F., and Martini, L., The 5alpha-reductase isozymes in the central nervous system, *Steroids*, 63, 246, 1998.

120. Kellog, C.K., Olson, V.G., and Pleger, G.L., Neurosteroid action at the GABA$_A$ receptor in fetal rat forebrain, *Dev. Brain Res.*, 108, 131, 1998.

121. Fernandez-Guasti, A. and Picazo, O., Sexual differentiation modifies the allopregnenolone anxiolytic actions in rats, *Psychoneuroendocrinology*, 24, 251, 1999.

122. Sapolsky, R.M., Packan, D.R., and Vale, W.W., Glucocorticoid toxicity in the hippocampus: *in vitro* demonstration, *Brain Res.*, 453, 369, 1988.

123. Behl, C., Lezoualch, F., Trapp, T., Widmann, M., Skutella, T., and Holsboer, F., Glucocorticoids enhance oxidative strees-induced cell death in hippocampal neurons *in vitro*, *Endocrinology*, 138, 101, 1997.

124. Mahendroo, M.S., Cala, K.M., and Russell, D.W., 5alpha-reduced androgens play a key role in murine parturition, *Mol. Endo.*, 10, 380, 1996.

125. Mahendroo, M.S., Cala, K.M., Landrum, C.P., and Russell, D.W., Fetal death in mice lacking 5alpha-reductase type 1 caused by estrogen excess, *Mol. Endo.*, 11, 917, 1997.

126. Naftolin, F. and Brawer, J.R., The effect of estrogens on hypothalamic structure and function, *Am. J. Obstet. Gynecol.*, 132, 758, 1978.

127. Faber, K.A. and Hughes, C.L., The effect of neonatal exposure to diethylstilbestrol, genistein, and zearalenone on pituitary responsiveness and sexually dimorphic nucleus volume in the castrated adult rat, *Biol. Reprod.*, 45, 649, 1991.

128. Gellert, R.J., Kepone, Mirex, Dieldrin and Aldrin: estrogenic activity and the induction of persistent vaginal estrus and anovulation in rats following neonatal treatment, *Environ. Res.*, 16, 131, 1978.

129. Kelce, W.R., Stone, C.R., Laws, S.C., and Gray, L.E., Jr., Kemppainen, J. A., Wilson, E. M., Persistent DDT metabolite *p, p'*-DDE is a potent androgen receptor antagonist, *Nature*, 375, 581, 1995.

130. Piva, F. and Martini, L., Neurotransmitters and the control of hypophyseal gonadal functions: possible implications of endocrine disruptors, *Pure Appl. Chem.*, 70, 1647, 1998.

131. Kelce, W.R., Monosson, E., Gamsik, M.P., Laws, S.C., and Gray, L.E., Jr., Environmental hormones disruptors: evidence that vinclozolin developmental toxicity is mediated by antiandrogenic metabolites, *Toxicol. Appl. Pharmacol.*, 126, 276, 1994.

132. Wong, C.I., Kelce, W.R., Sar, M., and Wilson, E.M., Androgen receptor versus agonist activities of the fungicide vinclozolin relative to hydroxyflutamide, *J. Biol. Chem.*, 270, 19998, 1995.

133. Fry, D.M. and Toone, C.K., DDT-induced feminization of gull embryos, *Science*, 213, 922, 1981.

134. Guillette, L.J., Gross, T.S., Masson, G.R., Matter, J.M., Percival, H.F., and Woodward, A.L., Developmental abnormalities of the gonad and abnormal sex hormone concentrations in juvenile alligators from contaminated and control lakes in Florida, *Environ. Health Perspect.*, 102, 680, 1994.

135. Guillette, L.G., Gross, T.S., Gross, D.A., Rooney, A.W., and Percival, H.F., Gonadal steroidogenesis *in vitro* from juvenile alligators obtained from contaminated or control lakes, *Environ. Health Perspect.*, 103, 31, 1995.

136. Gray, L.E., Jr., Ostby, J.S., and Kelce, W.R., Developmental effects of environmental antiandrogen: the fungicide vinclozolin alters sex differentiation of the male rat, *Toxicol. Appl. Pharmacol.*, 129, 46, 1994.

137. Mably, T.A., Moore, R.W., and Peterson, R.E., In utero and lactation exposure of male rats to 2, 3, 7, 8-tetrachlorodibenzo-*p*-dioxin. I. Effect on androgenic status, *Toxicol. Appl. Pharmacol.*, 114, 97, 1992.

138. Mably, T.A., Moore, R.W., Goy, R.W., and Peterson, R.E., In utero and lactation exposure of male rats to 2, 3, 7, 8-tetrachlorodibenzo-*p*-dioxin. II. Effect on sexual behavior and the regulation of luteinizing hormone secretion in adulthood, *Toxicol. Appl. Pharmacol.*, 114, 108, 1992.

139. Mably, T.A., Bjerke, D.L., Moore, R.W., Gendron-Fitzpatrick, A., and Peterson, R.E., In utero and lactation exposure of male rats to 2, 3, 7, 8-tetrachlorodibenzo-*p*-dioxin. III. Effect on spermatogenesis and reproductive capability, *Toxicol. Appl. Pharmacol.*, 114, 118, 1992.

140. Gray, L.E., Jr., Kelce, W.R., Monosson, E., Ostby, J.S., and Birnbaum, L.S., Exposure to TCDD during development permanently alters reproductive function in male Long Evans rats and hamsters: reduced ejaculated and epididymal sperm numbers and sex accessory gland weights in offspring with normal androgenic status, *Toxicol. Appl. Pharmacol.*, 131, 108, 1995.

141. Crews, D., Bergeron, J.M., and McLachlan, J.A., The role of estrogen in turtle sex determination and effect of PCBs, *Environ. Health Perpect.*, 103(suppl. 7), 73, 1995.

142. Korach, K.S., Sarver, P., Chae, K., McLachlan, J.A., and McKinney, J.D., Estrogen receptor-binding activity of polychlorinated hydroxybiphenils: conformationally restricted structural probes, *Mol. Pharmacol.*, 33, 120, 1998.

143. Hammer, R.P., Jr., The sexually dimorphic region of the preoptic area in rats contains denser opiate receptors binding sites in females, *Brain Res.*, 308, 172, 1984.

144. Hammer, R.P., Jr., The sex hormone-dependent development of opiate receptors in the rat medial preoptic area, *Brain Res.*, 360, 65, 1985.

145. Limonta, P., Dondi, D., Maggi, R., Martini, L., and Piva, F., Neonatal organization of the brain opioid systems controlling prolactin and luteinizing hormone secretion, *Endocrinology*, 124, 681, 1989.

146. Limonta, P., Dondi, D., Maggi, R., and Piva, F., Testosterone and postnatal ontogenesis of hypothalamic μ([³H]dihydromorphine) opioid receptors in the rat, *Dev. Brain Res.*, 62, 131, 1991.

147. Forest, M.G., David, M., and Morel, Y., Prenatal diagnosis and treatment of 21-hydroxylase deficiency, *J. Steroid Biochem. Mol. Biol.*, 45, 720, 1993.

148. Berenbaum, S.A. and Hines, M., Early androgens are related to childhood sex-typed toy preferences, *Psychol. Sci.*, 3, 203, 1992.
149. Carson, S.A. and Simpson, J.L., Virilization of female fetuses following maternal ingestion of progestational and androgenic steroids, in *Hirsutism and Virilism*, Mahesh, C.V.B. and Greenblatt, R.B., Eds., Butterworths, London, 1983, 177.
150. Lee, P.A., Brown, T.R., and La Torre, H., Diagnosis of the partial androgen insensitivity syndrome during infancy, *J. Am. Med. Assoc.*, 16, 2207, 1986.
151. Nagel, R.A., Lippe, B.M., and Griffin, J.E., Amdrogen resistance in the neonate: use of hormones of hypothalamic–pituitary–gonadal axis for diagnosis, *J. Pediatr.*, 109, 486, 1984.
152. Imperato-McGinley, J., Guerrero, L., Gautier, T., and Peterson, R.E., Steroid 5α-reductase deficiency in man: an inherited form of male pseudohermaphroditism, *Science*, 186, 1213, 1974.
153. Imperato-McGinley, J., Peterson, R.E., Gautier, T., and Sturla, E., Male pseudohermaphroditism secondary to 5α-reductase deficiency: a model for the role of androgens in both the development of the male phenotype and the evolution of a male gender identity, *J. Steroid Biochem.*, 11, 637, 1979.
154. Morishima, A., Grumbach, M.M., Simpson, E.R., Fisher, C., and Qin, K., Aromatase deficiency in male and female siblings caused by a novel mutation and the physiological role of estrogens, *J. Clin. Endocrinol. Metab.*, 80, 3689, 1995.
155. Bulun, S.E., Clinical review 78: aromarase deficiency in women and men: would you have predicted the phenotypes? *J. Clin. Endocrinol. Metab.*, 81, 867, 1996.
156. Carani, C., Qin, K., Simoni, M., Faustini-Fustini, M., Serpente, S., Boyd, J., Korach, K.S., and Simpson, E.R., Effect of testosterone and estradiol in a man with aromatase deficiency, *N. Engl. J. Med.*, 337, 91, 1997.
157. Granata, A.R.M., Rochira, V., Faustini-Fustini, M., Bevini, M., Madeo, B., and Carani, C., Sexual behavior in a man with aromatase deficiency: role of estrogen on male sexuality, *J. Endocrinol. Invest.*, 21(Suppl. 7), 17, 1998.

5

Estrogen in Gender-Specific Neural Differentiation

John B. Hutchison

CONTENTS

I. Introduction

During periods of perinatal brain development, there are phases of hormonal sensitivity when steroid sex hormones influence maturing neuronal mechanisms underlying both the neuroendocrine system and behavior.[1-3] However, the way in which these phasic effects of steroids interact with the changing fetal and neonatal environment to bring about behavioral and neuroendocrine development is still not understood. One of the major problems lies in trying to relate steroid action at a neuronal level to the complexity of behavioral and neuroendocrine mechanisms subject to genetically programmed brain development and rapidly changing external environmental effects.

This aspect has been studied most profitably by making use of the sexual differentiation of mammalian brain mechanisms of reproductive behavior. However, there have been major contributions made from research on other vertebrates, notably Aves and Reptilia (See Chapter 9 by Arnold and Chapter 8 by Crews).

There is little doubt now that the steroid hormones, androgens and estrogens, are involved in the sexual differentiation of behavioral development by direct action on the brain. However, as Beach[4] pointed out, steroid hormones also affect the development of peripheral sensory systems, for example, in the penis of the rat. It is important to emphasize that the effects of steroids on behavioral development depend to a large degree on changes in the sensory fields of cutaneous skin receptors[5] which are a consequence of functionally important fluctuations in seasonal stimuli such as temperature, humidity, and the effects of mutual stimulatory activity between male and female. Current ideas on the mode of action of steroids in the brain suggest that a sexually differentiated phenotype develops as a consequence of gonadal steroid action on an undifferentiated bipotential substrate. Theoretically, the distinction can be made between permanent "organizing" effects of steroids which are irreversible and the transient "activational" effects of steroids required for adult behavior.[16,17] Estrogens are well known from research in rodents to reproduce some of the effects of testosterone (T) on brain development.[8,9] Prenatal inhibition of brain estrogen formation by exogenous inhibitors impairs both reproductive behavior and the gonadotrophic feedback systems.[10] Estradiol-17β (E_2) also influences neuronal growth, and growth factors such as insulin-like growth factor I receptors and binding proteins in hypothalamic neurons.[11] Therefore, E_2 plays an important role in the early organization of both behavior and associated sexually dimorphic structures in the brain. There is also evidence that estrogen formation within the brain is important in androgen action.[12-14] E_2 is converted from circulating T within androgen target areas of the developing brain in all vertebrates examined so far. The conversion is catalyzed by an enzymatic complex consisting of cytochrome P-450$_{arom}$, expressed by the *CYP 19* gene, and NADPH reductase. The aromatase enzyme complex in question, which is a member of the P-450 superfamily,[15] shows how gene expression of a single brain enzyme early in development regulates the production of a single steroid, E_2, critical for the sexual differentiation of the gonads, the neuroendocrine system, and the brain as a whole.

The aim of this chapter is to discuss the role of aromatase in brain development. To understand the developmental role of estrogen formation within the brain as a dynamic process, it is necessary to relate aromatase activity in individual androgen target areas of the brain to steroid-dependent developmental changes in brain structures. Describing these relationships still presents a difficulty in view of the paucity of knowledge of how steroidogenic enzyme systems are regulated in the developing brain, and when physiologically active forms of these enzymes first appear in early development.

Moreover, both genetic and hormonal factors are involved in the sexual differentiation of the brain.[16]

II. Hormones in Brain Development — Conceptual Issues

Current views on the developmental actions of hormones on mammalian brain development and behavior are derived from embryological work.[17,18] There are three stages in sexual development from an undifferentiated primordium common to genetic males and females: (1) gonadal sex is determined by the presence or absence of the male testis-determining gene on the Y chromosome (Sry); (2) the reproductive tract and genital morphology are differentiated hormonally; (3) sexual differentiation of the brain and behavior occurs as the final stage, also under hormonal control. The general principle derived from Jost's embryological studies[17] is that in mammals the male phenotype develops when a specific hormonal signal (T) is present. In the absence of this hormonal signal, reproductive tissues differentiate into the female type. The best known example is the differentiation of the Müllerian (female) and Wolffian (male) ducts from undifferentiated primordia by the action of fetal testicular T. The Wolffian duct development requires the presence of T, whereas Müllerian duct development has no apparent hormonal requirement in the genetic female. Jost's work[17] also illustrates another important principle in that the development of the Müllerian ducts is suppressed in the male by a second hormone, the glycoprotein Müllerian inhibitory hormone (MIH) secreted by testicular Sertoli cells. Differentiation into a phenotypic male requires two processes acting on an undifferentiated primordium, masculinization and defeminization, involving suppression of female characteristics.

Initially, behavioral development in rodents appeared to differ from these somatic processes of duct differentiation. Earlier studies of mating behavior in the rat[6,19] and guinea pig[20] suggested that the postnatal action of T alone could account for behavioral masculinization without the accompanying action of a defeminizing factor. Behavioral differentiation was seen to occur along a masculine/feminine continuum that depended on T action. This idea led to the proposition of the classical organizational hypothesis that T, acting during the perinatal period, either before or immediately after birth, organizes the pattern of sexual behavior to the male type, irrespective of genetic sex. This differentiating effect, or "organization," was thought to be irreversible and to occur at a single "critical" or "time-limited" window in development. Both the suppression of feminine behavior and the enhancement of masculine behavior were assumed to occur as a consequence of the effects of T, suggesting a one-hormone process. The feminine behavioral type was assumed to require no hormonal effect. The second assumption of this classical hypothesis was that in adulthood, hormones appropriate to the sex of the individual "activated" the behavior.[6] Organization and activation not only occurred at different stages of life, but also were assumed to be different processes. The organization hypothesis has been extended to other vertebrates. In birds (specifically, Japanese quail), the action of steroids on reproductive behavior are the reverse of the effects in mammals. Ovarian E_2, acting in early embryonic development, is thought to irreversibly feminize the brain. As a corollary, therefore, the male can be demasculinized by E_2 during critical periods of behavioral differentiation. The female is the sex that is actively differentiated, and the male requires no differentiating process that depends on hormones.[21] This finding can be linked to the fact that in birds, unlike mammals, the female is the heterogametic sex. The organizational effect of hormones on sexual differentiation appears generally to be exerted on the heterogametic sex.[21] There is also an indication that the same hormones that are involved in the organization of sex-typical behavior later activate the behavior for the initiation of sexual interactions at puberty.[6]

There are a number of major problems with the organizational hypo-thesis; the first relates to the site of action of androgen. Does the hormone act exclusively on the brain during development? Hormones, including the sex steroids, affect sensory receptors and thereby modify afferent input to the brain.[2,4] For example, neonatal castration of a rat on Day 1 of life impairs penile development and the differentiation of sensory receptors (penile spines) and causes deficiencies in the organization of spinal reflex–control nuclei.[22] Therefore, the neonatally castrated male is deficient in both sensory and effector systems for sexual behavior later in adult life. The lack of the early effects of androgen on sexual differentiation can be explained by this peripheral effect, emphasizing Beach's[4] original prediction.

A second difficulty lies with the supposition that sexual differentiation is a unitary event involving a one-hormone T effect. The "orthogonal" model of sexual differentiation[23,24] established that in contrast to the classical theory, in which sexual differentiation occurs along a continuum, masculine and feminine aspects of sexual behavior are differentiated at separate stages in development, and probably independently. As has been pointed out by Yahr,[25] hormonal action that suppresses feminine behavior does not necessarily enhance masculine behavior. Important evidence for this model has been obtained using testicular feminized male Tfm rodents (rats and mice), which are mutants carrying the X-linked allele for testicular feminization. Such mutant mice and rats, which do not have androgen receptors in target tissues, show abnormally low levels of masculine components of mating behavior (e.g., mounting the female) despite treatment with testosterone in adulthood. However, if castrated as neonates, these Tfm mice do show feminine behavior (lordosis) as adults following estrogen and progesterone administration, suggesting that defeminization of coital behavior occurs in these androgen-insensitive males.[26] The terms *masculinization* and *defeminization* were introduced by Whalen[23] to emphasize the independence of the hormonal processes involved in the development of masculine and feminine behavior. These processes may occur at different stages of development (heterochrony, see Reference 25). Thus, the current idea of the sexual differentiation of behavior as applied to mammals, including primates, is quite similar to the original embryological hypothesis applied to the differentiation of the reproductive ducts by Jost. However, as both the classical theory and later modifications have been derived exclusively from studies of copulatory behavior in a few rodent species, more recent developmental work has been extended to other types of behavior and to other species such as a carnivore, the ferret.[27]

The third difficulty with the organizational hypothesis to be considered is whether hormonal "activation" of sexual behavior in adulthood, and "organization" during perinatal development are separable processes. The difference between these processes lies in the transient nature of steroid effects in the adult on mature brain mechanisms compared with the permanent organizational effects of hormones early in life. There are a number of possibilities that could, in theory, account for physiological differences. The sites of steroid action in the brain differ between the juvenile and adult; the cellular mechanisms, including steroid metabolism and receptor binding, differ; steroids exert separable actions depending on the developmental condition of the target areas in the brain at the time of steroid exposure, possibly in terms of both enzyme and steroid receptor constituents. It has been argued[28] that the distinction between organizational and activation processes is no longer tenable. This change in outlook has occurred because research on the avian brain provides evidence that does not agree with the idea that permanent organizing effects occur only during the critical sensitive period of early development. E_2 demasculinizes mechanisms underlying copulatory behavior in female Japanese quail embryos.[21] Therefore, females treated with E_2 as embryos do not show complete male copulatory behavior. However, if female Japanese quail are ovariectomized at hatching, a demasculinizing effect of E_2 can be demonstrated in adulthood.[29,30] This interesting finding suggests that in the absence of E_2 during the post-hatching period, the brain mechanisms underlying male sexual behavior remain receptive to the organizing effects of E_2 until maturity. Two aspects of the organizational hypothesis have to be reconsidered, therefore: (1) there appears to be no restricted time limit for the "critical period" for the differentiating effects of estrogen, and it is interesting to compare a similar lack of a rigidly fixed period for behavioral imprinting with hormonal imprinting in the brain; (2) the important conclusion can be drawn that, given the right conditions, mechanisms underlying behavior can be organized by steroids in the adult animal.[30,31] There is no fixed period for the hormonal imprinting effects on brain organization. The conclusion that steroid hormones can organize the adult brain has been supported by steroid effects involving the structural organization of the adult brain that previously have been thought to occur only in juveniles.

III. Estrogens in the Sexual Differentiation of the Mammalian Brain

There is increasing evidence that E_2 is involved in the organization of both the brain and reproductive behavior. As is well known, E_2, required for the process of differentiation, is available potentially to developing brain neurons from the aromatization of T locally in the hypothalamus and preoptic areas. Mutant Tfm rats and mice lacking the androgen receptor, mentioned above, have provided an insight into the role of E_2 in the sexual differentiation of behavior. Since Tfm male rats are relatively insensitive to T and show no female behavior as adults, the potential for female sexual behavior is probably suppressed in perinatal development by hormones other than androgen, notably E_2.[26] This discovery is of great interest because the hormonal mechanisms operating in development during masculinization and defeminization must differ, and E_2 appears to have different development effects from T. Support for this idea has also come from studies of the effects of nonaromatizable androgens such as 5α-dihydrotestosterone (DHT). This androgen is effective in masculinizing, but not defeminizing, copulatory behavior of the rat.[26] Second, the antiestrogen tamoxifen prevents behavioral defeminization by blocking E_2 action on brain cells. Aromatase inhibitors are also effective in preventing defeminization of coital behavior. There is, therefore, a great deal of evidence to suggest that aromatization and the production of E_2 in the neonatal rodent brain is involved in the suppression of the capacity to show female traits in male rats. Taken together, the existing evidence suggests that enzyme activity that forms E_2 in the brain is involved in behavioral defeminization during male development. The evidence is still not entirely conclusive. For example, events that accompany the use of a nonaromatizable androgen such as 5α-DHT can be difficult to interpret. The failure to masculinize behavior can be due to a reduction amount of T or DHT reaching the brain in view of extensive peripheral metabolism in the circulating blood or liver to catabolic metabolites.[13] Enzyme blockers such as 1, 4, 6-androstatrien-3, 17-dione (ATD) or the new nonsteroidal triazole aromatase inhibitors are not specific in their action on the aromatase system. They also interfere with the activity of enzymes that produce active metabolites of T such as DHT in the brain and may bind to steroid receptors. Gonadal steroids, particularly E_2,[32,33] also have a role in the regulation of the morphology of adult mammalian brain neurons. Dendritic spine density of rat hippocampus and in the ventromedial nucleus is influenced by E_2[34] and E_2 has synaptic effects.[35] There are also very rapid changes in both the organization of the neuronal membrane and synapses of the hypothalamus induced by E_2.[36,37] A role for the brain aromatase system has been suggested in the sexual differentiation of both the hippocampus and neocortex.[38] However, developmental neurogenesis and cell death related to estrogenic effects on behavioral mechanisms have not yet been established for the mammalian brain.

Estrogens have profound effects on the development of the rodent (rat) preoptic area. Thus, both T and E_2 influence development of the sexually dimorphic nucleus (SDN) to induce a larger volume and more constituent cells.[39,40] The Mongolian gerbil is also an interesting model for the study of sexual differentiation and bilateral asymmetry in brain development.[41] The male has an elaborate precopulatory sequence of sexual interaction which includes a stereotyped male ultrasonic vocalization. Our work shows that the volume of the sexually dimorphic area, pars caudalis, of the anterior hypothalamus (SDApc), is positively correlated with the emission rate of courtship ultrasonic vocalizations.[42] Since both vocal development and SDApc differentiation can be induced in females by treatment of neonates with a synthetic estrogen, diethylstilbestrol (DES), estrogen is involved specifically in brain organization after birth. Differentiation appears to be postnatal, but this does not preclude a fetal effect of hormones. The left, but not the right, SDApc is correlated with ultrasonic vocal emission,[42] providing the first indication of an asymmetric link among E_2 action, brain structure, and vocal behavior in the mammalian brain. Perception of auditory input in the zebra finch[43] and auditory input in mice[44] is also lateralized in the brain, suggesting that E_2 formation in early development has to be studied in the context of asymmetry in both motor and sensory brain functions.

IV. Control of Estrogen Formation in the Brain

Current ideas on the regulation of the brain aromatase system are largely derived from work on adult birds and rodents. Estrogens are formed locally as metabolites of T in hypothalamic cells of all vertebrates that have been studied so far.[14,45] Human placental, and rat, mouse and chicken ovarian aromatase genes have been cloned and sequenced.[46-48] However, the brain aromatase has not been characterized in any group apart from fish,[49] where there is a different gene expressing aromatase in the brain and retina to other tissues. However, the current view for mammals is that the aromatase protein is formed by a single gene producing a number of transcripts (mRNA), some of which are functional. The functional transcripts and how brain aromatase mRNA is translated are still largely unknown. Kinetic studies of aromatase activity in the ring dove reveal differences in K_m (substrate-binding affinity constant) between brain areas, between avian brain and ovary, and during development. What modulates aromatase gene expression in the dove or whether mRNAs for aromatases in different estrogen target sites are derived from different gene transcripts is still not clear. Adult dove preoptic aromatase activity can be increased markedly by sex steroid action. Exogenous

T increases the V_{max} of the preoptic enzyme fivefold in the dove without changing the K_m of the enzyme.[50] Induction of the enzyme involves increased aromatase gene expression and possible changes in the catalytic site. Formation and action of estrogen via its protein receptors occurs in the same areas of the canary brain, and in the same cells.[51] Colocalization of E_2 receptors and the aromatase has been shown to occur in developing rat and mouse brain neurons.[52-54]

Castration in the male ring dove drastically reduces aromatase activity, but does not affect the number of E_2 receptor cells in the preoptic area (POA) detected by immunocytochemistry.[55] The inductive effects of T on brain aromatase activity have been demonstrated in other avian[56,57] and mammalian species,[58,59] suggesting that steroid regulation of brain aromatase is widespread in vertebrates. T has recently been shown to increase aromatase mRNA in the rat [60] and mouse hypothalamus.[61,62] In both birds and mammals, there is synergism between E_2 and T in the induction of aromatase and its mRNA in the brain.[63,64] Rapid changes in environmental stimuli derived from sociosexual interaction have short-term effects on brain estrogen formation in the male dove. Visual stimuli in courtship interactions rapidly increase POA aromatase activity, an experimentally induced increase which parallels that seen during the normal male reproductive cycle.[65] Sociosexual stimuli have a major influence on T aromatization in the brain. Although species differ in the way in which aromatase activity is regulated, environmental stimuli have direct effects on steroid-metabolizing enzymes in the adult avian brain. Photoperiod also influences E_2 formation in the mammalian (hamster) brain. The anterior hypothalamus, which includes the photosensitive suprachiasmatic nucleus involved in the photoperiodic control of the testes and circadian rhythmicity, contains a functionally active aromatase system. This system is sensitive to photoinhibition independently of circulating androgen level in the hamster, whereas the POA associated with sexual behavior, which also contains relatively high aromatase activity, is insensitive to both photoinhibition and a reduction of circulating testosterone resulting from castration.[59] In addition to regulation mediated through changes in gene expression, endogenous brain-derived inhibitors of the aromatase (e.g., 5α-reduced androstanes) influence E_2 formation.[66] Other pathways of androgen metabolism (e.g., 5β-reduced androstanes) regulate brain aromatase activity by catabolic reduction of the substrate, T, and, therefore, the amount of E_2 formed as a metabolite.

To what extent do genetic and physiological factors regulate aromatase activity in early brain development? Prenatal sex differences in preoptic aromatase activity levels have been found in the rat[67,68] and ferret.[69,70] Similarly, a small sex difference in aromatase activity levels has been seen in pooled preoptic samples of embryonic Day 18 rats.[71] In the fetal rat[72] and ferret[27] exogenous sex steroids, including androgen introduced into the maternal circulation, do not appear to modify brain aromatase activity. However, fetal aromatase activity in the guinea pig is induced in the mediobasal hypothalamus, septum, and cortex by T or DHT given to the pregnant mother.[73]

Treatment of pregnant female ferrets with testosterone, DHT, or flutamide (an androgen-receptor blocker) do not influence fetal brain aromatase activity. The latter treatment suggests that androgen receptors are not involved in prenatal aromatase regulation. Neonatally, male rats have higher aromatase activity in the amygdala, temporal lobe, and anterior hypothalamus than females, whereas no sex differences have been found in the POA.[74] Aromatase activity is higher in the fetal male rat amygdala and anterior hypothalamus than in females,[75] but there are asymmetries in estrogen formation which complicate interpretation and attempts to localize the enzyme. Aromatase mRNA is expressed very early in prenatal ontogeny of the rat brain.[75] However, steroid effects on this early stage of brain aromatase development do not appear to have been tested. Neonatal gonadectomy and subsequent T treatment attenuate the sex difference in aromatase activity of adult rats in the POA.[76] It can be suggested that changes in aromatase activity are due in part to maturation of the androgen receptor system involved in regulation of the enzyme, although other factors such as neurotransmitters are undoubtedly involved. The current view from the literature on the mammalian brain suggests, therefore, that, unlike the adult, steroids do not directly modify developing brain aromatase activity before birth. There appears to be a fundamental difference between regulation of the brain aromatase systems in developing and adult mammalian brain. However, the results of fetal exposure to hormones via maternal treatment are difficult to interpret, because the metabolic potential of the feto-placental unit interferes with levels of exogenous steroid reaching the brain in mammals. There are, however, transient changes in the activity of the developing aromatase system at puberty,[72] suggesting that regulation of the enzyme can fluctuate according to stage of postnatal development.

Few studies have so far been able to demonstrate an inductive effect of T on the embryonic brain aromatase system of mammals. Initial work has been carried out in birds in view of the accessibility of the avian embryo for experimental alteration of steroid levels. The conclusions from work on the early embryogenesis of Japanese quail[77,78] are that (1) T influences E_2 formation in the developing brain, and (2) the sensitivity of the aromatase system is phasic in that cells containing the enzyme do not respond before a critical period of development. Thus, in quail, there is an abrupt appearance of sensitivity to androgen around embryonic Day 10 to 12.

In understanding the role of aromatase in the sexual differentiation of the central nervous system, the following questions have to be answered: (1) Is the enzyme localized in developing brain areas which later participate in the estrogenic action necessary to regulate the neuroendocrine system and behavior? (2) Which brain cells contain the aromatase and competing enzyme systems? (3) What regulates the brain aromatase? (4) Does E_2 formation reach a peak during specific steroid-sensitive periods of brain development? At present, the only way to begin to answer these questions, as they concern fetal mammalian brain development, is by using primary or organotype cell culture techniques in which cells containing the enzyme can be distinguished in the two sexes. At the same time a direct comparison has to be made with the intact developing brain using radiometric assay methods for enzyme actitivity, which have successfully demonstrated sex differences in the perinatal brain aromatase system previously.[67,74] The mouse is a good model, because stages of embryogenesis can be recognized following timed matings, and the aromatase gene has been cloned and sequenced.[48] The BALB/c mouse strain has an active and regionally localized aromatase system.[79] In addition, brain cell types can be identified using specific probes for neurons and various types of glia. Because there are sex differences in cultured hypothalamic neurons, identified for neurotransmitter systems,[80-84] the developing aromatase system can be related to the distribution of neuroactive catecholamines and peptides. For our studies, the activity of the developing aromatase system has been compared using tissue homogenates dissected from intact fetal and postnatal mouse brains, and cells derived from primary cell culture. Our objective has been to assess the development of this enzyme system in conditions matching as far as possible the *in vivo* situation. Using this design, we have compared hypothalamic and cerebral homogenates of fetuses of known sex. For example, tissue from embryonic day (ED) 13, 15, and 17 have been microdissected from fetuses and enzyme activity compared with neonates using a sensitive *in vitro* ^3H_2O aromatase microassay.[79] The developmental pattern of aromatase activity was traced in genderspecific neuronal cell cultures prepared from ED 13 and 15 mouse cerebral hemispheres and hypothalami. A regional distribution of aromatase activity exists which can be identified in samples from primary cell culture and also in micropunched samples taken *in vivo* from the intact brain. In the cortex, aromatase activity is present, but activity is low, and sex differences are difficult to measure in ED 15 mouse cultured cells or dissected cortex from fetuses or neonates.[85,86] However, sex differences in aromatase activity are found in hypothalamic cultured cells as early as ED 13. In ED 15 cultured hypothalamic cells from both sexes, aromatase activity is detectable after 3 days *in vitro* (DIV), but there are no sex differences. Moreover, sex differences appear after 6 DIV, and activity is significantly increased, indicating that maturation of the hypothalamic aromatase system occurs over time in culture and is accompanied by sexual differentiation of the aromatase system. Sexual differentiation is occurring in the absence of normal brain connectivity. Male mouse cultured hypothalamic cells always have higher aromatase activity than females, and the same may be true for the gerbil brain (Figure 5.1). Aromatase in the mouse brain is primarily neuronal, since treatment of ED 15 hypothalamic cultures with kainic acid results in a 70 to 80% decrease in aromatase activity compared with nontreated cultures, and the sex difference seen in hypothalamic cells is no longer present. Astroglial-enriched postnatal hypothalamic cultures exhibit very low aromatase activity after 6 DIV, and no sex differences are present, providing further evidence that aromatase in the fetal mouse brain is neuronal.[86]

Kinetic studies comparing aromatase activity in male and female hypothalamic cultured cells and homogenates from ED 17 fetuses indicate similar T

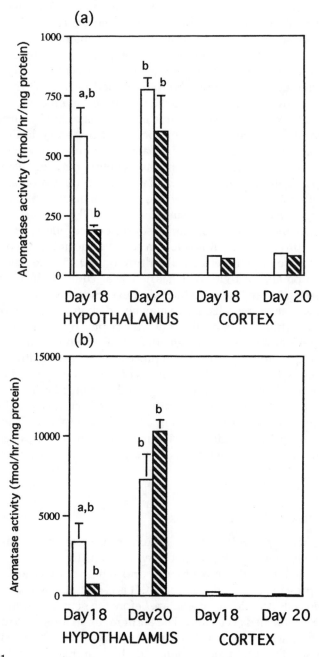

FIGURE 5.1
(a) Sex difference in aromatase activity on Day 18 but not on Day 20 of fetal development in gerbils. Hypothalamus and cortex were cultured for 6 days *in vitro.* [a]$p < 0.05$ male compared to female hypothalamus; [b]$p < 0.05$ male and female hypothalamus compared to cortex. (b) Microdissected brain samples from the intact brain showing a sex difference on Day 18 but not Day 20 of gestation. Open bars, male; cross-hatched bars, female.

substrate binding affinities (apparent K_m of about 40 nM), suggesting that cultured and intact neuronal aromatases have common catalytic properties.[86] Thus, the mouse primary cell culture work demonstrates that the sexually dimorphic aromatase activity in the embryonic brain is neuronal rather than astroglial. There is regional specificity in aromatase activity within the brain, and sex differences are apparent both in the perinatal intact hypothalamus and ED 15 cultured neurons. Since ED 17 hypothalamic neurons show a more-marked sex difference in enzyme activity than ED 15 neurons, a rapid maturation of the aromatase system occurs at this stage in development which appears to match a peak in circulating T in male fetuses.

The developmental profile of aromatase activity in hypothalamic cultures is also of particular interest. After 3 DIV in ED 15 cells, aromatase activity does not differ significantly between the sexes but exhibits a different developmental pattern (male greater than female) by 6 DIV. At present, we cannot determine whether this developmental sex difference reflects a male-specific, cell-intrinsic program, or if it is due to a cell-specific induction of the developing aromatase system before cell culture on ED 15. The developmental increase in aromatase activity, which is neuronal, could be due either to more cells expressing aromatase or to increased expression within an existing set of aromatase-containing cells. There may be a dichotomy in the source of the enzyme due to translation of different transcripts. This possibility depends on whether the aromatase cells seen after 3 DIV are shown to be neuroblasts or developing neuronal soma without processes. Since the cultures are grown in a medium with undetectable amounts of sex steroids, sex differences in aromatase-containing neurons of the early embryonic brain develop in the absence of T or E_2.[87] However, hypothalamic cells cultured on ED 13 or 15 may be influenced at earlier stages in development by circulating steroids. Recently, cell type and region-specific mRNA expression has been demonstrated in mouse cultured brain cells,[88] but sex differences were not examined in this study.

A. Aromatase in Fetal Brain Cells

Aromatase activity in cultured cells and microdissected brain samples taken from the developing hypothalamus provide a sensitive measurement in trying to establish sexual dimorphisms, and whether there are changes in enzyme expression during development. It is essential, however, to establish which brain cells contain the aromatase before the neuroanatomical distribution can be established accurately. The adult rat brain aromatase distribution has been identified using polyclonal antibodies.[89] However, there is disagreement over whether brain areas containing high aromatase activity, such as the POA, contain a large network of immunoreactive cells. In our studies we have developed a polyclonal, mouse-specific aromatase antibody, deduced from a cDNA sequence of mouse ovarian P-450 aromatase.[48] Use of this antibody has allowed us to identify aromatase-containing cells and their morphological characteristics in embryonic cultures (Figure 5.2). The anti-body recognizes a single protein band with an apparent molecular weight of approximately 55 kDa which corresponds to previously reported data for the aromatase protein in rat ovaries,[89] human placenta,[90] and ovaries and hypo-thalamus.[91] The antibody also inhibits aromatase activity in male mouse hypothalamic culture homogenates.[92] Aromatase immunoreactivity is local-ized in neurons of the intact developing mouse hypothalamus and cortex, but is not seen in astroglial cells or oligodendrocytes. Consequently, these obser-vations confirm that aromatase in the brain and sex differences in androgen aromatization are restricted to neurons. The neuronal aromatase is colocal-ized in different cells with both androgen and estrogen receptors. In contrast to the aromatase, glial cells appear to be involved in other pathways of andro-gen metabolism, for example, 5α-reduction and androstanedione-3α-diol formation.[93] The neuronal localization of aromatase in mouse cultures con-trasts with data on the avian zebra finch brain in which cultured glial cells express relatively high levels of aromatase activity and mRNA signal.[94,95]

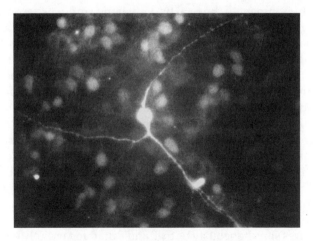

FIGURE 5.2
Aromatase neuron with dendrites and varicosities stained immunocytochemically with anti-P-450$_{AROM}$ using four expressed codons of the P-450 aromatase gene (*CYP19*)

Hypothalamic cells cultured from ED 13 mouse fetuses contain only a few aromatase-immunoreactive (AR-IR) cells, and there are no immunoreactive processes. However, cultures of ED 15 hypothalamic cells form a dense net-work of microtubuli-associated protein (MAP2)-positive neurons (60 to 70% of cells) consisting of neural aggregates of clumped soma and processes.[86] Cultures also contain AR-IR cell aggregates and interconnecting processes which also show immunoreactivity. Double immunofluorescent staining for neurons (using MAP2) and aromatase indicates that the AR-IR cells are neu-rons, probably linked by AR-IR-positive processes. These processes do not appear in ED 13 cultured neurons even when the culture period is extended from 6 to 10 days. There is a direct relationship between the number of cells in hypothalamic culture and the enzyme activity measured by radiometric assay. A pronounced increase in aromatase activity between ED 13 and 15 is correlated with the progressive increase in AR-IR cell number. A regional difference between embryonic cortex and hypothalamic aromatase activity matched by the AR-IR neurone number in each area can be measured.[87] The number of AR-IR cells is low in the cortex compared with the hypothalamus. Significantly higher numbers of AR-IR neurons which costain with MAP2 occur in male cultures compared with females. However, the aromatase activity expressed per single AR-IR neurone in ED 15 cultures does not differ between male and female (capacity for estrogen formation: 0.084 ± 0.01 and 0.080 ± 0.01 fmol/h/mg protein, respectively). Therefore, there are likely to be more aromatase neurons in the male hypothalamus than the female, rather than more aromatase activity in fetal male neurons.

B. Regulation of the Fetal Brain Aromatase by Steroids

Does the T environment of the fetal hypothalamus influence aromatase activity in this region? Brain aromatase in the intact mouse and rat brain is clearly neuronal; therefore, the effects of manipulation of steroid level can be studied experimentally using a primary cell culture system. Interference by the feto-placental unit, maternal metabolism, and afferents extrinsic to the fetal hypothalamus can be avoided. When the cultures are treated with T for the entire culture period (10^{-8} M T for 6 DIV), the steroid treatment doubles the aromatase activity relative to controls. This increase is specific to the hypothalamic cultures and is not seen in cultures of cortex cells. Female cultured hypothalamic neurons are also steroid-sensitive and show increased aromatase activity in response to T.[91] Therefore, inductive effects of T are not specific to male neurons. By using the aromatase antibody, it is possible to identify hormonal effects on AR-IR cell number. T increases the number of AR-IR neurons by at least 70% in cultures of male hypothalamic cells. Treatment of cultures simultaneously with T and flutamide entirely eliminates the T effect. This action of flutamide suggests that the androgen receptor is involved in the induction of neuronal fetal aromatase.[66,91] We have no evidence so far that T induces proliferation of neuroblasts and neurons expressing aromatase, because MAP2-IR neurons in cultures containing AR-IR neurons do not stain for PCNA which is expressed in proliferating cells undergoing cell division. T, therefore, is likely to increase aromatase gene expression in mouse hypothalamic neurons. However, one possibility that cannot be excluded is that T may also increase the number of cells expressing aromatase by decreasing cell death in the neuronal population of the male hypothalamus. Steroids, for example, the glucocorticoids, are known to influence both neurogenesis and cell death in the brain.[96] This androgen-induced decrease in apoptosis could be more prevalent in the male than the female brain. The effects of T may also be mediated indirectly via catecholamines for example.[97] Cyclic AMP is also involved.[72,98] Neurosteroids synthesized within the brain may also modulate aromatase activity.[99,100]

C. Effects of an Aromatase Inhibitor

The effects of a triazole aromatase inhibitor (Arimidex) on the development of behavior, the brain, and the reproductive endocrine system were tested; 4 h after a single injection of Arimidex in adult gerbils aromatase activity was reduced in the POA, but not in other areas of the brain (Figure 5.3). Another

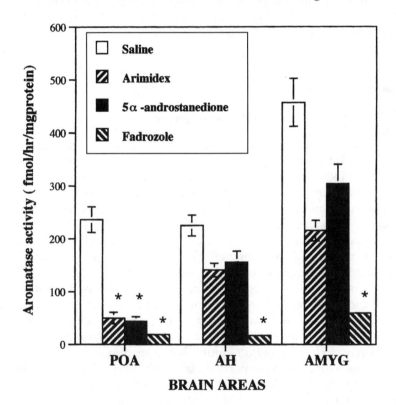

FIGURE 5.3
Aromatase inhibitor effects in adult gerbil brains 4 h after treatment with a single injection. *Each group compared to saline, $p < 0.05$.

aromatase inhibitor (Fadrozole) reduced aromatase activity in all areas of the brain without exception. Fadrozole is toxic to the central nervous system in that it acts nonspecifically throughout the brain. 5α-Androstanedione which is an endogenous inhibitor of brain aromatase, produced in the brain from T, is effective in the POA but not in other areas.[87] Therefore, Arimidex matches the results obtained with the natural inhibitor. A single injection of Arimidex reduces aromatase activity significantly in both the male and the female fetus when given to the mother at Day 18 of gestation in the gerbil (Figure 5.4a). This is the period when there is a sex dimorphism in aromatase activity in the brain during early development (see Figure 5.1).

Arimidex in the gerbil also has a long-term action if given on Day 18 of gestation to the mother. This is an effect on the endocrine system. Thus Arimidex is highly effective in reducing anogenital distance in the male, suggesting that aspects of testicular descent are controlled by estrogen. Arimidex also has long-term behavioral effects in that the reproductive behavior of the male is selectively influenced by Arimidex. In males treated with Arimidex, the latency to the first litter is extended after only a single injection to the mother of Arimidex on Day 18 of gestation (Figure 5.4b). Undoubtedly, for behavioral effects Arimidex must influence the developing hypothalamus.

V. Conclusions

Steroids have functional effects on the differentiation of sexual dimorphisms in the brain. Estrogens, notably E_2 derived by aromatization of T within brain neurons are likely to be effective in differentiating behavioral and neuroendocrine mechanisms in male birds and mammals, including primates such as humans. In the adult brain, the P-450 enzyme, aromatase, converts T to E_2. The enzyme is expressed genetically (*CYP 19* gene) and is present throughout the neurone, including dendritic and axonal terminals. The product, E_2, is probably transynaptic and may affect synaptic transmission. In the adult brain, circulating steroid levels and environmental factors, such as sociosexual and seasonal stimuli, affect the formation of E_2 from T by aromatase activity in brain areas related to behavior. Questions which still have to be resolved are (1) What determines changes in aromatase activity effective for the differentiation of the sexually dimorphic brain? (2) When does this critical regulation occur? There is little doubt that in mammals the control of aromatase gene expression is important both in the earliest fetal and postnatal brain development and later in puberty. The limits of the sensitive periods for brain aromatase regulation are still virtually unknown. In the *in vivo* fetal mouse brain, E_2 formation is neuronal rather than glial.

There are more neurons containing aromatase activity developing in the embryonic male than in the female, and this sex difference exists at early stages of embryonic development. However, the brain aromatase does not appear to be regulated by T during this very early embryonic period. Regulation is genetic during the earliest stages of brain development. It is only later in development that the brain aromatase system is controlled by T and possibly other steroids. Development of the androgen receptor may have a role in determining when steroid-sensitive phases of aromatase regulation are initiated. Although T is important in the control of E_2 formation in the brain, the exact mechanism of T action on the aromatase gene is still

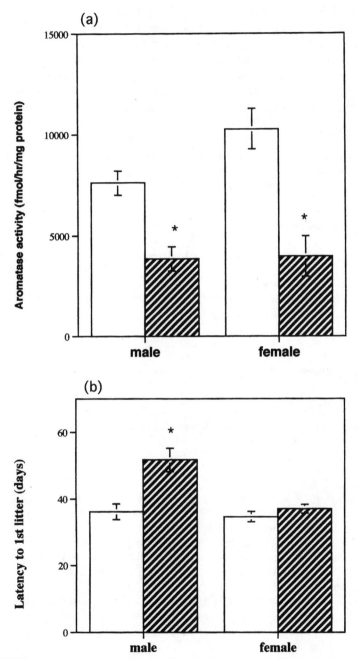

FIGURE 5.4

(a) Effects of Arimidex on hypothalamic aromatase activity of gerbil fetuses. Pregnant females were injected with Arimidex at the onset of the period of sexual dimorphism in aromatase activity, on Day 17, and killed at embryonic Day 20. (b) Long-term effect on latency to first litter of male and female gerbils treated with Arimidex on embryonic Day 18. *p, 0.05. Open bars, saline control; cross-hatched bars, Arimidex.

unknown. Factors that promote the development of connectivity in steroid-sensitive neurons, and whether, as is likely, there is a steroid-independent phase in neuronal ontogeny, are also unknown. However, developing aromatase-containing neuroblasts probably form processes that connect to other aromatase cells. Since the sex of individuals can be identified from an early embryonic age, it should be possible in the future to determine the relative contribution of genetic and hormonal factors in brain differentiation. The estrogen-forming capacity of the developing male hypothalamus has the characteristics and plasticity necessary for provision of effective E_2 for brain differentiation. The aromatase activity required develops at critical stages in prenatal ontogeny.

Acknowledgments

I am grateful to Dr. Rosemary E. Hutchison for her comments on the manuscript and for her contributions to this work. The contributions of Dr. Andrew Wozniak are also greatly appreciated.

References

1. Arnold, A.P., Developmental plasticity in neural circuits controlling bird song: sexual differentiation and the neural basis of learning, *J. Neurobiol.*, 23, 1506, 1992.
2. Breedlove, S.M., Sexual dimorphism in the vertebrate nervous system, *J. Neurosci.*, 12, 4133, 1992.
3. Hutchison, J.B. and Hutchison, R.E., Sexual development at the neurohormonal level: the role of androgens, in *Pedophilia: Biosocial Dimensions*, Feierman, J., Ed., Springer-Verlag, Berlin, 1990, 510.
4. Beach, F.A., Hormonal factors controlling the differentiation, development and the display of copulatory behavior in the Ramstergig and related species, in *Biopsychology of Development*, Aronson L. and Tobach, E., Eds., Academic Press, New York, 1971, 249.
5. Komisaruk, B.R., Adler, N.T., and Hutchison, J.B., Genital sensory field: enlargement by estrogen treatment in female rats, *Science*, 178, 1295, 1972.
6. Goy, R.W. and McEwen, B.S., *Sexual Differentiation of the Brain*, MIT Press, Cambridge, MA, 1980.
7. Arnold, A.P. and Gorski, R.A., Gonadal steroid induction of structural sex differences in the central nervous system, *Annu. Rev. Neurosci.*, 7, 413, 1984.
8. McEwen, B.S., Lieberburg, I., Chaptal, C., and Krey, L., Aromatization: important for sexual differentiation of the neonatal rat brain, *Horm. Behav.*, 9, 249, 1979.

9. MacLusky, N.J. and Naftolin, F., Sexual differentiation of the central nervous system, *Science*, 211, 1294, 1981.

10. Choate, J.V.A. and Resko, J.A., Prenatal inhibition of aromatase activity affects luteinizing hormone feedback mechanisms and reproductive behaviors of adult guinea pigs, *Biol. Reprod.*, 51, 1273, 1994.

11. Pons, S. and Torres-Aleman, I., Estradiol modulates insulin-like growth factor I receptors and binding proteins in neurons from the hypothalamus, *J. Neuroendocrinol.*, 5, 267, 1993.

12. Naftolin, F., Ryan, K.J., Davies, I.J., Reddy, V.V., Flores, F., Petro, Z., Kuhn, M., White, R.J., Takaoka, Y., and Wolin, L., Formation of estrogens by central neuroendocrine tissues, *Recent Prog. Horm. Res.*, 31, 295, 1975.

13. Hutchison, J.B. and Steimer Th., Androgen metabolism in the brain: behavioral correlates, *Prog. Brain Res.*, 61, 23, 1984.

14. Naftolin, F., Brain aromatization of androgens, *J. Reprod. Med.*, 39, 257, 1994.

15. Simpson, E.R. , Zhao, Y., Agarwal, V.R. , Dodson, M., Bulun, S.E., Hinshelwood, M.M. , Graham-Lawrence, S., Sun, T., Fisher, C.R., Qin, K., and Mendelson, C.R. Aromatase expression in health and disease, *Recent Prog. Horm. Res.*, 52, 185, 1997.

16. Pilgrim, C. and Hutchison, J.B., Developmental regulation of sex differences in the brain: can the role of gonadal steroids be re-defined? *Neuroscience*, 60, 843, 1994.

17. Jost, A.A., A new look at the mechanisms controlling sex differentiation in mammals, *Johns Hopkins Med. J.*, 130, 38, 1972.

18. Jost, A.A. and Magre, S., Testicular development phases and dual hormonal control of sexual organogenesis, in *Sexual Differentiation: Basic and Clinical Aspects*, Serio, M., Motta, M., Aznisi, M., and Martini, L., Eds., Raven Press, New York, 1984, 112.

19. Harris, G.W. and Levine, S., Sexual differentiation of the brain and its experimental control, *J. Physiol.*, 181, 379, 1965.

20. Phoenix, Ch.H., Goy, R.W., Gerall, A.A., and Young, W.C., Organizing action of prenatally administered testosterone propionate on the tissues mediating mating behavior in the female guinea pig, *Endocrinology*, 55, 369, 1959.

21. Adkins, E.K., Hormonal basis of sexual differentiation in the Japanese quail, *J. Comp. Physiol. Psychol.*, 89, 61, 1975.

22. Breedlove, S.M. and Arnold, A.P., Hormonal control of a developing neuromuscular system. II. Sensitive periods for the androgen-induced masculinization of the rat spinal nucleus of the bulbocavernosus, *J. Neurosci.*, 3, 424, 1983.

23. Whalen, R.E., Sexual differentiation: models, methods and mechanisms, in *Sex Differences in Behavior*, Friedman, R.C., Richart, R.M., and Van de Wiele, R.L., Eds., Wiley, New York, 1974, 467.

24. Yahr, P., Searching for the neural correlates of sexual differentiation in a heterogeneous tissue, in *Neurobiology*, Gilles, R. and Balthazart, J., Eds., Springer-Verlag, Berlin, 1985, 180.

25. Yahr, P., Sexual differentiation of behavior in the context of developmental psychobiology, in *Handbook of Behavioral Neurobiology*, Vol. 9, Blass, E.M., Ed., Plenum Press, New York, 197, 1988.

26. Olsen, K.L., Genetic determinants of sexual differentiation, in *Hormones and Behavior in Higher Vertebrates*, Balthazart, J., Pröve, E., and Gilles, R., Eds., Springer-Verlag, Berlin, 1985, 138.

27. Baum, M.J., Carroll, R.S., Cherry, J.A., and Tobet, S.A., Steroidal control of behavioral, neuroendocrine and brain sexual differentiation: studies in a carnivore, the ferret, *J. Neuroendocrinol.*, 2, 401, 1990.

28. Arnold, A.P. and Breedlove, S.M., Organizational and activational effects of sex steroids on brain and behavior: a reanalysis, *Horm. Behav.*, 19, 469, 1985.

29. Hutchison, R.E., Hormonal differentiation of sexual behavior in Japanese quail, *Horm. Behav.*, 11, 363, 1978.

30. Schumacher, M. and Balthazart, J., Sexual differentiation is a biphasic process in mammals and birds, in *Neurobiology*, Gilles, R. and Balthazart, J., Eds., Springer-Verlag, Berlin, 1985, 203.

31. Hutchison, J.B. and Hutchison, R.E., Phasic effects of hormones in the avian brain during behavioral development, in *Neurobiology*, Gilles, R. and Balthazart, J., Eds., Springer-Verlag, Berlin, 1985, 105.

32. Garcia-Segura, L.M., Baetens, D., and Naftolin, F., Synaptic remodelling in arcuate nucleus after injection of estradiol valerate in adult female rats, *Brain Res.*, 366, 131, 1986.

33. Nishizuka, M. and Arai, Y., Organizational action of estrogen on synaptic pattern in the amygdala: implications for sexual differentiation of the brain, *Brain Res.*, 213, 422, 1981.

34. Gould, E., Woolley, C.S., Frankfurt, M., and McEwen, B.S., Gonadal steroids regulate dendritic spine density in hippocampal pyramidal cells in adulthood, *J. Neurosci.*, 10, 1286, 1990.

35. Garcia-Segura, L.M., Olmos, G., Tranque, P., and Naftolin, F., Rapid effects of gonadal steroids upon hypothalamic neuronal membrane ultrastructure, *J. Steroid Biochem.*, 27, 615, 1987.

36. Matsumoto, A., Synaptogenic action of sex steroids in developing and adult neuroendocrine brain, *Psychoneuroendocrinology*, 16, 25, 1991.

37. Naftolin, F., Garcia-Segura, L.M., Keefe, D., Leranth, C., MacLusky, N.J., and Brawer, J.R., Estrogen effects on the synaptology and neural membranes of the rat hypothalamic arcuate nucleus, *Biol. Reprod.*, 42, 21, 1990.

38. MacLusky, N.J., Clark, A.S., Naftolin, F., and Goldman-Rakic, P.S., Estrogen formation in the mammalian brain: possible role of aromatase in sexual differentiation of the hippocampus and neocortex, *Steroids*, 50, 459, 1987.

39. Gorski, R.A., Gordon, J.H., Shryme, J.E., and Southam, A.M., Evidence for a morphological sex difference within the medial preoptic area of the rat brain, *Brain Res.*, 148, 333, 1978.

40. Kawata, M., Roles of steroid hormones and their receptors in structural organization in the nervous system, *Neurosci. Res.*, 24, 1, 1995.

41. Yahr, P. and Stevens, D.R., Hormonal control of sexual and scent marking behaviors of male gerbils in relation to the sexually dimorphic area of the hypothalamus, *Horm. Behav.*, 21, 331, 1987.

42. Holman, S.D. and Hutchison, J.B., Differential effects of neonatal castration on the development of sexually dimorphic brain areas in the gerbil, *Dev. Brain Res.*, 61, 147, 1991.

43. Nottebohm, F., Alvarez-Buylla, A., Cynx, J., Kirn, J., Ling, C.-Y., Nottebohm, M., Suter, R., Tolles, A., and Williams, H., Song learning in birds: the relationship between perception and production, in *Philosophical Transactions: Biological Sciences*, Krebs, J.R. and Horn, G., Eds., Royal Society of London, London, 329, 115, 1990.

44. Ehret, G., Left hemisphere advantage in the mouse brain for recognizing ultrasonic communication calls, *Nature*, 325, 249, 1987.
45. Callard, G.V. and Pasmanik, M., The role of estrogen as a parahormone in brain and pituitary, *Steroids*, 50, 475, 1987.
46. McPhaul, M.J., Noble, J.F., Simpson, E.R., Mendelson, C.R., and Wilson, J.D., The expression of a functional cDNA encoding the chicken cytochrome P-450arom (Aromatase) that catalyses the formation of estrogen from androgen, *J. Biol. Chem.*, 263, 16358, 1988.
47. Simpson, E.R., Regulation of expression of the genes encoding steroidogenic enzymes, *J. Steroid Biochem. Mol. Biol.*, 40, 45, 1991.
48. Terashima, M., Toda, K., Kawamoto, T., Kuribayashi, I., Ogawa, Y., Maeda, T., and Shizuta, Y., Isolation of a full-length cDNA encoding mouse aromatase P450, *Arch. Biochem. Biophys.*, 285, 231, 1991.
49. Callard, G.V., Drygas, M., and Gelinas, D., Molecular and cellular physiology of aromatase in the brain and retina, *J. Steroid Biochem. Mol. Biol.*, 44, 541, 1993.
50. Steimer, Th. and Hutchison, J.B., Androgen increases formation of behaviorally effective estrogen in the dove brain, *Nature*, 292, 345, 1981.
51. Fusani, L., Action of Oestrogen on Brain Mechanisms of Behavior in the Ring Dove (*Streptopelia risoria*) and the canary (*Serinus canaria*), Ph.D. dissertation, University of Cambridge, U.K., 1999.
52. Tsuruo, Y., Ishimura, K., Fujita, H., and Osawa, Y., Immunocytochemical localization of aromatase-containing neurons in the rat brain during pre- and postnatal development, *Cell Tissue Res.*, 278, 29, 1994.
53. Tsuruo, Y., Ishimura, K., and Osawa, Y., Presence of estrogen-receptors in aromatase-immunoreactive neurons in the mouse-brain, *Neurosci. Lett.*, 195, 49, 1995.
54. Tsuruo, Y., Ishimura, K., Hayashi, S., and Osawa, Y., Immunohistochemical localization of estrogen-receptors within aromatase-immunoreactive neurons in the fetal and neonatal rat-brain, *Anat. Embryol.*, 193, 115, 1996.
55. Gahr, M. and Hutchison, J.B., Behavioral action of estrogen in the male dove brain: area differences in codistribution of aromatase activity and estrogen receptors are steroid-dependent, *Neuroendocrinology*, 56, 74, 1992.
56. Schumacher, M. and Balthazart, J., Testosterone-induced brain aromatase is sexually dimorphic, *Brain Res.*, 370, 285, 1986.
57. Balthazart, J., Foidart, A., and Harada, N., Immunocytochemical localization of aromatase in the brain, *Brain Res.*, 514, 327, 1990.
58. Roselli, C.E., Ellinwood, W.E., and Resko, J.A., Regulation of brain aromatase activity in rats, *Endocrinology*, 14, 192, 1984.
59. Hutchison, R.E., Hutchison, J.B., Steimer, Th., Steel, E., Powers, J.B., Walker, A.P., Herbert, J., and Hastings, M.H., Brain aromatization of testosterone in the male Syrian hamster: Effects of androgen and photoperiod, *Neuroendocrinology*, 53, 194, 1991.
60. Abdelgadir, S.E., Resko, J.A., Ojeda, S.R., Lephart, E.D., McPhaul, M.J., and Roselli, C.E., Androgens regulate aromatase cytochrome P450 messenger ribonucleic acid in rat brain, *Endocrinology*, 135, 395, 1994.
61. Yamada, K., Harada, N., Tamaru, M., and Takagi, Y., Effects of changes in gonadal hormones on the amount of aromatase messenger RNA in mouse brain diencephalon, *Biochem. Biophys. Res. Commun.*, 195, 462, 1993.

62. Yamada, N.M., Hirata, S., and Kato, J., Distribution and postnatal changes of aromatase mRNA in the female rat brain, *J. Steroid Biochem. Mol. Biol.*, 48, 529, 1994.

63. Harada, N. and Yamada, K., Ontogeny of aromatase messenger ribonucleic acid in mouse brain: fluorometrical quantitation by polymerase chain reaction, *Endocrinology*, 131, 2306, 1992.

64. Harada, N., Abe-Dohmae, S., Loeffen, R., Foidart, A., and Balthazart, J., Synergism between androgens and estrogens in the induction of aromatase and its messenger RNA in the brain, *Brain Res.*, 622, 243, 1993.

65. Hutchison, J.B., How does the environment influence the behavioral action of hormones? in *The Development and Integration of Behavior*, Bateson P.P.G., Ed., Cambridge University Press, Cambridge, U.K., 1991, 149.

66. Hutchison, J.B., Wozniak, A., Beyer, C., and Hutchison, R.E., Regulation of sex-specific formation of estrogen in brain development, endogenous inhibitors of aromatase, *J. Steroid Biochem. Mol. Biol.*, 56, 201, 1996.

67. George, F.W. and Ojeda, S.R., Changes in aromatase activity in the rat brain during embryonic, neonatal and infantile development, *Endocrinology*, 111, 522, 1982.

68. Weisz, J., Brown, B.L., and Ward, I.L., Maternal stress decreases steroid aromatase activity in brains of male and female rat fetuses, *Neuroendocrinology*, 35, 374, 1982.

69. Krohmer, R.W. and Baum, M.J., Effect of sex, intrauterine position and androgen manipulation on the development of brain aromatase activity in fetal ferrets, *J. Neuroendocrinology*, 1, 265, 1989.

70. Weaver, C.E. and Baum, M.J., Differential regulation of brain aromatase by androgen in adult and fetal ferrets, *Endocrinology*, 128, 1247, 1991.

71. Tobet, S.A., Baum, M.J., Tang, H.B., Shim, J.S., and Canick, J.A., Aromatase activity in the perinatal rat forebrain: effects of age, sex and intra-uterine position, *Dev. Brain Res.*, 23, 171, 1985.

72. Lephart, E.D. and Ojeda, S.R., Hypothalamic aromatase activity in male and female rats during juvenile peripubertal development, *Neuroendocrinology*, 51, 385, 1990.

73. Connolly, P.B., Roselli, C.E., and Resko, J.A., Aromatase activity in developing guinea pig brain: ontogeny and effects of exogenous androgens, *Biol. Reprod.*, 50, 436, 1994.

74. MacLusky, N.J., Philip, A., Hurlburt, C., and Naftolin, F., Estrogen formation in the developing rat brain: sex differences in aromatase activity during early post-natal life, *Psychoneuroendocrinology*, 10, 355, 1985.

75. von Ziegler, N.I. and Lichtensteiger, W. Asymmetry of brain aromatase activity: region- and sex-specific developmental patterns, *Neuroendocrinology*, 55, 512, 1992.

76. Steimer, Th. and Hutchison, J.B., Is androgen-dependent aromatase activity sexually differentiated in the rat and dove preoptic area? *J. Neurobiol.*, 21, 787, 1990.

77. Schumacher, M. and Hutchison, J.B., Testosterone induces hypothalamic aromatase during early development in quail, *Brain Res.*, 377, 63, 1986.

78. Schumacher, M., Hutchison, R.E., and Hutchison, J.B., Ontogeny of testosterone-inducible brain aromatase activity, *Brain Res.*, 441, 98, 1988.

79. Wozniak, A., Hutchison, R.E., and Hutchison, J.B., Localisation of aromatase activity in androgen target areas of the mouse brain, *Neurosci. Lett.*, 146, 191, 1992.

80. Beyer, C., Epp, B., Fassberg, J., Reisert, I., and Pilgrim, C., Region- and sex-specific differences in maturation of astrocytes in dissociated cell cultures of embryonic rat brain, *Glia*, 3, 55, 1990.

81. Beyer, C., Pilgrim, C., and Reisert, I., Dopamine content and metabolism in mesencephalic and diencephalic cell cultures: sex differences and effects of sex steroids, *J. Neurosci.*, 11, 1325, 1991.

82. Beyer, C., Kolbinger, W., Froehlich, U., Pilgrim, C., and Reisert, I., Sex differences of hypothalamic prolactin cells develop independently of the presence of sex steroids, *Brain Res.*, 593, 253, 1992.

83. Beyer, C., Eusterschulte, B., Pilgrim, C., and Reisert, I., Sex steroids do not alter sex differences in tyrosine hydroxylase activity of dopaminergic neurons *in vitro*, *Cell Tissue Res.*, 270, 547, 1992.

84. Kolbinger, W., Trepel, M., Beyer, C., Pilgrim, Ch., and Reisert, I., The influence of genetic sex on sexual differentiation of diencephalic dopaminergic neurons *in vitro* and *in vivo*, *Brain Res.*, 544, 349, 1991.

85. Beyer, C., de la Silva, C., and Hutchison, J.B., Kainic acid has cell-specific effects on survival of hypothalamic mouse neurons *in vitro*, *NeuroReport*, 4, 547, 1993.

86. Beyer, C., Wozniak, A., and Hutchison, J.B., Sex-specific aromatization of testosterone in mouse hypothalamic neurons, *Neuroendocrinology*, 58, 673, 1993.

87. Hutchison, J.B., Beyer, C., Green, S., and Wozniak, A., Brain formation of estrogen in the mouse: sex dimorphism in aromatase development, *J. Steroid Biochem. Mol. Biol.*, 49, 407, 1994.

88. Abe-Dohmae, S., Tanaka, R., and Harada, N., Cell type- and region-specific expression of aromatase mRNA in cultured brain cells, *Mol. Brain Res.*, 24, 153, 1994.

89. Sanghera, M.K., Simpson, E.R., McPhaul, M.J., Kozlowski, G., Conley, A.J., and Lephart, E.D., Immunocytochemical distribution of aromatase cytochrome P450 in the rat brain using peptide-generated polyclonal antibodies, *Endocrinology*, 129, 2834, 1991.

90. Harada, N., Novel properties of human placental aromatase as cytochrome P-450: purification and characterization of a unique form of aromatase, *J. Biochem.*, 103, 106, 1988.

91. Beyer, C., Green, S.J., and Hutchison, J.B., Androgens influence sexual differentiation of embryonic mouse hypothalamic aromatase neurons *in vitro*, *Endocrinology*, 135, 1220, 1994.

92. Beyer, C., Green, S.J., Barker, P.J., Huskisson, N.S., and Hutchison, J.B., Aromatase-immunoreactivity is localised specifically in neurons in the developing mouse hypothalamus and cortex, *Brain Res.*, 638, 203, 1994.

93. Melcangi, R.C., Celotti, F., Castano, P., and Martini, L., Differential localization of the 5alpha-reductase and the 3alpha-hydroxysteroid dehydrogenase in neuronal and glial cultures, *Endocrinology*, 132, 1252, 1993.

94. Arnold, A.P., Amur-Vmarjec, S., Campagnoni, A., and Schlinger, B.A., Glia express high levels of aromatase in cultures of developing zebra finch telencephalon, *Soc. Neurosci. Abstr.*, 231, 1992.

95. Schlinger, B.A., Amur-Umarjee, S., Shen, P., Campagnoni, A.T., and Arnold, A.P., Neuronal and non-neuronal aromatase in primary cultures of developing zebra finch telencephalon, *J. Neurosci.*, 14, 7541, 1994.
96. Gould, E. and McEwen, B.S., Neuronal birth and death, *Curr. Opin. Neurobiol.*, 3, 676, 1993.
97. Canick, J.A., Tobet, S.A., Baum, M.J., Vaccaro, D.E., Ryan, K.J., Leeman, S.E., and Fox, T.O., Studies on the role of catecholamines in the regulation of the developmental pattern of hypothalamic aromatase, *Steroids*, 50, 509, 1987.
98. Lephart, E.D., Simpson, E.R., and Ojeda, S.R., Effects of cyclic AMP and androgens on in vitro brain aromatase enzyme activity during prenatal development in the rat, *J. Neuroendocrinology*, 4, 29, 1991.
99. Tsutsui, K. and Yamazaki, T., Avian neurosteroids. I. Pregnenolone biosynthesis in the quail brain, *Brain Res.*, 678, 1, 1995.
100. Usui, M., Yamazaki, T., Kominami, S., and Tsutsui, K. Avian neurosteroids. II. Localization of a cytochrome P450scc-like substance in the quail brain, *Brain Res.*, 678, 10, 1995.

6

Glial Cells are Involved in Organizational and Activational Effects of Sex Hormones in the Brain

Luis Miguel Garcia-Segura, Gloria Patricia Cardona-Gomez, Jose Luis Trejo, Maria Carmen Fernandez-Galaz, and Julie Ann Chowen

CONTENTS

I. Introduction

Brain development and activity depend on coordinated functional interactions between glial cells and neurons. For many years glial cells were not considered to be relevant components in the process of sexual differentiation of brain structures. However, recent evidence has demonstrated that glial cells are affected by gonadal steroids and are actively involved in the organizational and activational effects of these hormones.[1,2] Astrocytes and oligodendrocytes in culture express receptors for sex steroids,[2,3] suggesting that at least in early stages of brain development glial cells may be a direct target of these hormones. Furthermore, estrogen receptor-immunoreactive glia have been detected *in situ* in the medial preoptic area and median eminence.[4] Sex hormones affect myelination by acting on Schwann cells[5] and modulate the response of nerve tissue to injury by acting on microglia and astroglia.[1] In addition to being a target for sex steroids, glial cells are also involved in their metabolism[6,7] and participate in the organizational and activational effects of sex steroids on synapse formation[8] and synaptic plasticity.[9]

Among glial cells, astroglia play a critical role in providing metabolites, trophic factors, and neuromodulators to neurons and in the regulation of extracellular ion concentrations and local cerebral blood flow. Furthermore, recent studies have shown that astroglia have a fundamental role in neural signaling by modulating synaptic transmission and synaptic plasticity. Therefore, astroglia are a crucial element to consider in order to understand the cellular mechanisms involved in the action of sex hormones on brain development and function. Furthermore, it is important to keep in mind the participation of astroglia in the response to injury, in regard to the therapeutical implications of sex steroids as promoters of neural regeneration and prevention of neuronal death. Here we review evidence for the sex hormone regulation of astroglial development and function. In addition, the role of astroglia in the interaction of sex hormones and growth factors is discussed.

II. Sex Hormones Promote Astroglia Differentiation and Plasticity

Several studies have shown that testosterone and estradiol promote astroglia differentiation and growth of astroglia processes *in vitro* and regulate the expression of the specific astrocytic marker glial fibrillary acidic protein (GFAP).[10-14] Effects of gonadal steroids on astroglia have also been demonstrated *in situ*.[14-21] For instance, administration of estradiol to adult ovariectomized rats, castration of newborn males, and testosterone administration to newborn females result in significant changes in the immunohistochemical distribution of GFAP in the striatum,[15,16] while castration of adult male rats results in elevated levels of GFAP mRNA in the hippocampus.[18] Furthermore, sex differences in astrocytic immunoreactivity have been detected in the hippocampus, striatum, and cerebellum.[16,22] In adult females, the surface density of GFAP-immunoreactive cells in the hilus of the dentate gyrus decreases after removal of circulating gonadal steroids by ovariectomy and increases after the pharmacological administration of either 17β-estradiol or progesterone, but not 17α-estradiol, to ovariectomized animals.[23] Furthermore, GFAP-immunoreactivity fluctuates during the estrous cycle following the physiological variations in the circulatory levels of ovarian hormones.[23]

These latter changes in immunoreactivity may reflect a cytoplasmic redistribution of GFAP, since neither GFAP transcription nor messenger RNA change in the hilus of the dentate gyrus during the estrous cycle.[14]

The focus of most studies on the effects of gonadal steroids on glia has been the neuroendocrine hypothalamus of rodents.[8-12,17,20,21,24-30] Sex differences in astroglia morphology and in the expression of astroglia cell markers have been described in several hypothalamic regions.[8,20,24,27,29] Furthermore, estradiol and testosterone regulate GFAP expression in hypothalamic astroglia in developing and adult animals.[8-12,14,17,20,21,24,25] Changes in GFAP immunoreactivity have also been described during the estrous cycle in the arcuate nucleus of the hypothalamus.[9,30] In contrast to what has been observed in the hilus of the dentate gyrus, changes in GFAP immunoreactivity in the arcuate nucleus are accompanied by modifications in the transcription of GFAP.[14,28]

III. Astroglia are Involved in Synaptic Effects of Sex Hormones

Studies conducted in the rat hypothalamic arcuate nucleus indicate that astroglia are involved in synaptic effects of gonadal hormones. Matsumoto and Arai, in a series of pioneering studies,[31,35] showed the existence of sex differences in synaptic connectivity in this nucleus and defined the role of estrogen and androgens in the genesis of sexually dimorphic synaptic contacts. More recent studies suggest that astroglia may be involved in the organizational effects of sex steroids on arcuate neurons. It has been proposed that testosterone or its metabolite estradiol may regulate the generation of the sexually dimorphic pattern of synaptic contacts by affecting the growth of astroglial processes on neuronal surfaces and the amount of neuronal membrane available for the establishment of synaptic contacts.[8] The genesis of the sexually dimorphic pattern of synaptic connectivity is accompanied by sex differences in the levels of GFAP, the number of astroglial cell processes and the amount of neuronal membrane surface covered by glia.[8,20,29] These sex differences are dependent on perinatal androgens. Administration of testosterone to newborn females increases GFAP and its mRNA levels, the number of astroglial cell processes, and the proportion of neuronal membrane covered by glia, while decreasing the final number of axosomatic synapses per neuron to male levels. Castration of newborn males results in the opposite outcome.[8,20]

In addition to the organizational effect of gonadal steroids, estrogen promotes synaptic plasticity in the arcuate nucleus. In postpuberal female rats, there is an estrogen-induced transient disconnection of GABAergic inputs to the somata of arcuate neurons during the preovulatory and ovulatory phases of the estrous cycle. This synaptic remodeling is blocked by progesterone and begins with the onset of female puberty.[9,30] Astrocytes appear to play a significant role in these synaptic changes as well. In the afternoon of proestrus, estradiol induces synthesis of GFAP and the growth of glial processes which ensheathe the neuronal membrane and displace the synaptic terminals. Glial processes retract and synapses reform in the afternoon of estrus.[9,30] Hormonally induced changes in the astrocytic ensheathing of hypothalamic neurons are not a peculiarity of the rat. These changes are also associated with estrogen-mediated modifications in the number of synaptic inputs to hypothalamic neurons in primates.[36,37]

IV. Glial Cells are Involved in the Interaction of Sex Hormones and Growth Factors

The complexity of gonadal steroid action on the brain is not only a consequence of the complexity of cellular interactions in neural tissue. Interactions between the signaling cascades of membrane receptors for peptide factors and intracellular steroid hormone receptors should also be taken into consideration. Sex hormones may interact with neurotrophins[38-41] and growth factors[42-48] to exert their effects on neurons and glial cells. These factors which include insulin-like growth factor-I (IGF-I), basic fibroblast growth factor (bFGF), and transforming growth factors α(TGF-α) and β(TGF-β), are involved in the maturation and regulation of the neuroendocrine hypothalamus.[42-47] For instance, estrogen regulates the production of TGF-α by hypothalamic glia[43] and this factor regulates luteinizing hormone-releasing hormone (LHRH) release via a glia to neuron signaling pathway acting on prostaglandin E_2 receptors in LHRH neurons.[46]

In the mediobasal hypothalamus and median eminence TGF-α is expressed by astrocytes and tanycytes.[43] Tanycytes are specialized glial cells that have many ultrastructural and immunological similarities to astrocytes, while preserving a radial shape characteristic of the astroglia of submammalian vertebrates. It has been proposed that tanycytes of the arcuate nucleus and median eminence may be involved in endocrine regulation.[1] Tanycytes are in close contact with LHRH processes and may create a focal concentration of factors, such as TGF-α, that may affect neurohormone release. In addition, tanycyte end feet encapsulate neurosecretory terminals in the median eminence and may regulate neuronal contact with portal capillaries.

Tanycytes in the rat arcuate nucleus and median eminence are immunoreactive for IGF-I as well.[49] IGF-I immunoreactive levels in tanycytes show sex differences in the rat arcuate nucleus, with adult females showing significantly lower IGF-I levels than males of the same age. This sex difference is abolished by early postnatal androgenization of females[49] suggesting that it may be dependent on the perinatal burst of androgen which is produced by the testis of developing male rats. Furthermore, IGF-I levels in tanycytes increase in both male and female rats at the time of puberty. Females show an abrupt increase in IGF-I immunoreactive levels in tanycytes between the morning and the afternoon of the first proestrus. Henceforth, IGF-I immunoreactivity fluctuates according to the different stages of the estrus cycle. IGF-I immunoreactive levels are high in the afternoon of proestrus, after the peak of estrogen in plasma, remain elevated in the morning of the following day and then decrease to basal conditions by the morning of metestrus.[49] In addition, IGF-I levels decrease in tanycytes when gonadal steroid levels are reduced by ovariectomy and increase in a dose-dependent manner when ovariectomized rats are injected with 17β-estradiol.[49]

An interesting question in regard to the hormonal-induced changes in IGF-I immunoreactivity in tanycytes is the source of IGF-I, since tanycytes do not express mRNA for IGF-I.[49] The modifications in IGF-I levels may result from hormonal modulation of its accumulation from blood or cerebrospinal fluid. To test this possibility, IGF-I was labeled with digoxigenin and injected intravenously or in the lateral cerebral ventricle. In both cases we found that various subsets of neurons and glial cells throughout the central nervous system, including tanycytes, specifically accumulate the labeled IGF-I.[50] The accumulation of IGF-I was specific since it was substantially decreased by the administration of unlabeled IGF-I or unlabeled insulin, which also acts on IGF-I receptors, and was blocked by a specific IGF-I receptor antagonist.[50]

Brawer et al.[51] reported ultrastructural modifications in the ventricular surface of tanycytes in the arcuate nucleus of female rats at different stages of the estrous cycle. The number of microvilli increased during proestrus, remained elevated during estrus, and decreased in metestrus. The significance of these changes is still unknown, but may well be related to the uptake of substances from the cerebrospinal fluid. By using colloidal gold ultrastructural immunolocalization techniques, we have recently observed that IGF-I receptors are enriched in the microvilli of tanycytes (Garcia-Segura et al., unpublished). Therefore, changes in the extension of microvilli may be related to modifications in the number of IGF-I receptors exposed to the lumen of the third ventricle, and this, in turn, may influence the uptake of IGF-I by tanycytes. In order to test whether IGF-I uptake by tanycytes fluctuates during the different stages of the estrous cycle, we injected IGF-I labeled with digoxigenin in the lateral cerebral ventricle of cycling female rats. The number of tanycytes labeled with digoxigenin showed prominent changes associated with the different phases of the estrous cycle, suggesting that tanycytes may mediate hormonal effects on the levels of IGF-I in the mediobasal hypothalamus.[52]

V. Estrogen Activation of Hypothalamic Astroglia and Neuro-Glia Plasticity Depend on IGF-I Receptors

IGF-I levels fluctuate in the rat hypothalamic arcuate nucleus during the estrous cycle in parallel with synaptic and astroglia plastic changes. Furthermore, estrogen-induced increases in IGF-I immunoreactivity in the arcuate nucleus are associated with axosomatic synaptic disconnection, the growth of astroglial processes, and increased GFAP expression. Since coactivation of estrogen and IGF-I receptors is necessary for the action of both factors on hypothalamic neuronal differentiation and survival *in vitro*,[53] and since IGF-I is involved in synaptic plasticity in other neuronal systems, we decided to test whether IGF-I is involved in the glial and synaptic plastic changes of the arcuate nucleus.

One of the most prominent effects of estradiol on the arcuate nucleus of adult rats is its induction of an increase in GFAP levels in the afternoon of proestrus.[9,28,30] This estrogen-induced increase in GFAP protein and mRNA levels during the afternoon of proestrus is associated with the redistribution of astroglia cytoskeletal components, the growth of astrocyte processes, the ensheathing of neuronal somata by glial processes, and the transient disconnection of inhibitory GABAergic synapses from neuronal somata by the interposed glial processes.[9] These changes are also elicited by the administration of estradiol to adult ovariectomized rats.[9,30] To test whether or not astrocyte activation by estrogen in the arcuate nucleus is dependent on IGF-I, hypothalamic tissue fragments from ovariectomized rats, which contained the arcuate nucleus and the median eminence, were incubated in an artificial cerebrospinal fluid in the presence or absence of estradiol.[54] Surprisingly, the hormone did not induce a significant increase in GFAP immunoreactive levels, which is in contrast to what is observed *in vivo*. The effect of the hormone was observed, however, in the presence of insulin, although insulin alone had no effect on GFAP immunoreactivity. Furthermore, the effect of estradiol in the presence of insulin was abolished when the fragments were incubated with a specific IGF-I receptor antagonist peptide. This suggests that the effect of estradiol on arcuate nucleus astrocytes, in the presence of insulin, depends on the activation of IGF-I receptors. Since insulin, at concentrations that act on IGF-I receptors, was unable to increase GFAP immunoreactive levels, the activation of IGF-I receptors alone is not enough to stimulate glial cells. This supports the existence of a coordinated cross-talk mechanism between the IGF-I and estradiol signaling pathways to activate astroglia in the hypothalamus.

The role of IGF-I in estradiol-induced glial activation and synaptic plasticity *in situ* was assessed by the intracerebroventricular administration of a specific IGF-I receptor antagonist to cycling female rats. In agreement with previous findings, the number of synaptic inputs to arcuate neuronal somata in control rats showed a significant decrease between the morning of proestrus and the morning of estrus. This decline in synaptic inputs, as well as the accompanying increase in glial ensheathing of neuronal somata, was blocked by the IGF-I receptor antagonist. In contrast, the IGF-I receptor antagonist did not affect the basal number of synapses, nor the morphology of synaptic terminals or length of the synaptic contacts.[55] Furthermore, the administration of the IGF-I receptor antagonist to ovariectomized females injected with estradiol blocked the effect of estradiol on synapses and astroglia. These findings indicate that IGF-I receptor activation may be involved in the hormonally induced remodeling of arcuate nucleus synapses and astroglia during the estrous cycle.

Further studies are needed to determine the precise mechanism involved in the coordinated action of IGF-I and estrogen in the regulation of synaptic plasticity. Arcuate neurons express IGF-I receptors[56] and may be, therefore, a direct target for IGF-I. IGF-I may affect pre- and/or postsynaptic mechanisms, since ultrastructural studies have shown that IGF-I receptors are present both in a subset of axosomatic presynaptic terminals and a subset of neuronal somata of the rat arcuate nucleus.[56] Arcuate astrocytes are another possible cellular target for IGF-I since they also express IGF-I receptors.[56] Tanycytes express IGF-I receptors as well[56] and, as has been mentioned before, can accumulate IGF-I from extrahypothalamic sources and show changes during the estrous cycle in IGF-I immunoreactivity and IGF-I accumulation. Therefore, tanycytes may participate in the regulation of synaptic plasticity by regulating IGF-I levels in the arcuate nucleus. All of these possibilities reflect the complex cellular and molecular interactions that should be taken into account when explaining the actions of sex hormones in the central nervous system.

VI. Conclusions

Astrocytes and tanycytes are targets for estrogen and testosterone and are apparently involved in the action of sex steroids in the brain. Sex hormones induce changes in the expression of GFAP, the growth of astrocytic processes and the degree of apposition of astroglial processes on neuronal membranes, and these changes are linked to modifications in the number of synaptic inputs. These findings suggest that astrocytes may participate in the genesis of androgen-induced sex differences in synaptic connectivity and in estrogen-induced synaptic plasticity in the adult brain. Astrocytes and tanycytes may also participate in the cellular effects of sex steroids by releasing neuroactive substances and by regulating the local accumulation of specific growth factors, such as IGF-I, that are involved in estrogen-induced synaptic plasticity.

References

1. Garcia-Segura, L.M., Chowen, J.A., and Naftolin, F., Endocrine glia: roles of glial cells in the brain actions of steroid and thyroid hormones and in the regulation of hormone secretion, *Front. Neuroendocrinol.* 17, 180–211, 1996.

2. Jung-Testas, I. and Baulieu, E.E., Steroid hormone receptors and steroid action in rat glial cells of the central and peripheral nervous system, *J. Steroid Biochem. Mol. Biol.*, 65, 243–251,1998.

3. Santagati, S., Melcangi, R.C., Celotti, F., Martini, L., and Maggi, A., Estrogen receptor is expressed in different types of glial cells in culture, *J. Neurochem.*, 63, 2058–2064, 1994.

4. Langub, M.C. and Watson, R.E., Estrogen receptor-immunoreactive glia, endothelia, and ependyma in guinea pig preoptic area and median eminence: electron microscopy, *Endocrinology*, 130, 364–372, 1992.

5. Koenig, H.L., Schumacher, M., Ferzaz, B., Do Thi, A.N., Ressouches, A., Guennoun, R., Jung-Testas, I., Robel, P., Akwa, Y., and Baulieu, E.E., Progesterone synthesis and myelin formation by Schwann cells, *Science*, 268, 1500–1503, 1995.

6. Melcangi, R.C., Celotti, F., Castano, P., and Martini, L., Differential localization of the 5α-reductase and the 3α-hydroxysteroid dehydrogenase in neuronal and glial cultures, *Endocrinology*, 132, 1252–1259, 1993.

7. Melcangi, R.C., Poletti, A., Cavarretta, I., Celotti, F., Colciago, A., Magnaghi, V., Motta, M., Negri-Cesi, P., and Martini, L., The 5α-reductase in the central nervous system: expression and modes of control, *J. Steroid Biochem. Mol. Biol.*, 65, 295–299, 1998.

8. Garcia-Segura, L.M., Dueñas, M., Busiguina, S., Naftolin, F., and Chowen, J.A., Gonadal hormone regulation of neuronal-glial interactions in the developing neuroendocrine hypothalamus, *J. Steroid Biochem. Mol. Biol.*, 53, 293–298, 1995.

9. Garcia-Segura, L.M., Chowen, J.A., Parducz, A., and Naftolin, F., Gonadal hormones as promoters of structural synaptic plasticity: cellular mechanisms, *Prog. Neurobiol.*, 44, 279–307, 1994.

10. Garcia-Segura, L.M., Torres-Aleman, I., and Naftolin, F., Astrocytic shape and glial fibrillary acidic protein immunoreactivity are modified by estradiol in primary rat hypothalamic cultures, *Dev. Brain Res.*, 47, 298–302, 1989.

11. Torres-Aleman, I., Rejas, M.T., Pons, S., and Garcia-Segura, L.M., Estradiol promotes cell shape changes and glial fibrillary acidic protein redistribution in hypothalamic astrocytes *in vitro*: a neuronal-mediated effect, *Glia*, 6, 180–187, 1992.

12. Garcia-Segura, L.M., Cañas, B., Parducz, A., Rougon, G., Theodosis, D., Naftolin, F., and Torres-Aleman, I., Estradiol promotion of changes in the morphology of astroglia growing in culture depends on the expression of polysialic acid on neuronal membranes, *Glia*, 13, 209–216, 1995.

13. Del Cerro, S., Garcia-Estrada, J., and Garcia-Segura, L.M., Neuroactive steroids regulate astroglia morphology in hippocampal cultures from adult rats, *Glia*, 14, 65–71, 1995.

14. Stone, D.J., Song, Y., Anderson, C.P., Krohn, K.K., Finch, C.E., and Rozovsky, I., Bidirectional transcription regulation of glial fibrillary acidic protein by estradiol *in vivo* and *in vitro*, *Endocrinology*, 139, 3202–3209, 1998.

15. Tranque, P. A., Suarez, I., Olmos, G., Fernandez, B., and Garcia-Segura, L.M., Estradiol-induced redistribution of glial fibrillary acidic protein immunoreactivity in the rat brain, *Brain Res.*, 406, 348–351, 1987.

16. Garcia-Segura, L.M., Suarez, I., Segovia, S., Tranque, P.A., Cales, J.M., Aguilera, P., Olmos, G., and Guillamon, A., The distribution of glial fibrillary acidic protein in the adult rat brain is influenced by the neonatal levels of sex steroids, *Brain Res.*, 456, 357–363, 1988.

17. Schipper, H.M., Lechan, R.M., and Reichlin, S., Glial peroxidase activity in the hypothalamic arcuate nucleus: effects of estradiol valerate-induced persistent estrus, *Brain Res.*, 507, 200–207, 1990.

18. Day, J.R., Laping, N.J., McNeil, T.H., Schreiber, S.S., Pasinetti, G., and Finch, C.E., Castration enhances expression of glial fibrillary acidic protein and sulfated glycoprotein-2 in the intact and lesion-altered hippocampus of the adult male rat, *Mol. Endocrinol.*, 4, 1995–2002, 1990.

19. Day, J.R., Laping, N.J., Lampert-Etchells, M., Brown, S.A., O'Callaghan, J.P., McNeill, T.H., and Finch, C.E., Gonadal steroids regulate the expression of glial fibrillary acidic protein in the adult male rat hippocampus, *Neuroscience*, 55, 435–443, 1993.

20. Chowen, J.A., Busiguina, S., and Garcia-Segura, L.M., Sexual dimorphism and sex steroid modulation of glial fibrillary acidic protein messenger RNA and immunoreactive levels in the rat hypothalamus, *Neuroscience*, 69, 519–532, 1995.

21. Stone, D.J., Rozovsky, I., Morgan, T.E., Anderson, C.P., Hajian, H., Finch, C.E., Astrocytes and microglia respond to estrogen with increased apoE mRNA *in vivo* and *in vitro*, *Exp. Neurol.*, 143, 313–318, 1997.

22. Suárez, I., Bodega, G., Rubio, N., and Fernández, B., Sexual dimorphism in the hamster cerebellum demonstrated by glial fibrillary acidic protein (GFAP) and vimentin immunoreactivity, *Glia*, 5, 10–16, 1992.

23. Luquín, S., Naftolin, F., and Garcia-Segura, L.M., Natural fluctuation and gonadal hormone regulation of astrocyte immunoreactivity in dentate gyrus, *J. Neurobiol.*, 24, 913–924, 1993.

24. Tobet, S.A. and Fox, T.O., Sex- and hormone-dependent antigen immunoreactivity in developing rat hypothalamus, *Proc. Natl. Acad. Sci. U.S.A.*, 86, 382–386, 1989.

25. Toran-Allerand, C.D., Neurite-like outgrowth from CNS explants may not always be of neuronal origin, *Brain Res.*, 513, 353–357, 1990.

26. McQueen, J.K., Wright, A.K., Arbuthnott, G.W., and Fink, G., Glial fibrillary acidic protein (GFAP)-immunoreactive astrocytes are increased in the hypothalamus of androgen-insensitive testicular feminized (Tfm) mice, *Neurosci. Lett.*, 118, 77–81, 1990.

27. Suárez, I., Bodega, G., Rubio, N., and Fernández, B., Sexual dimorphism in the distribution of glial fibrillary acidic protein in the supraoptic nucleus of the hamster, *J. Anat.* (London), 178, 79–82, 1991.

28. Kohama, S.G., Goss, J.R., McNeill, T.H., and Finch, C.E., Glial fibrillary acidic protein mRNA increases at proestrus in the arcuate nucleus of mice, *Neurosci. Lett.*, 183, 164–166, 1995.

29. Mong, J.A., Kurzweil, R.L., Davis, A.M., Rocca, M.S., and McCarthy, M.M., Evidence for sexual differentiation of glia in rat brain, *Horm. Behav.*, 30, 553–562, 1996.

30. Garcia-Segura, L.M., Luquin, S., Parducz, A., and Naftolin, F., Gonadal hormone regulation of glial fibrillary acidic protein immunoreactivity and glial ultrastructure in the rat neuroendocrine hypothalamus, *Glia*, 10, 59–69, 1994.

31. Matsumoto, A. and Arai, Y., Developmental changes in synaptic formation in the hypothalamic arcuate nucleus of female rats, *Cell Tissue Res.*, 169, 143–56, 1976.

32. Arai, Y. and Matsumoto, A., Synapse formation of the hypothalamic arcuate nucleus during post-natal development in the female rat and its modification by neonatal estrogen treatment, *Psychoneuroendocrinology*, 3, 31–45, 1978.

33. Matsumoto, A. and Arai, Y., Synaptogenic effect of estrogen on the hypothalamic arcuate nucleus of the adult female rat, *Cell Tissue Res.*, 198, 427–433, 1979.

34. Matsumoto, A. and Arai, Y., Sexual dimorphism in "wiring pattern" in the hypothalamic arcuate nucleus and its modification by neonatal hormonal environment, *Brain Res.*, 190, 238–242, 1980.

35. Matsumoto, A. and Arai, Y., Effect of androgen on sexual differentiation of synaptic organization in the hypothalamic arcuate nucleus: an ontogenetic study, *Neuroendocrinology*, 33, 166–169, 1981.

36. Witkin, J. W., Ferin, M., Popilskis, S. J., and Silverman, A. J., Effects of gonadal steroids on the ultrastructure of GnRH neurons in the Rhesus monkey: synaptic input and glial apposition, *Endocrinology*, 129, 1083–1092, 1991.

37. Naftolin, F., Leranth, C., Perez, J., and Garcia-Segura, L. M., Estrogen induces synaptic plasticity in adult primate neurons, *Neuroendocrinology*, 57, 935–939, 1993.

38. Berg von der Emde, K., Dees, W. L., Hiney, J. K., Hill, D. F., Dissen, G. A., Costa, M. E., Moholt-Siebert, M., and Ojeda, S. R., Neurotrophins and the neuroendocrine brain: different neurotrophins sustain anatomically and functionally segregated subsets of hypothalamic dopaminergic neurons, *J. Neurosci.*, 15, 4223–4237, 1995.

39. Toran-Allerand, C.D., Miranda, R.C., Bentham, W., Sohrabji, F., Brown, E.J., Hochberg, R.B., and MacLusky, N.J., Estrogen receptors colocalize with low-affinity NGF receptors in cholinergic neurons of the basal forebrain, *Proc. Natl. Acad. Sci. U.S.A.*, 89, 4668–4672, 1992.

40. Toran-Allerand, C.D., Mechanisms of estrogen action during neural development: mediation by interactions with the neurotrophins and their receptors? *J. Steroid Biochem. Mol. Biol.*, 56, 169–178, 1996.

41. Sohrabji, F., Miranda, R.C., and Toran-Allerand, C.D., Identification of a putative estrogen receptor element in the gene encoding brain-derived neurotrophic factor, *Proc. Natl. Acad. Sci. U.S.A.*, 92, 11110–11114, 1995.

42. Ojeda, S.R., Dissen, G.A., and Junier, M.P., Neurotrophic factors and female sexual development, *Front. Neuroendocrinol.*, 13, 120–162, 1992.

43. Ma, Y.J., Junier, M.P., Costa, M.E., and Ojeda, S.R., Transforming growth factor-α gene expression in the hypothalamus is developmentally regulated and linked to sexual maturation, *Neuron*, 9, 657–670, 1992.

44. Melcangi, R.C., Galbiati, M., Messi, E., Piva, F., Martini, L., and Motta, M., Type 1 astrocytes influence luteinizing hormone-releasing hormone release from the hypothalamic cell line GT1-1: Is transforming growth factor β the principle involved?, *Endocrinology*, 136, 679–686, 1995.

45. Wetsel, W.C., Hill, D.F., and Ojeda, S.R., Basic fibroblast growth factor regulates the conversion of pro-luteinizing hormone-releasing hormone (Pro-LHRH) to LHRH in immortalized hypothalamic neurons, *Endocrinology*, 137, 2606–2616, 1996.

46. Rage, F., Lee, B.J., Ma, Y.J., and Ojeda, S.R., Estradiol enhances prostaglandin E2 receptor gene expression in luteinizing hormone-releasing hormone (LHRH) neurons and facilitates the LHRH response to PGE2 by activating a glia-to-neuron signaling pathway, *J. Neurosci.*, 17, 9145–1956, 1997.

47. Hiney, J.K., Srivastava, V., Nyberg, C.L., Ojeda, S.R., and Les Dees, W., Insulin-like growth factor I of peripheral origin acts centrally to accelerate the initiation of female puberty, *Endocrinology*, 137, 3717–3728, 1996.

48. Ma, Z.Q., Santagati, S., Patrone, C., Pollio, G., Vegeto, E., and Maggi, A., Insulin-like growth factors activate estrogen receptor to control the growth and differentiation of the human neuroblastoma cell line SK-ER3, *Mol. Endocrinol.*, 8, 910–918, 1994.

49. Dueñas, M., Luquin, S., Chowen, J.A., Torres-Aleman, I., Naftolin, F., and Garcia-Segura, L.M., Gonadal hormone regulation of insulin-like growth factor-I-like immunoreactivity in hypothalamic astroglia of developing and adult rats, *Neuroendocrinology*, 59, 528–538, 1994.

50. Fernandez-Galaz, M.C., Torres-Aleman, I., and Garcia-Segura, L.M., Receptor-mediated internalization of insulin-like growth factor-I in neurons and glia of the central nervous system of the adult rat, *Eur. J. Anat.*, 2, 147–158. 1998.

51. Brawer, J.R., Lin, P.S., and Sonnenschein, C., Morphological plasticity in the wall of the third ventricle during the estrous cycle in the rat: a scanning electron microscopic study, *Anat. Rec.*, 179, 481–489, 1974.

52. Fernandez.-Galaz, M.C., Torres-Aleman, I., and Garcia-Segura, L.M., Endocrine-dependent accumulation of IGF-I by hypothalamic glia, *NeuroReport*, 8, 373–377, 1996.

53. Dueñas, M., Torres-Aleman, I., Naftolin, F., and Garcia-Segura, L.M., Interaction of insulin-like growth factor-I and estradiol signalling pathways on hypothalamic neuronal differentiation, *Neuroscience*, 74, 531–539, 1996.

54. Fernandez-Galaz, M.C., Morschl, E., Chowen, J.A., Torres-Aleman, I., Naftolin, F., and Garcia-Segura, L.M., Role of astroglia and insulin-like growth factor-I in gonadal hormone-dependent synaptic plasticity, *Brain Res. Bull.*, 44, 525–531, 1997.

55. Fernandez-Galaz, M.C., Naftolin, F., and Garcia-Segura, L.M., Phasic synaptic remodeling of the rat arcuate nucleus during the estrous cycle depends on insulin-like growth factor-I receptor activation, *J. Neurosci. Res.*, 55, 286–292, 1999.

56. Garcia-Segura, L.M., Rodriguez, J.R., and Torres-Aleman, I., Localization of the insulin-like growth factor I receptor in the cerebellum and hypothalamus of adult rats: an electron microscopic study, *J. Neurocytol.*, 26, 479–490, 1997.

7

Sex Steroids and Weakly Electric Fish: A Model System for Activational Mechanisms of Hormone Action

Harold H. Zakon

CONTENTS

I. Introduction: Organization, Activation, and Fish

In 1959 Phoenix et al. published a landmark paper on the development of sex behavior in guinea pigs which first distinguished between organizational and activational effects of steroid hormones on behavior and, by inference, the brain. In their view, depending on the perinatal hormonal environment, a fundamentally "bisexual" brain becomes canalized by the survival of circuits underlying either male or female behavior, and the suppression or elimination of the circuits causing the behavior of the other sex; this is organization. The circuits that remain are triggered by steroids in adulthood to produce the appropriate behavior; this is activation. Thus, for example, because female circuits are perinatally suppressed in male rodents, hormonal treatments that elicit robust lordosis in adult females do not in males. Conversely, because the circuits that control male behavior are perinatally suppressed in females, hormones that activate mounting in males cannot in females.

Despite elaborations and modifications of this original idea,[1] this distinction has been a useful guiding concept for studies of hormone action on the central nervous system (CNS). This formulation is essentially correct for mammals and birds in the case of many behaviors and brain structures.[2,3] A case can be made, however, that this scenario does not hold true for teleost (bony) fish. Teleost fish comprise the largest number of vertebrate species (over 20,000) with the greatest diversity of reproductive strategies.[4,5] While androgens may have organizational effects on peripheral structures such as bone or muscle in some species with male-specific intromittent organs (gonopodia), fin rays, or swim-bladder musculature,[6-8] there is little evidence for an organizational role of steroids in the central control of reproductive behaviors.

Like birds and mammals, the teleost brain is initially bisexual. However, the fish brain never becomes canalized and is, therefore, capable of generating behaviors of either sex well into adulthood under the right hormonal conditions. Fish of one sex will display behaviors or neuroendocrine reflexes of the opposite sex when hormonally treated in the laboratory.[9,10] Some fish can be made behaviorally hermaphroditic: when hormonally primed and presented with both a courting male and a gravid female, a goldfish will alternate between male- and female-typical behaviors in a single session.[11] This even occurs naturally: some species of fish are sequentially, and some even simultaneously, hermaphroditic in their reproduction and reproductive behaviors.[12,13] In addition, in a genetically all-female species, when treated with androgens, females still show male sex behavior.[14] Thus, we can conlcude that the brain of fish does not undergo organization and remains neurally bisexual, that the dominant mode of action of hormones in fish is activational, and that it is the presence of particular hormones that activate either male or female circuits in adults.

II. Why Study Electric Fish?

We study the activational effects of sex steroid hormones on the electrical activity of central and peripheral circuits that generate and sense a sexually dimorphic communication signal in weakly electric fish. Electric fish are good experimental animals to study the actions of steroids on electrical activity for a number of reasons. First, their hormone-dependent communication signals are stereotyped, easily measured, and quantifiable. Second, the neural structures that generate these signals are simple, each cell type is defined, they are hierarchically-organized, their synaptic connections are known, and they are accessible for electrophysiological observations, in particular, for biophysical analyses of their ionic currents. Third, there are interesting species differences in these communication signals and in the influence of steroids on their signals.[5,15] Fourth, the time course of steroid effects in these fish is slow (days for initiation and weeks for completion). While many steroid actions are more rapid, such as those controlling reproductive behaviors in rodents, others, such as androgen-induced aggression in female mice[16] or, on the cellular level, corticosteroid-induced changes in ion channel composition in adrenal chromaffin cells,[17] show a similar slow time course. Therefore, they will be a good system to study slow presumably genomic effects of steroids on electrical activity.

III. The Electric Organ Discharge (EOD) as a Communication Signal

Weakly electric fish produce electric fields from an electric organ and sense electric fields with specialized sensory receptor cells.[18] They sense the distortions of their *own* EODs imposed by nearby objects to locate and identify those objects, a process termed *electrolocation*. In addition, they sense the EODs of *other* electric fish and from sometimes subtle differences in their EOD waveforms, can extract information on the species, sex, and individual identity of the sender.[19-21]

Some species generate pulsatile EODs of variable rate, while others produce extremely regular sinusoidal discharges. In this chapter, we will focus on the species *Sternopygus macrurus* which produces a sinusoidal EOD (Figure 7.1). In this species the frequency at which each fish discharges depends on age and sex. As a whole, the species discharges from about 50 to 200 Hz: mature males discharge from 50 to 90 Hz, mature females from 110 to 200 Hz, and juveniles discharge at intermediate and overlapping frequencies.[20-24] Behavioral studies show that fish can detect and respond to the EODs of conspecifics. In particular, males respond to female, but not male, EOD frequency mimics with courtship displays. Interestingly, the EOD frequencies of presumed mated pairs of fish are an octave apart, with the female being at twice the EOD frequency of the male.[21]

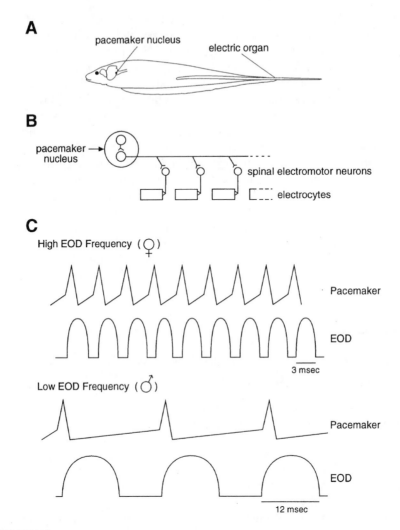

FIGURE 7.1
Schematic illustration of the electromotor system in the species *Sternopygus macrurus*. (A)
The EOD is controlled by the pacemaker nucleus, a midline nucleus in the ventral medulla.
(B) Pacemaker neurons synapse on relay neurons and these send their axons down the spinal
cord to innervate spinal electromotor neurons. The electromotor neurons then synapse on
the posterior end of the electrocytes, the cells of the electric organ. (C) The frequency of the
EOD is controlled by the firing frequency of the pacemaker neurons, while the duration of
each EOD pulse is determined by the membrane properties of the electrocytes. Note that
the EOD pulse has a fourfold variation in duration when a female with a high EOD frequency
is compared with a male with a low EOD frequency.

IV. The EOD-Generating Circuitry: The Motor Output

The EOD is controlled by a small number (~100) of intrinsic pacemaking neurons located in a midline nucleus, called the pacemaker nucleus (PMN), in the ventral medulla (see Figure 7.1). Pacemaker neurons, are endogenously active oscillatory neurons that are chemically and electrotonically coupled to each other and a second group of about 50 neurons, called relay cells.[25-27] The relay neurons send large-diameter axons out of the nucleus and down the spinal cord to innervate electromotoneurons (EMNs) which are found along much of the length of the cord.[28] The EMNs then innervate the electrocytes, the myogenically derived cells of the electric organ. The rapid conduction of the large-diameter relay axons ensures that the electrocytes discharge simultaneously and their action potentials add to form each EOD pulse.[29]

The resting frequency of the EOD is determined by the firing frequency of the pacemaker neurons, while the shape of each EOD pulse is dictated by the membrane properties of the electrocytes. In other words, to produce a sinusoidal EOD, a fish discharging at a low frequency, such as a sexually mature male, must make long-duration electric organ pulses, while a fish firing at a high frequency, such as a sexually mature female, must make narrower pulses. EOD frequency and EOD pulse duration are highly correlated.[30] The importance of this observation is that the electrical activities of two distinct cell types must be coordinated with remarkable precision in order to maintain a sinusoidal waveform. Furthermore, this precision is all the more remarkable because the sensory receptors of each fish are also sharply tuned to its own EOD frequency.

V. Tuning of Electroreceptors: The Sensory Link

The fish's own electric fields and those of other fish are sensed by a specific class of sensory receptors called tuberous receptors (Figure 7.2). These are hair-cell-like receptors innervated by the afferent fibers of the lateral line. Tuberous electroreceptors are located in capsules in the epidermis all over the body surface, most densely over the head and anterior end of the fish.[18] There are about 20 receptors per capsule.

One extremely interesting aspect of these sensory receptors is that they are tuned and the receptors of each fish are best tuned to its own EOD frequency.

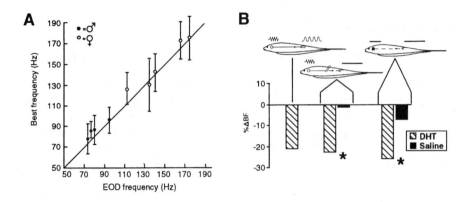

FIGURE 7.2

Electroreceptors are tuned to each fish's EOD frequency and this tuning is influenced by androgens. (A) A graph of sensory receptor tuning for four male and five female *Sternopygus*. In all, 20 receptor afferent fibers were recorded for each fish and the mean and standard deviation of the best frequency — the frequency to which the receptor is most sensitive — is plotted against each fish's EOD frequency. Note that the best frequency of the electroreceptors is matched to each fish's EOD frequency, and is also sexually dimorphic. (B) The extent to which electroreceptor best frequency (BF) changes after a variety of treatments. On the left is illustrated a 20% lowering of receptor BF after DHT treatment. DHT still lowers receptor tuning by a comparable amount when the spinal cord is cut, thereby interrupting the EOD (middle fish). DHT still significantly lowers receptor BF when the PMN is lesioned both eliminating the EOD and abolishing any putative PMN calibration signal (fish on right). Asterisks indicate significance at <0.5.

This is evident in recordings made from afferent fibers which are most sensitive to frequencies at or close to the fish's own EOD frequency[31-34] (see Figure 7.2). It can be seen additionally from recordings made directly from the receptor organs themselves. When a tuned element such as a tuning fork is struck suddenly, it rings; that is, it oscillates at its resonant frequency. An electrical resonator behaves similarly. When a brief electrical pulse is applied to the opening of a tuberous electroreceptor, it generates an oscillation, the frequency of which is close to the fish's EOD frequency.[35]

Since electric fish use this sensory system to detect objects in their vicinity, it is most efficacious if each fish has a different EOD frequency, and each fish's receptors are narrowly tuned to its own EOD frequency. On the other hand, since the EOD is also used as a communication signal, receptors must be capable of sensing the EODs of other fish.[33] In other words, the EODs of other individuals function as noise during electrolocation and signal during communication. Because of these competing demands, the receptors are tuned broadly enough to be sensitive to the EODs of other fish, including members of the opposite sex, while still being most sensitive to their own EODs.[34]

While the mechanisms underlying receptor tuning are not understood, it is believed that they are similar to those of electrically tuned inner ear hair cells of other species including fish, birds, and reptiles.[36]

VI. The EOD and Electroreceptor Tuning are Modified by Gonadal Steroids

One might suspect that a sexually dimorphic signal may be under the influence of sex steroids. Indeed, this is the case. Analysis of blood samples taken from *Sternopygus* in the field just prior to the onset of the breeding season show that while levels of testosterone (T) are similar in both sexes, males have higher levels (~1 to 5 ng/ml) of 11-ketotestosterone (11KT — the dominant piscine androgen) than females, which have vanishingly low levels.[23] T levels are often similar in the two sexes of fish and high levels of 11KT are associated with the occurrence of sexually dimorphic coloration, secondary structures, and behaviors in males in many species.[4] EOD frequency in male *Sternopygus* is inversely correlated with androgen levels both in field-captured specimens and in fish treated in the laboratory with human chorionic gonadotropin to induce androgen secretion.[22,23] Thus, EOD frequency varies in a graded way with plasma androgen level.

Treatment of fish with DHT (dihydrotestosterone) or T lowers, that is masculinizes, EOD frequency over the course of 2 weeks (Figure 7.3). This is a potent effect and occurs in gonadectomized or gonadally intact juvenile or adult fish of both sexes.[24,30] However, if the frequency (determined by the pacemaker neurons) is changed, then each pulse (determined by the electrocytes) must change in its duration by a comparable amount in order to preserve the sinusoidal form of the EOD. Correspondingly, androgens broaden the EOD pulse (see Figure 7.3), which is the summed action potentials of the electrocytes, as well as the action potentials recorded intracellularly from single electrocytes[24,30] (see Figure 7.6A and C). In a remarkable feat of biophysical coordination, androgens also shift the tuning of a fish's electroreceptors to lower frequencies to keep them sensitive to the fish's new EOD frequency.[31,35]

Do female fish have higher EOD frequencies, etc. than males simply because androgens lower the male's from an intermediate frequency range, or is there an active hormonal action moving female EOD to higher values? Plasma estrogen levels do not correlate with EOD frequency in females.[23] Nevertheless, treatment of fish with E_2 (5 to 12 ng/ml) raises, or feminizes, EOD frequency and shortens the EOD pulse with about the same time course.[24,37] It is not yet known whether E_2 shifts the tuning of receptors in the opposite direction; this seems likely however.

It may appear difficult to reconcile the lack of correlation between circulating E_2 and EOD frequency with the effects of E_2 treatment on EOD frequency on fish. The answer might be that plasma levels of E_2 are a misleading indicator and that a better correlation might occur between local levels of E_2 within each tissue that are determined by local aromatization within the tissue. Fish are notable for high levels of aromatase in their CNS.[38] It is currently

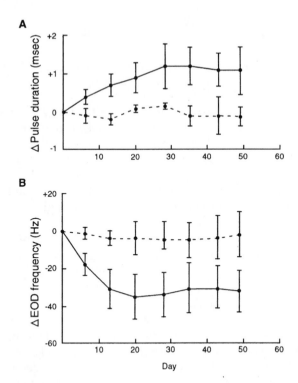

FIGURE 7.3

DHT implants (solid lines), but not empty capsules (dotted lines), increase EOD pulse duration (A) and lower EOD frequency (B) over 2 to 3 weeks. (From Mills, A.C. and Zakon, H.H., Coordination of EOD frequency and pulse duration in a weakly electric wave fish: the influence of androgous, *J. Comp. Physiol.*, 161, 417, 1987. With permission.)

unkown how much aromatase activity occurs in the electrosensory or motor nuclei of any electric fish.

VII. How are EOD Frequency, EOD Pulse Duration, and Receptor Tuning Shifted in Tandem by Steroids?

The precise regulation of electrical activity in three tissues — pacemaker, electrocytes, and electroreceptors — raises the question of how the actions of steroids are coordinated across tissues. The most parsimonious mechanism would be that hormones act directly on the pacemaker and then the pacemaker serves as a "calibration signal" which retunes the receptors and changes the duration of the electrocyte action potential (Figure 7.4A). A second hypothesis is that the hormones act independently on each cell type but

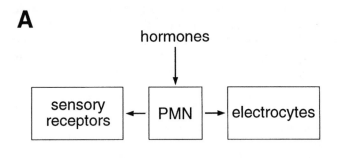

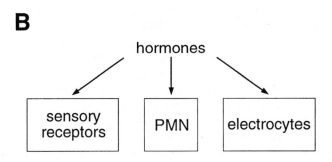

FIGURE 7.4

Steroid hormones could influence the PMN, sensory receptors, and electrocytes and keep them in register either by acting *only* on the PMN, which would then calibrate receptor tuning and electrocyte action potential duration (A), or they could act independently on each tissue with no communication among them (B).

quantitatively regulate the electrical activity of each to keep them in register (Figure 7.4B).

All of the evidence supports the latter hypothesis. First, electroreceptors become tuned to lower frequencies in DHT-treated fish with spinal cord transections (this interrupts the descending input from the pacemaker to the spinal motoneurons and silences the EO) showing that electroreceptors do not need to sense the fish's own EOD to become retuned[39] (see Figure 7.2B). Second, when spinally transected fish are treated with DHT and stimulated with mimic electric fields that go up in frequency rather than down, as DHT would normally cause, the electroreceptors "ignore" the electrical stimulus and decrease in their tuned frequency. Third, if the pacemaker nucleus is lesioned, the electroreceptors still come down in frequency when treated by DHT (see Figure 7.2B).[40] In all of the above cases, controls implanted with empty silastic capsules or injected with saline show no change in receptor tuning.

Electrocyte action potential duration is probably also independent of pacemaker activity. At least in the short term, anesthetization of fish results in a lowering of EOD frequency. Yet, when EO pulse duration is tracked for up to an hour, there is no broadening of the pulse duration.[30] On the other hand, small implants of androgens in the EO broaden the EO pulse. In these experiments EOD frequency remains constant so that the EO pulse changes despite the EO being driven at a constant frequency.[41]

VIII. Steroid Receptors are Localized in Most Hormone-Sensitive Tissues

If hormones act directly on the electroreceptors and electrocytes, then they ought to possess hormone receptors. When reacted for an antibody (Ab) to the androgen receptor (AR) (PG-21),[42] electroreceptors and electrocytes from mature males with low EOD frequencies show label in their nuclei[43] (Figure 7.5). Tissues taken from females and reacted similarly do not have labeled nuclei (Figure 7.5A). When a second sample of these tissues is taken from the same females 3 h after an injection of DHT, PG-21 now labels their nuclei (Figure 7.5B). This effect is specfic since neither E_2 nor progesterone injections cause female tissues to be labeled with PG-21. This shows that both sexes possess ARs, that ARs are occupied and translocated to the nucleus in mature males but not in females, and that they can be easily moved to the nucleus in females injected with androgen and not other steroids.

An interesting paradox is that while the PMN is a presumed site of hormone action, PG-21 does not label any cells within it. This is true even when ARs can be visualized in other cells in the section. The possibilities are that there are membrane receptors, that there are distinct nuclear receptors in pacemaker neurons that are not recognized by PG-21, or that the signal is below the level of detection.

Additionally, electrocytes have been shown to label with an Ab to the E_2 receptor (Hs222).[37] We have not yet examined whether other cells in the electrosensory system label with this Ab.

IX. Biophysical Mechanisms Underlying Action Potential Broadening in the Electrocytes

The mechanisms by which hormones work independently in these different tissues to maintain correlated electrical activity is still a mystery. One necessary step to finding an answer to this question is to know what ion currents are affected in each tissue type. We have done this so far only for the electrocytes and this has provided additional surprises.

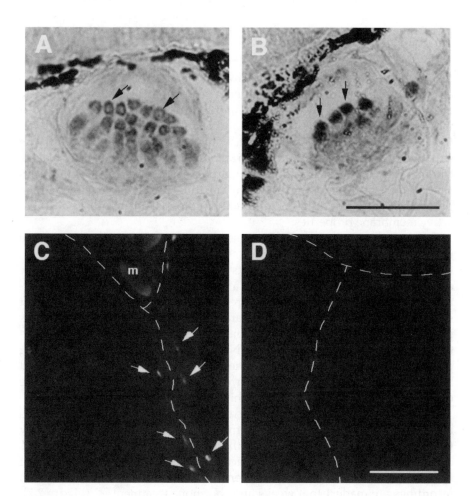

FIGURE 7.5

The electroreceptors and the electrocytes have androgen receptors. (A) Cross section of a tuberous electroreceptor organ of a female fish reacted with the antibody PG-21 against mammalian androgen receptor. The arrows indicate the rings around the nuclei due to artifactual labeling of the secondary antibody; however, the nuclei are unlabeled. (B) After 3 h that same female was injected with DHT, another piece of skin was removed and reacted. This time the nuclei are darkly labeled by PG-21 (arrows). Fish injected with saline do not show labeling (data not shown). (C) Cross section of the electric organ of a male fish labeled with PG-21 and a fluorescent secondary antibody. The dashed lines denote the borders of the electrocytes, (m) indicates a fascicle of muscle fibers. The arrows point to the labeled nuclei at the margin of the electrocytes. (D) An electrocyte treated with PG-21 and the peptide against which the PG-21 was generated does not label. This illustrates specific block of the antibody. Calibration bars = 50 μm.

Differences in action potential shape could arise from differences in passive membrane properties, such as membrane capacitance or resistance, or active properties, the voltage-dependent ion currents of the membrane. The size of the cell may influence the passive membrane responses insofar as the total membrane capacitance or number of channels may change with changes in cell surface area.

The electrocytes of *Sternopygus* are long, cigar-shaped cells, a few millimeters in length and a few hundred microns in diameter. There is no sex difference in the size of the electrocytes either in field-caught sexually mature male and female fish, or in DHT-implanted and control fish in the laboratory, even when the EOD frequencies of males and females or control and hormone-treated fish are significantly different.[44] This observation contrasts with the striking sex difference in electrocyte size found in some other species of electric fish.[45-47] In addition, injection of hyperpolarizing current pulses into *Sternopygus* electrocytes did not reveal any sex difference in passive membrane properties. Thus, sex differences in action potential shape must be due to active membrane properties.

The electrocytes are innervated on their posterior ends and all of their active ion channels are at that end as well.[48,49] Using the voltage-clamp technique, three voltage-dependent ionic currents have been identified in the posterior end of the *Sternopygus* electrocyte: a Na^+ current, an inward-rectifying K^+ current, and an outward-rectiyfing K^+ current [50]. The inward-rectifying K^+ current sets the electrocyte resting potential and shows little variation in any of its properties from individual to individual nor does it seem to be steroid sensitive. As in many other cells types, the Na^+ current is responsible for the rising phase of the action potential. The duration of the action potential is determined both by the rate at which the Na^+ current shuts off and by the onset of the delayed rectifier potassium current, which helps to repolarize the membrane.

The Na^+ current of the *Sternopygus* electrocyte activates and inactivates more rapidly in fish with higher EOD frequencies and slowly in fish with low EOD frequencies[51] (Figure 7.6B). Furthermore, treatment of fish with DHT causes the current to activate and inactivate more slowly, and treatment of fish with E_2 causes these voltage-dependent kinetics to become faster when recordings are made 1 to 3 weeks after hormone treatment[37,51] (Figure 7.7). The delayed rectifying K^+ currents also show individual variation in their activation kinetics; it is not yet known whether the kinetics of the K^+ current are affected by hormones, but this seems likely.

These results raise some interesting questions about how the ionic currents of a cell can be so precisely regulated by steroids. Given the slow time course for steroid-induced changes in ionic currents and the presence of nuclear androgen and estrogen receptors in the electrocytes, our working hypothesis is that these are "classical" transcriptional effects. First, steroids may regulate genes for the ion channels themselves. This would presume that different isoforms of the channel (either different genes or splice products of one or more genes) would produce ion channels with different kinetic properties. We have partially cloned two Na^+ channel genes from the *Sternopygus* electric organ. These are *SKM1* and *SKM2* which are homologues of mammalian muscle Na^+ channel genes (Lopreato, personal communication). Interestingly, the currents generated by these different Na^+ channel genes in mammals have different activation and inactivation kinetics. At this point we do

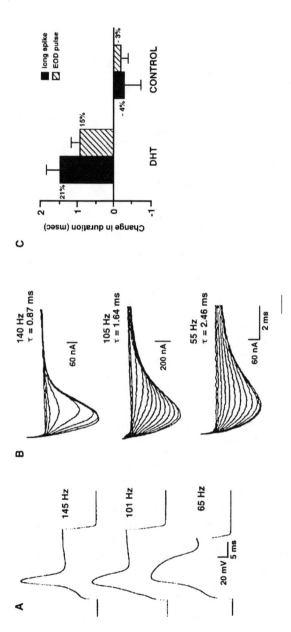

FIGURE 7.6

Electrocyte action potential duration and Na⁺ current is individually distinct and influenced by hormones. (A) Intracellular recordings made in electrocytes of high, middle, and low EOD frequency juvenile *Sternopygus*. Current was injected with one electrode and the voltage responses of the membrane were recorded with another. The number to the right of each trace gives each fish's EOD frequency. Note the systematic individual variation in action potential duration. (B) Voltage clamp recordings from electrocytes from fish with comparable EOD frequencies as in A show that the Na⁺ current activates and inactivates at different rates. EOD frequency and the exponential time constant of decay of the peak current (its rate of inactivation) is given to the right of each family of current traces. (C) Androgens broaden the EOD pulse and the electrocyte action potential (labeled long spike). In this experiment recordings of these two parameters are made before and after treatment with DHT or empty capsules. Note that the control fish show no change in either parameter, but the hormone-treated fish show an increase in both. (A and B from Ferrari, M.B., McAnelly, M.L., and Zakon, H.H., Individual variation in and androgen-modulation of the sodium current in electric organ, *J. Neurosci.*, 15, 4023, 1995; C from Mills, A. and Zakon, H.H., Chronic androgen treatment increases action potential duration in the electric organ of *Strenopygus*, *J. Neurosci.*, 11, 2349, 1991. With permission.)

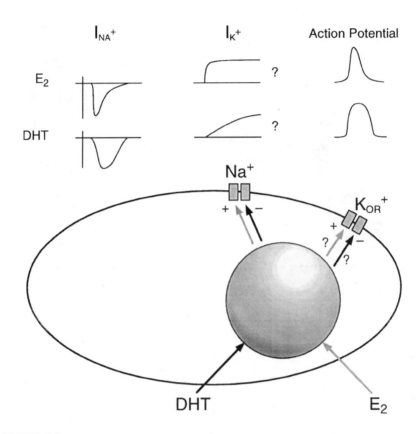

FIGURE 7.7
Schematic drawing of an electrocyte illustrating the opposing actions of DHT and estrogen (E_2) on the kinetics of the Na^+ current, their likely actions on the kinetics of the K^+ current, and their effect on action potential duration. The currents on the top of the figure are schematic voltage-clamp recordings of the Na^+ and K^+ currents. The dark arrows in the electrocyte indicate DHT action, the gray arrows indicate the action of E_2. These arrows project to the nucleus (large round gray circle) to suggest a nuclear site of action of the hormones. (From Zakon, H.H., The effects of steroid hormones on electrical activity of excitable cell, *Trends Neurosci.*, 5, 21, 202–207, 1998. With permission.)

not yet know whether they are differentially expressed in electrocytes of male and female fish or whether their kinetics also differ in fish.

A second possibility is that the genes that are affected are those that code for proteins that modify the function of the ion channel. These would include so-called beta subunits, which are proteins that bind to the channel and alter their levels of expression or kinetics. While beta subunits are known to associate with Na^+ channels in mammalian CNS and muscle, there is no evidence for their presence in the eel EO using low-stringency *in situ* hybridization with probes against the mammalian beta subunits.[52] Other possible targets are proteins known to serve as an "adaptor" between the Na^+ channel and the submembranous cytoskeleton, such the dystrophins.[53,54]

Third, hormones might regulate the levels of enzymes that modify the kinetics of the channels, such as kinases. The amino acid sequence of the Na^+ channel from eel EO possesses consensus sequences for phosphorylation by protein kinase A as well as sites for other post-translational modifications such as glycosylation.[55-57] We examined whether activation of protein kinase A in electrocytes influences Na^+ current kinetics. We find that it enhances the magnitude of the current greatly, but has no influence on the kinetics of the current.[58] So that, while the magnitude of the EOD is likely to be regulated by this pathway, the duration of the EOD pulse is not. Another intriguing target is nitric oxide, which is known to act on Na^+ currents.[59] NO synthase is present in the electrocytes (Smith and Unguez, personal communication).

X. Conclusion

Understanding how sex steroids modify the electrical activity of cells is a major challenge in the field of behavioral neuroendocrinology. The electrosensory and motor system of weakly electric fish has provided a good model system to address this question. We are just begining a biophysical analysis of the pertinent cells, and a molecular analysis is in its infancy. It is our hope that this system will contribute to an understanding of activational mechanisms of steroid hormones in all vertebrates.

References

1. Arnold, A. and Breedlove, S., Organizational and activational effects of sex steroid hormones on vertebrate behavior: a reanalysis, *Horm. Behav.*, 19, 469, 1985.
2. Arnold, A. and Gorski, R., Gonadal steroid induction of structural sex differences in the central nervous system, *Annu. Rev. Neurosci.*, 7, 423, 1984.
3. Matsumoto, A. and Arai, Y., Sexual differentiation of neuronal circuitry in the neuroendocrine hypothalamus, *Biomed. Rev.*, 7, 5, 1997.
4. Brantley, R., Wingfield, J., and Bass, A.H., Sex steroid levels in *Porichthys notatus*, a fish with alternative reproductive tactics, and a review of the hormonal bases for male dimorphism among teleost fishes, *Horm. Behav.*, 27, 332, 1993.
5. Dunlap, K., Thomas, P., and Zakon, H.H., Diversity of sexual dimorphism in electrocommunication signals and its androgen regulation in a genus of electric fish, *Apteronotus*, *J. Comp. Physiol.*, 183, 77, 1998.
6. Brantley, R., Marchaterre, M., and Bass, A.H., Androgen effects on vocal muscle structure in a teleost fish with inter- and intra-sexual dimorphism, *J. Morphol.*, 216, 305, 1993.

7. Herfeld, S. and Moller, P., Effects of 17a methyltestosterone on sexually dimorphic characters in the weakly discharging electric fish, *Brienomyrus niger* (Gunther, 1866) (mormyridae): electric organ discharge, ventral body wall indentation, anal-fin ray bone expansion, *Horm. Behav.*, 34, 303, 1998.

8. Rosa-Molinar, E., Fritzsch, B., and Hendricks, S., Organizational-activational concept revisited: sexual differentiation in an Atherinomorph teleost, *Horm. Behav.*, 30, 563, 1996.

9. Kobayashi, M., Furukawa, K., Kim, M., and Aida, K., Induction of male-type gonadotropin secretion by implantation of 11-ketotestosterone in female goldfish, *Gen. Comp. Endocrinol.*, 108, 434, 1997.

10. Landsman, R., David, L., and Drews, B., Effects of 17a-methyltestosterone and mate size on sexual behavior in *Poecilia reticulata*, *Proc. Third Int. Symp. Reproductive Physiology of Fish*, St. Johns, Newfoundland, 133, 1987.

11. Stacey, N., and Kobayashi, M., Androgen induction of male sexual behaviors in female goldfish, *Horm. Behav.*, 30, 434, 1996.

12. Shapiro, D.Y., Differentiation and evolution of sex changes in fishes, *BioScience*, 37, 490, 1987.

13. Francis, R., Sexual lability in teleosts: developmental factors, *Q. Rev. Biol.*, 67, 1, 1992

14. Schlupp, I., Parzefall, J., Epplen, J., Nanda, I., Schmid, M., and Schartl, M., Pseudomale behaviour and spontaneous masculinizaton in the all-female teleost *Poecilia formosa* (teleostei: Poeciliidae), *Behaviour*, 122, 88, 1992.

15. Dunlap, K., Thomas, P., and Zakon, H.H., Diversity of sexual dimorphism in electrocommunication signals and its androgen regulation in a genus of electric fish, *Apteronotus*, *J. Comp. Physiol. A*, 183, 77, 1998.

16. Simon, N., McKenna, S., Lu, S.-F., and Cologer-Clifford, A., Development and expression of hormonal systems regulating aggression, in *Understanding Aggressive Behavior in Children*, New York Academy of Sciences, New York, 1996, 8.

17. Xie, J. and McCobb, D., Control of alternative splicing of potassium channels by stress hormones, *Science*, 280, 443, 1998.

18. Zakon, H.H., The electroreceptive periphery, in *Electroreception*, Bullock, T.H. and Heiligenberg, W., Eds., Wiley, New York, 1986, 103.

19. McGregor, P.K. and Westby, G.W.M., Discrimination of individually characteristic electric organ discharges by a weakly electric fish, *Anim. Behav.*, 43, 977, 1992.

20. Hopkins, C.D., Sex differences in electric signaling in an electric fish, *Science*, 176, 1035, 1972.

21. Hopkins, C.D., Electric communication in the reproductive behahavior of *Sternopygus macrurus* (Gymnotoidei), *Z. Tierpsychol.*, 35, 518, 1974.

22. Zakon, H.H., Yan, H.-Y., and Thomas, P., Human chorionic gonadotropin-induced shifts in the electrosensory system of the weakly electric fish, *Sternopygus*, *J. Neurobiol.*, 21, 826, 1990.

23. Zakon, H.H., Thomas, P., and Yan, H.Y., Electric organ discharge frequency and plasma sex steroid levels during gonadal recrudescence in a natural population of the weakly electric fish *Sternopygus macrurus*, *J. Comp. Physiol. A*, 169, 493, 1991.

24. Meyer, J.H., Steroid influences upon the discharge frequencies of a weakly electric fish, *J. Comp. Physiol. A*, 153, 29, 1983.

25. Elekes, K. and Szabo, T., Comparative synaptology of the pacemaker nucleus in the brain of the weakly electric fish (gymnotidae), in *Advances in Physiological Sciences: Sensory Physiology of Aquatic Lower Vertebrates*, Szabo, T. and Czeh, G., Eds., Pergamon Press, Oxford, 1981, 107.

26. Dye, J. and Heiligenberg, W., Intracellular recording in the medullary pacemaker nucleus of the weakly electric fish, *Apteronotus*, during modulatory behaviors, *J. Comp. Physiol. A*, 161, 187, 1987.

27. Moortgat, K.T., Keller, C.H., Bullock, T.H., and Sejnowski, T., Submicrosecond pacemaker precision is behaviorally modulated: the gymnotiform electromotor pathway, *Proc. Natl. Acad. Sci. U.S.A.*, 95, 4684, 1998.

28. Bennett, M.V.L., Pappas, G.D., Gimenez, M., and Nakajima, Y., Physiology and ultrastructure of electrotonic junctions. IV. Medullary electromotor nuclei in gymnotid fish, *J. Neurophysiol.*, 30, 236, 1967.

29. Assad, C., Rasnow, B., Stoddard, P.K., and Bower, J.M., The electric organ discharges of the gymnotiform fishes: II. *Eigenmannia*, *J. Comp. Physiol. A*, 183, 419, 1998.

30. Mills, A.C. and Zakon, H.H., Coordination of EOD frequency and pulse duration in a weakly electric wave fish: the influence of androgens, *J. Comp. Physiol.*, 161, 417, 1987.

31. Zakon, H.H. and Meyer, J.H., Plasticity of electroreceptor tuning in the weakly electric fish, *Sternopygus macrurus*, *J. Comp. Physiol. A*, 153, 477, 1983.

32. Hopkins, C.D., Stimulus filtering and electroreception: tuberous electroreceptors in three species of gymnotoid fish, *J. Comp. Physiol.*, 111, 171, 1976.

33. Fleishman, L., Communication in the weakly electric fish *Sternopygus macrurus* I. The neural basis of conspecfic EOD detection, *J. Comp. Physiol. A*, 170, 335, 1992.

34. Fleishman, L., Zakon, H., and Lemon, W., Communication in the weakly electric fish *Sternopygus macrurus* II. Behavioral test of conspecfic EOD detection ability, *J. Comp. Physiol. A*, 170, 349, 1992.

35. Meyer, J.H. and Zakon, H.H., Androgens alter the tuning of electroreceptors. *Science*, 217, 635, 1982.

36. Wu, Y.-C., Art, J., Goodman, M. and Fettiplace, R., A kinetic description of the calcium-activated potassium channel and its application to electrical tuning of hair cells, *Prog. Biophys. Mol. Biol.*, 1995, 63, 131–158.

37. Dunlap, K., McAnelly, M., and Zakon, H., Estrogen modifies an electrocommunication signal by altering the electrocyte sodium current in an electric fish, *Sternopygus*, *J. Neurosci.*, 17, 2869–2875, 1997.

38. Callard, G.V., Aromatization in brain and pituitary: an evolutionary perspective, in *Metabolism of Hormonal Steroids in the Neuroendocrine Structures*, Norris, P.O. and Jones, R.E., Ed., Raven Press, New York, 1984, 79.

39. Keller, C.H., Zakon, H.H., and Sanchez, D.Y., Evidence for a direct effect of androgens upon electroreceptor tuning, *J. Comp. Physiol. A*, 158, 301, 1986.

40. Ferrari, M.B. and Zakon, H.H., The medullary pacemaker nucleus is unnecessary for electroreceptor tuning plasticity in Sternopygus, *J. Neurosci.*, 9, 1354, 1989.

41. Few, W.P. and Zakon, H.H., Androgens act locally to broaden electric organ pulse duration, in *Abstr. Fifth Int. Congress of Neuroethology*, 323, 1998.

42. Prins, G.S., Birch, L., and Greene, G.L., Androgen receptor localization in different cell types of the adult rat prostate, *Endocrinology*, 129, 3187, 1991.

43. Gustavson, S., Zakon, H., and Prins, G., Androgen receptors in the brain, electroreceptors, and electric organ of a wave-type electric fish, *Abstr. Soc. Neurosci.*, 19, 371, 1994.

44. Mills, A., Zakon, H.H., Marchaterre, M.A., and Bass, A.H., Electric organ morphology of *Sternopygus macrurus*, a wave-type, weakly electric fish with a sexually dimorphic EOD, *J. Neurobiol.*, 23, 920, 1992.

45. Bass, A.H. and Hopkins, C.D., Hormonal control of sexual differentiation: changes in electric organ discharge waveform, *Science*, 220, 971, 1983.

46. Bass, A.H. and Volman, S.F., From behavior to membranes: Testosterone-induced changes in action potential duration in electric organs, *Proc. Natl. Acad. Sci. U.S.A.*, 84, 9295, 1987.

47. Hopkins, C.D., Comfort, N.C., Bastian, J., and Bass, A.H., Functional analysis of sexual dimorphism in an electric fish, *Hypopomus pinnicaudatus*, Order Gymnotiformes, *Brain Behav. Evol.*, 35, 350, 1990.

48. Mills, A. and Zakon, H.H., Chronic androgen treatment increases action potential duration in the electric organ of *Sternopygus*, *J. Neurosci.*, 11, 2349, 1991.

49. Bennett, M.V.L., Modes of operation of electric organs, *Ann. N.Y. Acad. Sci.*, 54, 458, 1961.

50. Ferrari, M.B., Zakon, H.H., Conductances contributing to the action potential of *Sternopygus* electrocytes, *J. Comp. Physiol. A*, 173, 281, 1993.

51. Ferrari, M.B., McAnelly, M.L., and Zakon, H.H., Individual variation in and androgen-modulation of the sodium current in electric organ, *J. Neurosci.*, 15, 4023, 1995.

52. Isom, L.L., De Jonhg, K.S., Patton, D.E., Reber, B.F.X., Offord, J., Charbonneau, H., Walsh, K., Goldin, A.L., and Catterall, W.A., Primary structure and functional expresson of the β_1 subunit of the rat brain sodium channel, *Science*, 256, 839, 1992.

53. Gee, S., Madhavan, R., Levinson, S., Caldwell, J., Sealock, R., and Froehner, S., Interaction of muscle and brain sodium channels with multiple members of the syntrophin family of dystrophin-associated proteins, *J. Neurosci.*, 18, 128, 1998.

54. Morris, G., Sedgwick, S., Ellis, J., Pereboev, A., Chamberlain, J., and Nguyen, T., An epitope structure for the C-terminal domain of dystrophin and utrophin, *Biochemistry*, 37, 11117, 1998.

55. Recio-Pinto, E., Thornhill, W., Duch, D., Levinson, S., and Urban, B., Neuraminidase treatment modifies the function of electroplax sodium channels in planar lipid bilayers, *Neuron*, 5, 675, 1990.

56. Bennett, E., Urcan, M., Tinkle, S., Koszowski, A., and Levinson, S., Contribution of sialic acid to the voltage dependence of sodium channel gating. A possible electrostatic mechanism, *J. Gen. Physiol.*, 109, 327, 1997.

57. Emerick, M.C. and Agnew, W.S., Identification of phosphorylation sites for adenosine 3',5'-cyclic phosphate dependent protein kinase on the voltage-sensitive sodium channel from *Electrophorus electricus*, *Biochemistry*, 28, 8367, 1989.

58. McAnelly, M.L. and Zakon, H.H., Protein kinase A activation increases sodium current magnitude in the electric organ of *Sternopygus*, *J. Neurosci.*, 16, 4383, 1996.

59. Li, Z., Chapleau, M., Bates, J., Bielefeldt, K., Lee, H., and Abboud, F., Nitric oxide as an autocrine regulator of sodium currents in baroreceptor neurons, *Neuron*, 20, 1039, 1998.

8

Evolution of Brain Mechanisms Controlling Sexual Behavior

David Crews and Jon Sakata

CONTENTS

I. Introduction

The comparative perspective (i) offers insight into the rules, principles, or generalizations that govern the structure and function of the brain and (ii) illuminates the roots from which they emerged.[1,2] For example, it is reasonable to assume that traits or characteristics shared by many different vertebrates are evolutionarily more ancient, and hence more fundamental, than those traits less widely shared and hence less fundamental. This is known as the biogenetic law[3] and is central to the theoretical insights of the great neurologist J. Hughlings Jackson.[4] Information on what is common (fundamental) and what is uncommon (recently derived) can best be gathered by studying a variety of species, including nonmammalian species. Further, atypical organisms can be especially useful, for their unusual adaptations illustrate alternative solutions to particular problems. Also, they "often force one to abandon standard methods and standard points of view" with the result that "in trying to comprehend their special and often unusual adaptation, one often serendipitously stumbles on new insights."[5] Finally, studying reptiles, the present-day representatives of the ancestors of mammals and birds, yields insight into the origin and adaptation of neuroendocrine mechanisms of species-typical behavior in the higher vertebrates.

The primary question addressed in my (D.C.) work concerns how brain mechanisms controlling behavior might evolve. That is, how has the brain come to exploit specific external and internal stimuli so that they serve as triggers for adaptive responses? A basic tenet of neuroethology is that the structures and functions of the central nervous system are adaptational responses to the environment. While we know that the neural mechanisms underlying behavior can be modified by mutation, evolutionary selection pressures, or by hybridization of closely-related species, we know very little about *how* brain-behavior relationships evolved. This is of considerable interest and fundamental importance to our understanding of the neural control of complex behaviors, but it is difficult to address for three reasons:

1. To demonstrate that microevolutionary changes in the neural mechanisms controlling a specific behavior are a result of selection, it is necessary to establish both that individuals inherently differ in their performance of the behavior, and that these differences in behavior lead to differential reproductive success. Only when these facts have been established can the issue of differences in mechanism be meaningfully addressed.

2. We rarely know the exact phylogenetic relations among the species at hand. Even though closely-related species may be compared, the common ancestor to these species usually no longer exists, and further, there is no way of determining the exact number of intervening species since the original divergence.

3. Behavior-genetic analyses such as screening for mutants or selective breeding reveal the potential for the brain to change in response to artificial selection, not how the brain responds to the selection pressures present in nature.

The end result is that we often are only able to make assumptions regarding the patterns of evolutionary change.

II. Initial Studies (1978–1987)

In 1978, David Crews and Kevin Fitzgerald independently observed two liz-ards mating. Ordinarily this would not be a notable event. It was serendipi-tous however because we knew that both animals were females and, further, that the species consisted entirely of females, reproducing by obligate parthe-nogenesis. What was remarkable was that the behaviors exhibited by the par-thenogenetic whiptail were identical to the courtship and copulatory behavior of its direct sexual ancestor. On seeing the animals engaged in this behavior, since termed pseudosexual behavior, it struck one of us (D.C.) that this display of both male- and female-like "sexual" behaviors *alternately by a single individual* demonstrates perfectly the fundamental bisexuality of the vertebrate brain. Simultaneously hermaphroditic species make the same point in that individuals alternate in their behavioral roles, although in such species the gonad is an ovotestis that releases both sperm and eggs.

Since that time the Crews Lab has been studying whiptail lizards (genus *Cnemidophorus*) because they afford a particularly good opportunity to inves-tigate the evolution of the neuroendocrine mechanisms underlying repro-duction. (In this section references to individual facts are not cited; rather we refer the reader to a review[6] for specific citations). This is because a direct ancestor-descendant phylogeny is present and two different forms of repro-duction exist, sexual and asexual. Approximately one-third of extant whiptail lizard species are all-female (parthenogenetic) species that resulted from hybrid unions of sexual species. For example, the parthenogenetic desert-grasslands whiptail (*Cnemidophorus uniparens*) descended from a hybridiza-tion event between two sexually reproducing species, the rusty rumped whiptail (*C. burti*) and the little striped whiptail (*C. inornatus*); two-thirds of the triploid genome of the descendant parthenogenetic species is derived from the little striped whiptail, the maternal ancestral species.

Although genetically similar, the desert-grasslands whiptail (hereafter the parthenogenetic or descendant whiptail) and the little striped whiptail (here-after the sexual ancestor whiptail) differ in several aspects of their reproduc-tive biology. For example, (i) estradiol (E_2) concentrations in reproductively active parthenogenetic whiptails are approximately five-fold lower than in reproductively active female sexual whiptails, and (ii) while the sexual ances-tral species display the typical vertebrate pattern in that the male mounts and intromits and the female is receptive to the male's courtship and copulation, individual parthenogenetic whiptails alternate between displaying male-like pseudocopulatory behaviors and female-like receptive behaviors depending upon the stage of follicular development (Figure 8.1).

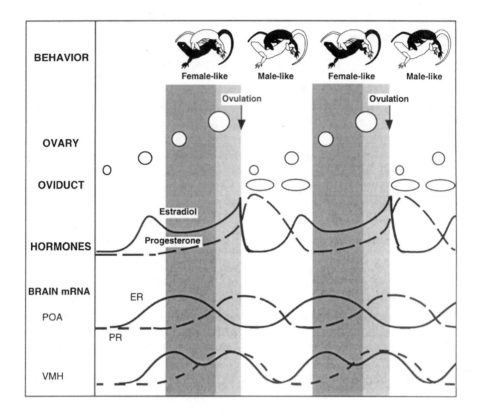

FIGURE 8.1

Relation among male-like and female-like pseudosexual behavior, ovarian state and circulating levels of E_2 and P during different stages of the reproductive cycle of the parthenogenetic whiptail lizard. The transition from receptive to mounting behavior occurs at the time of ovulation (arrow). Also shown are the relative changes in abundance of the gene transcripts coding for ER and PR in the POA and the VMH, brain areas which are involved in the regulation of male- and female-typical sexual behaviors. (Redrawn from Crews, D., Unisexual organisms as model systems for research in the behavioral neurosciences, in *Evolution and Ecology of Unisexual Vertebrates*, Dawley, R.M. and Bogart, J.P., Eds., New York State Museum, Albany, New York, 1989, 132. With permission.)

Our early experiments indicated that engaging in pseudosexual behavior stimulates ovarian growth in the parthenogenetic whiptail *Cnemidophorus uniparens* just as male courtship stimulates ovarian growth in its sexual ancestor, *C. inornatus*. In both species, the time to the first ovulation is decreased significantly if sexual (or pseudosexual) behavior is present. Indeed, females of the sexual species will only lay eggs if a sexually active male is present; females housed with a male that fails to court because he has been castrated neither ovulate nor lay eggs. Similarly, in the parthenogenetic whiptail, engaging in pseudosexual behavior increases the likelihood of ovulation. Over the course of a reproductive season this effect can be substantial. Isolated parthenogens will eventually ovulate, but it is rare that they will produce no more than one clutch. Two intact parthenogens housed together will each produce usually two or more clutches. If a parthenogen is caged with another parthenogen who has been ovariectomized and hormonally-treated so as to only exhibit male-like behavior, the intact individual will produce two and sometimes three clutches (this is the same number of clutches produced in nature). Thus, study of sexual and unisexual whiptail lizards meets two major challenges in evolutionary studies — namely, demonstration of differential reproductive success and presence of the ancestral species.

Pseudosexual behavior in parthenogenetic whiptails is related to the ovarian cycle (Figure 8.1). Female-like receptive behavior is limited to the preovulatory stage of the follicular cycle, whereas the expression of male-like mounting behavior occurs most frequently during the postovulatory stages of the cycle. Differences in the behavioral roles during pseudocopulations are paralleled by differences in the circulating levels of sex steroid hormones. That is, individuals show primarily female-like behavior during the preovulatory stage when E_2 concentrations are relatively high and progesterone (P) levels are relatively low; in contrast, individuals display male-like behavior in the postovulatory phase when concentrations of E_2 are low and P levels are high.

In the sexual ancestral species, sexual activity in males is dependent on testicular androgen, and administration of exogenous androgen stimulates male-like pseudosexual behavior in the unisexual whiptail. This led to the expectation that androgen levels would be elevated during the male-like phase of the ovarian cycle. However, it was a surprise to find that androgens remained uniformly undetectable in the parthenogen throughout the reproductive cycle. In addition, the nature and pattern of sex steroid hormone secretions in the parthenogens are virtually identical to the females of the sexual species. Together these data indicate that the evolution of parthenogenesis and pseudosexual behavior have *not* been accompanied by an alteration of the female-typical pattern of endocrine changes.

Changes in behavior commonly occur at transitions in circulating levels of hormones. The close parallel between the transition from female- to male-like pseudosexual behavior and from E_2 dominance to P dominance in the circulation at ovulation (Figure 8.1) suggested that this shift in hormone concentrations may play a crucial role in controlling the expression of male-like pseudosexual behavior. To test this hypothesis, parthenogens had their ovaries removed and were then given a P or E_2 implant or a blank capsule. Animals were paired with another similarly ovariectomized parthenogen that had, or had not, received hormone treatment. The results were clear-cut. Pseudocopulations occurred only in pairs in which both individuals were hormone-treated in a complementary fashion (e.g., in pairings of E_2- and P-treated individuals). Further, in all pseudocopulations, the P-treated parthenogen assumed the male-like role while animals treated with E_2 exhibited the female-like role. In the absence of the appropriate hormones, pseudosexual behavior was never exhibited. Thus, it appears that the postovulatory surge in P has been exploited as the hormonal cue triggering male-like pseudosexual behaviors.

III. Recent Studies (1988–1998)

During the last decade the focus of the laboratory has been on individual variation in the capacity of P to induce male-typical behavior (= P-sensitivity), the evolution of underlying mechanisms, and structure-function relationships in brain and behavior. (A) It was first documented that there exists individual variation in sensitivity to P in males of the sexual ancestral species. This work suggests that individual variation in P-sensitivity in the sexual ancestor served as the substrate for the evolution of P-activation of male-like pseudosexual behavior in the descendant unisexual species. Work with mammalian models (rats and mice) revealed that this role of P in modulating sexual behavior of males may be widespread. (B) To study the evolution of the molecular neuroendocrine events regulating sexual and pseudosexual behavior, the neuroanatomical distribution of estrogen receptor (ER), progesterone receptor (PR), and androgen receptor (AR) mRNA, the sensitivity to circulating hormones, and the regulation of ER and PR mRNA by hormones in the sexual and unisexual species were characterized. (C) Finally, the potential for structure-function relationships between sexual dimorphisms in behavior and sexual dimorphisms in the hypothalamus was investigated. Detailed reviews of these findings along with specific citations can be found in Godwin and Crews[7] and Young and Crews.[8]

A. Individual Variation in Progesterone Sensitivity

Given the cost that the evolutionary loss of a central nervous system structure would seem to entail,[9] it is easier to simply evolve another system of controls. That is, the structure remains, but the agent activating that structure changes. In terms of the male-like pseudosexual behavior in parthenogens, this would mean that a female physiology must become capable of stimulating specific brain areas to express male-like behaviors at the appropriate time. For this to happen, there must exist a predisposition for such novel functional relationships in the ancestral species.

Earlier experiments established that mating behavior in males of the ancestral sexual species is dependent upon testicular androgens (see Reference 6). How could an androgen-dependent male-typical behavior of the sexual ancestral species evolve to become a P-dependent male-like behavior in the unisexual descendant species? Existing features can be produced by two distinct historical processes.[10] One is adaptation, or the gradual selection of traits resulting in improved functions. Some traits, however, evolved from features that served other roles, or had no function at all, and were co-opted for their current role because they enhance fitness. This latter process may be termed *exaptation*. In adaptation, traits are constructed by selection for their present functions, while exaptations are co-opted for a new use.

In the present case, variation in P-sensitivity among male whiptails of the ancestral sexual species appears to be the substrate on which selection operated, resulting in the novel hormone-brain-behavior relationship observed in the parthenogen. That is, it was found that in approximately one-third of the males of the sexual ancestral species exogenous P was capable of stimulating or maintaining sexual behavior; further, the majority of males that were vigorous courters were P-sensitive, whereas none of the low courting males were P-sensitive (see Reference 6). Assuming that a P-sensitive male was involved in the hybridization process, the postovulatory surge in P presents a reliable stimulus that, given the low circulating concentrations of androgens, was co-opted to trigger mounting behavior in the parthenogen.

Named for its central role in female reproduction, P traditionally has been thought to have little or no function in the control of sexual behavior in males. Indeed, early experiments indicated that administration of P to male birds and rats will inhibit their sexual behavior.[11-14] In fact, this viewpoint has become so entrenched that it serves as a rationale for the use of progestins in the "chemical castration" of sex-offenders.[15,16] However, the physiology of P secretion reveals a marked diurnal rhythm in male rats[17] and humans[18-20] that positively correlates with periods of sexual behavior. Further, work with several species of reptiles has demonstrated that exogenous P, whether administered systemically or directly into the brain, will stimulate courtship and copulatory behavior in castrated males,[21-25] that P is acting in its native form and not via its conversion to androgens,[23] and that dihydrotestosterone (DHT) and P can synergize in stimulating sexual behavior in males[22-24, 26] much as E_2 and P synergize in stimulating sexual behavior in female rats.[27] These data prompted a reassessment of the evidence gleaned from mammalian work, and it was discovered that most of these data were derived from pharmacological dosages of P or from the administration of synthetic progestins that have anti-androgenic properties (see Reference 28).

Recent studies with rats demonstrate that P administered both systemically to produce physiological titers or directly into the POA stimulates the expression of sexual behavior of intact and castrated males[29,30] and, as in the lizard studies, T and P treatments synergize to stimulate sexual behavior in castrated males.[28] This work was extended recently using progesterone receptor knockout (PRKO) mice to confirm the facilitatory role of PR in sexual behavior in males. These experiments demonstrate that males with targeted disruption in the PR show a rapid loss in sexual behavior following castration relative to wildtype (WT) males and a reduced responsiveness to T replacement therapy (Figure 8.2).[31] This finding parallels that observed in whiptail lizards; males sensitive to P are also sensitive to DHT.[23] Finally, P may also be important in the sexual differentiation of rats. Wagner et al.[32] have recently discovered that in rats the number of PR immunoreactive cells is high in the medial POA of males, but virtually absent in females, from late in gestation (embryonic day 20) until 10 days after birth. This is a period corresponding to the surge in T in males, and prenatal treatment with T will masculinize females and induce PR expression to levels similar to that of nor-

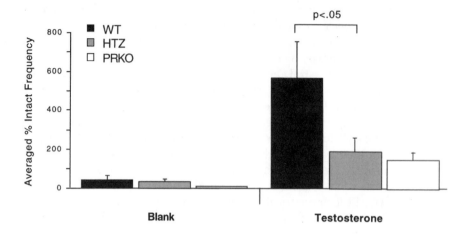

FIGURE 8.2

Average frequency of male-typical sexual behaviors in WT, PRKO, and heterozygote (HTZ) mice. Values are averages of three frequency measures (mounts, intromissions, and ejaculations) and are expressed as a percentage of levels performed while intact. (Redrawn from Phelps, S.M., Lydon, J.P., O'Malley, B.W., and Crews, D., Progesterone receptor potentiates induction of male-typical sexual behavior by testosterone in male transgenic mice, *Horm. Behav.*, 34, 294, 1998. With permission.)

mal males. Interestingly, neonatal treatment with P directs the development of a particular reproductive strategy in male tree lizards.[33]

B. Evolution of the Molecular Neuroendocrine Events Regulating Sexual and Pseudosexual Behavior

That sex steroid hormones and their receptors in the brain modulate the expression of sexual behavior is another tenet of behavioral neuroendocrinology. Although certain aspects of the structure and expression of sex steroid receptors are remarkably conserved across all vertebrate classes, sexual behaviors and their attendant physiology vary widely among species. Comparison of species with different hormone-brain-behavior relationships reveals three aspects of sex steroid receptor gene expression which may underlie species differences in endocrine physiology and behavior[8] (1) neuroanatomical distribution of sex steroid receptors, (2) sensitivity to sex steroid hormones, and (3) variation in the magnitude of sex steroid hormone receptor gene expression in response to hormones.

1. *The distribution of PR, AR, and ER in the brain is evolutionarily conserved.* Gonadal steroid hormones act upon specific areas of the vertebrate brain to affect the reproductive physiology and behavior of the animal. Steroid receptors are members of a superfamily of transcription factors that mediate the effects of steroid hormones by modulating gene expression in the cells containing the receptors. The neuroanatomical distributions of steroid hor-

mone receptor-containing cells have been described for several species using steroid autoradiography, immunocytochemistry, and more recently, *in situ* hybridization. The polymerase chain reaction was used to amplify and clone fragments of the PR, AR, and the ER genes of whiptail lizards. These clones were then used to synthesize probes for use in *in situ* hybridization assays and to map the neuroanatomical distribution of all three sex steroid hormone receptors in the brains of parthenogenetic and sexual whiptail lizards. The distribution of receptor-containing cells in the whiptail lizard is in agreement with previous reports in other species, with receptor-containing cells concentrated in septal, amygdaloid, cortical, preoptic and hypothalamic nuclei. Studies with mammals using a variety of methods indicate that the anterior hypothalamus (AH) and preoptic area (POA) are heterogeneous structures, and the patterns of steroid receptor gene expression in the whiptail brain are consistent with this interpretation.

2. *Species differences in the sensitivity to sex steroid hormones have a specific relationship with sex steroid levels and sex steroid hormone receptor expression in the brain.* Circulating concentrations of gonadal steroid hormones and reproductive behavior in female vertebrates vary as a function of ovarian state. Steroids secreted by the ovary, specifically E_2 and P, influence the expression of behaviors associated with reproduction by interacting with sex steroid receptors located in specific regions of the brain. Comparison of females of the ancestral sexual species to the parthenogen reveals that the circulating E_2 concentrations in the parthenogen are approximately five-fold lower than in female sexual whiptails, though the display of receptive behaviors do not differ between the species. To assess whether this species difference is linked to gene expression of sex steroid receptors, ER and PR mRNA expression were analyzed in several brain regions of ovariectomized, vitellogenic, and postovulatory individuals from the sexual and unisexual species using *in situ* hybridization. The regulation of sex steroid receptor gene expression is region specific, and, furthermore, species differences exist in the level of sex steroid receptor gene expression in specific regions. Specifically, E_2 increases the abundance of ER mRNA in the ventromedial hypothalamus (VMH) in both females of the sexual species and the parthenogen, but the magnitude of the increase is greater in the unisexual whiptail. Because of the evolutionarily conserved role of the VMH in the expression of female-like receptive behaviors, this species difference in ER mRNA expression may account for the increased sensitivity to E_2 in the parthenogen.

A recent study has established that the ancestral sexual and descendant parthenogen species differ also in the estrogenic regulation of PR mRNA in the POA (Figure 8.3).[34] While E_2 treatment does not increase PR mRNA expression in the periventricular region of the POA (PvPOA) of females of the sexual species, E_2 does increase PR mRNA expression in the PvPOA of the parthenogen. This finding suggests a possible proximate mechanism important in the evolution of pseudocopulatory behavior. Thus, the current model of the origin of pseudosexual behavior in the parthenogenetic whiptail lizard postulates that E_2 secreted during the preovulatory phase of the ovarian cycle

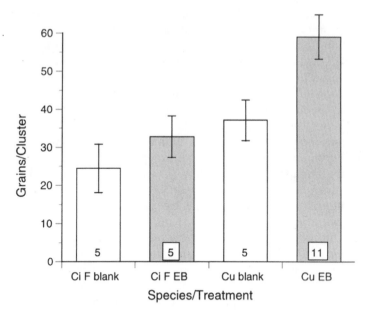

FIGURE 8.3
PR mRNA abundance in the PvPOA for females of the ancestral sexual species (*C. inornatus* or Ci F) and the descendant parthenogenetic species (*C. uniparens* or Cu) given either blank or estradiol benzoate (EB) injections. Depicted is the abundance of PR mRNA measured as average number of silver grains per cluster (mean ± SEM) in the PvPOA of the ancestral sexual species and the descendant parthenogenetic species. (Redrawn from Goodwin, J. and Crews, D., Progesterone receptor mRNA expression in the hypothalamus of whiptail lizards: Regional and species differences, *J. Neurobiol.*, 39, 287, 1999. With permission.)

upregulates ER and PR mRNA in the VMH and stimulates the expression of female-like receptive behavior. This pattern is similar to that in females of the ancestral sexual species. The rise in E_2 also stimulates increases in PR mRNA in the PvPOA, sensitizing this brain region to the surge in P which follows ovulation. This report and previous work demonstrating higher PR mRNA levels in the PvPOA of the parthenogen compared to females of the sexual species over the course of the ovarian cycle suggest that a species difference in sensitivity to P in this brain region underlies the observed species differ-ence in the display of pseudocopulatory behavior.

3. *Female rats and mice, which have an abbreviated follicular phase and brief peri-ods of estrus, differ in their reproductive physiology compared to other vertebrates, like whiptail lizards, which have extended follicular phases and prolonged periods of estrus.* While much is known about how estrogens regulate sex steroid recep-tor expression in rats and mice, little is known about the effects of E_2 on sex steroid receptor expression in lizards. To better understand the molecular mechanisms involved in the control of receptive behavior in whiptail lizards, the effects of exogenous E_2 on the regulation of ER and PR gene expression in several brain regions were investigated. First, after determining a dosage of estradiol benzoate which reliably induced receptive behavior in ovariecto-

mized parthenogens, *in situ* hybridization was used to examine the effects of that dosage on ER and PR mRNA expression in the brain 24 h after injection. Estrogen treatment results in a significant upregulation of ER mRNA expression in the VMH, downregulation in the lateral septum, and no change in the dorsal hypothalamus and PvPOA. The same dosage results in increased PR mRNA expression in the PvPOA, but no significant changes in PR mRNA expression are observed in the periventricular nuclei of the hypothalamus or the torus semicircularis. The upregulation of ER gene expression by E_2 in the VMH of lizards is opposite to that reported in female rats in which E_2 downregulates ER expression in the ventromedial nucleus of the hypothalamus (VMN). Differences in reproductive physiology between rats and mice and other vertebrates may be related to these neural differences.

C. Sexual Dimorphisms in Behavior vs. Sexual Dimorphisms in the Hypothalamus

The large literature on sex differences in the vertebrate brain will not be reviewed here. Rather, we focus on sexual dimorphisms in the reptilian brain, and in particular in the whiptail lizard (see Reference 7). Unisexual vertebrates enable us to address from a new perspective two fundamental questions (1) Are there neural circuits for both male- and female-typical sexual behaviors? and (2) Are structural differences in brain areas related to the frequency and intensity of these behaviors?

1. Dual Neural Circuits Subserving Sexual Behavior

Are there dual neural circuits in the vertebrate brain, one mediating mounting and intromission behavior (including the AH and POA), the male-typical mating pattern, and the other mediating receptive behavior (including the VMH), the female-typical mating pattern? Although researchers have often commented on males that exhibit female-typical sexual behaviors or, conversely, females that exhibit male-typical sexual behaviors, the bulk of modern research has focused on the neuroendocrine mechanisms controlling homotypical behaviors, namely mounting behavior in gonadal males and receptive behavior in gonadal females. In other words, each neural circuit has been studied extensively, but almost always in isolation of its complement.

In males of the ancestral sexual whiptail species and in the parthenogen, intracranial implantation of androgens into the preoptic area-anterior hypothalamus (POAH) continuum elicits mounting behavior. Androgen implants into the VMH not only fail to elicit mounting behavior but also fail to affect receptive behaviors. Conversely, implantation of E_2 into the VMH activates receptivity both in the females of the sexual species and in the parthenogen, while E_2 implants into the POAH continuum have no effect on receptive or mounting behavior. Lesions of the dorsolateral VMH, an area containing high concentrations of ER mRNA in the whiptail, inhibit receptive behavior.

Lesions in the POAH continuum impair courtship both in males of the sexual species and in the parthenogen. These results highlight the conservation in vertebrates of the VMH as a brain area critical for the expression of female-typical sexual behavior and of the POAH continuum as an area integral for the expression of male-typical sexual behavior.

In the sexual whiptail the POAH is larger in males, while the VMH is larger in females.[35,36] During hibernation and following castration, the POAH shrinks while the VMH enlarges (i.e., these areas become female-like). Golgi studies indicate that soma size of neurons in these areas follows the same pattern. [37] In other words, the somata of neurons in the POAH continuum are larger in males than in females, while the somata of neurons in the VMH are larger in females. These studies indicate that sexual dimorphisms in the sexual species are seasonally plastic in the adult and sensitive to testicular androgens.

While the parthenogen displays both male-typical and female-typical sexual behaviors, its brain is not morphologically bisexual. Rather, the POAH and VMH are similar in size to those of *females* of the ancestral sexual species, even in parthenogens exhibiting male-typical behaviors. This same relationship holds for neuronal somata size.[37] Even if parthenogens are treated with androgens so that they exhibit male-like copulatory behaviors and coloration, brain morphology remains unchanged and feminine. Despite this lack of apparent structural difference in these brain areas, the bisexual nature of the brain is revealed by patterns of metabolic activity.[38] That is, 2-fluorodeoxyglucose is concentrated in the POAH of parthenogens exhibiting male-like mounting behavior and concentrated in the VMH of individuals exhibiting female-like receptive behavior. Because the parthenogens are genetically identical and of the same sex, the confounds of gender and genetic differences do not exist. Thus, the parthenogen provides a unique model for the study of neural circuits underlying sex-typical sexual behaviors in the same brain.

Aside from the structural dimorphisms between brains of male and female sexual whiptails, another dimorphic trait concerns the estrogenic regulation of ER and PR mRNA in discrete brain areas.[39] *In situ* hybridization analysis has revealed sex and regional differences in estrogenic effects on ER and PR mRNA abundance in the ancestral sexual species. Females but not males respond to E_2 treatment with increases in ER and PR mRNA expression in the VMH. The VMH sex differences described here are similar to those in rats in that females exhibit estrogenic regulation of ER and PR mRNA while males do not, suggesting that this pattern is evolutionarily conserved. Neither sex nor estrogen effects have been definitively shown for ER or PR mRNA abundance in the POAH. Sex differences in the response to E_2 in the VMH may therefore underlie sex differences in the display of receptive behavior. Indeed, the parthenogen demonstrates the PR mRNA increase and displays female-like receptive behavior in response to E_2 treatment.

However, recent experiments indicate that the sexual dimorphism in the ability of the VMH to respond to E_2 is not irreversibly differentiated. In comparison to males castrated for 1 week, males castrated and maintained for 6 weeks will show a significant upregulation of PR mRNA in the VMH after E_2 treatment as compared to control males.[40] In fact, they do not differ from similarly treated long-term ovariectomized females, which raises the possibility that long-term castrate males might also respond behaviorally to E_2 treatment. Preliminary evidence, however, indicates this is not the case. Male whiptails are never sexually receptive to another male, even when their brains resemble those of females morphologically (as during hibernation) and in terms of steroid hormone receptor gene expression (as after long-term castration).

The expression of three sexually dimorphic behavioral and neural traits in the ancestral sexual species has been characterized. Females, but not males, will display receptive behavior in response to exogenous E_2. In addition, exogenous E_2 treatment will greatly increase PR mRNA in the VMH of females but not of males. Finally, females have smaller POAH volumes and larger VMH volumes than males,[35] and this neurophenotypic trait appears sensitive to androgen levels in males but not in females. After castration, males have POAH and VMH volumes similar to conspecific females.[41] Androgen-replacement therapy reinstates male-like morphology in males, but does not masculinize neural morphology in females. The phenotype of the descendant parthenogen is very similar to that of females of the ancestral sexual species. Exogenous E_2 treatment induces receptive behavior and increases PR mRNA in the VMH of the parthenogen. The volumes of the POAH and VMH are comparable to females of the sexual species and are not affected by either ovariectomy or androgen administration. However, androgen treatment effectively induces male-typical pseudosexual behavior in the parthenogen. Thus, gross morphological changes in hypothalamic and preoptic areas are neither necessary nor sufficient for the expression of heterotypical sexual behaviors.

Such findings raise interesting questions as to the meaning of sexual dimorphisms in the vertebrate brain. The parthenogen clearly retains the ability to express male-like behaviors. But it does so not because it has developed a morphologically masculinized POAH, but because it has co-opted the naturally-occurring P surge to trigger the masculine behavioral potential that remains in a feminized brain. Thus, research on the parthenogen suggests that behavioral differences need not be paralleled by structural differences in the brain. Indeed, structural differences do not exist in brain area volumes between courting and non-courting male whiptails during the breeding season or between courting and non-courting males castrated and given identical T treatment. Taken together these studies suggest that behavioral differences do not necessarily imply gross structural differences in the brain.

2. Genetic vs. Hormonal Organization of the Brain

A strong test of the hypothesis that structural dimorphisms in the brain need not underlie behavioral dimorphisms would be to examine "male" parthenogens. However, males have never been found in this all-female species naturally, and administration of T or DHT before and/or after hatching has consistently failed to alter ovarian development. One of us (D.C.) had come to assume that the parthenogen had lost the gene(s) required for testes development. However, it was found that if eggs are treated up to 12 days but not 20 days after oviposition with as little as 1 μg of Fadrozole, an aromatase inhibitor, all hatchlings will have fully developed testes and vasa deferentia, and lack any signs of ovarian tissue or oviducts![7,42-44] This suggests that gonadal structures become irreversibly determined between embryonic Days 12 and 20, the time during which the gonad first becomes histologically distinct as an ovary in unmanipulated hatchlings. As adults, these "created males" exhibit masculinized coloration, possess male-typical accessory sex structures, and display only male-typical copulatory behaviors.

The paradigm established from studies with mammals is that the sexual differentiation of the brain is determined by hormones secreted by genetically determined differentiated gonads. Recent work in birds and mammals has suggested however that some sexually differentiated traits can be determined directly by genetic factors, independent of hormonal input.[45] Specifically, the size of sexually dimorphic brain nuclei in the zebra finch[46] and sexually dimorphic sensitivity of specific neurons to androgen treatment in mice[47] are affected by genotype. Establishing the generality of this phenomenon across taxa is imperative, and the "created male" parthenogen enables us to address this question.

Recent studies with "created male" parthenogens indicate that sexual dimorphisms in the brain may be determined directly by genetic factors rather than by hormones. Two different measures of sexual dimorphisms in the brain, volume of the POA and estrogen regulation of PR mRNA in the VMH, are female-like in the brains of "created males," indicating that the dimorphisms observed in the ancestral sexual species are likely due to genotype and not testicular hormones.[44] Taken together, these data suggest that there is both a developmentally organized as well as genetically determined difference between male and female brains.

This work with "created males" is significant for two reasons. First, it indicates that despite the loss of males and lack of dependence upon sperm to activate development, the genes regulating testis development have been retained on the autosomes in the parthenogenetic whiptail, a finding consistent with reports that *SRY*, the trigger for testis development in eutherian mammals, is not specific to males in nonmammals.[48,49] Second, it enables us to test in a novel way the recent hypothesis that *SRY*-like genes may have direct organizational effects on the brain independent of sex steroid hormones.[45] The fact that male parthenogens can be created allows investigation of the independent contributions of gonadal and genetic sex. For example, the hypothesis that estrogenic regulation of PR mRNA in the VMH is primarily controlled by gonadal sex can be rigorously tested by comparing "created male" parthenogens to unmanipulated parthenogens. Because all parthenogens are genetically identical, the influence of testicular development on sexual differentiation without the confound of genetic differences can be assessed.

Acknowledgments

We thank Sarah Woolley, Steven Phelps, and Cindy Gill for their comments on the manuscript. Research reported herein was supported by grants from the National Institute of Mental Health (MH R37 41770 and MH K05 00135).

References

1. Bullock, T.H., Why study fish brains? Some aims of comparative neurology today, in *Fish Neurobiology and Behavior,* Davis, R.E. and Northcutt, R.G., Eds., Univ. of Michigan Press, Ann Arbor, 1983, 361.
2. Bullock, T.H., Comparative neuroscience holds promise for quiet revolutions, *Science,* 225, 473, 1984.
3. Gould, S.J., *Ontogeny and Phylogeny,* Belknap Press of Harvard University, Cambridge, MA, 1977.
4. Jackson, J.H., Croonian lectures, *Science,* 1, 501, 1884.
5. Bartholomew, G.A., Scientific innovation and creativity: a zoologist's point of view, *Amer. Zool.,* 22, 227, 1982.
6. Crews, D., Unisexual organisms as model systems for research in the behavioral neurosciences, in *Evolution and Ecology of Unisexual Vertebrates,* Dawley, R.M. and Bogart, J.P., Eds., New York State Museum, Albany, New York, 1989, 132.
7. Godwin, J. and Crews, D., Sex differences in the nervous system of reptiles, *Cell. Mol. Neurobiol.,* 17, 649, 1997.
8. Young, L.J. and Crews, D., Comparative neuroendocrinology of steroid receptor gene expression and regulation: relationship to physiology and behavior, *Trends Endocrinol. Metab.,* 6, 317, 1995.
9. Kavanau, J.L., Conservative behavioural evolution, the neural substrate, *Anim. Behav.,* 39, 758, 1990.
10. Gould, S.J., and Vrba, E.S., Exaptation—a missing term in the science of form, *Paleobiology,* 8, 4, 1982.
11. Erickson, C.J., Bruder, R.H., Komisaruk, B.R., Lehrman, D.S., Selective inhibition of androgen-induced behavior in male ring doves (*Streptopelia risoria*), *Endocrinol.,* 81, 39, 1967.
12. Erpino, M.J., Hormonal control of courtship behavior in the pigeon (*Columba livia*), *Anim. Behav.,* 17, 401, 1969.

13. Erpino, M.J., Temporary inhibition by progesterone of sexual behavior in intact male mice, *Horm. Behav.*, 4, 335, 1973.

14. Bottoni, L., Lucini, V., and Massa, R., Effect of progesterone on the sexual behavior of the male Japanese quail, *Gen. Comp. Endocrinol.*, 57, 345, 1985.

15. Bradford, J.M.W., Treatment of sexual offenders with cyproterone acetate, in *Handbook of Sexology*, Vol. 6: *The Pharmacology and Endocrinology of Sexual Function*, Sitse, J.M.A., Ed., Elsevier Publishers, New York, 1988, 526.

16. Lehne, G.K., Treatment of sex offenders with medroxyprogesterone acetate, in *Handbook of Sexology*, Vol. 6: *The Pharmacology and Endocrinology of Sexual Function*, Sitse, J.M.E., Ed., Elsevier, New York, 1988, 516.

17. Kalra, P.S. and Kalra, S.P., Circadian periodicities of serum androgens, progesterone, gonadotropins and luteinizing hormone-releasing hormone in male rats: the effects of hypothalamic deafferentation, castration and adrenalectomy, *Endocrinology*, 101, 1821, 1977.

18. Vermueulen, A. and Verdonck, L., Radioimmunoassay of 17b-hyroxy-5a-androstan-3-one, 4-androstene-3, 17-dione, dehydroepiandrosterone, 17-hydroxyprogesterone and progesterone and its application to human male plasma, *J. Steroid Biochem.*, 7, 1, 1976.

19. Kage, A., Fenner, A., Weber, B., and Schoneshofer, M., Diurnal and ultradian variations of plasma concentrations of eleven adrenal steroid hormones in human males, *Klin. Wochenschr.*, 60, 659, 1982.

20. Opstad, K., Circadian rhythm of hormones is extinguished during prolonged physical stress, sleep, and energy deficiency in young men, *Eur. J. Endocrinol.*, 13, 56, 1994.

21. Lindzey, J. and Crews, D., Hormonal control of courtship and copulatory behavior in male *Cnemidophorus inornatus*, a direct ancestor of a unisexual, parthenogenic lizard, *Gen. Comp. Endocrinol.*, 64, 411, 1986.

22. Lindzey, J. and Crews, D., Psychobiology of sexual behavior in a whiptail lizard, *Cnemidophorus inornatus*, *Horm. Behav.*, 22, 279, 1988.

23. Lindzey, J. and Crews, D., Interactions between progesterone and androgens in the stimulation of sex behavior in male little striped whiptail lizards, *Cnemidophorus inornatus*, *Gen. Comp. Endocrinol.*, 86, 52, 1992.

24. Young, L.J., Greenberg, N., and Crews, D., The effects of progesterone on sexual behavior in male green anole lizards (*Anolis carolinensis*), *Horm. Behav.*, 25, 477, 1991.

25. Crews, D., Godwin, J., Hartman, V., Grammer, M., Prediger, E.A., and Sheppherd, R., Intrahypothalamic implantation of progesterone in castrated male whiptail lizards (*Cnemidophorus inornatus*) elicits courtship and copulatory behavior and affects androgen receptor- and progesterone receptor-mRNA expression in the brain, *J. Neurosci.*, 16, 7314, 1996.

26. Lindzey, J. and Crews, D., Effects of progesterone and dihydrotestosterone on stimulation of androgen-dependent sex behavior, accessory sex structures, and *in vitro* binding characteristics of cytosolic androgen receptors in male whiptail lizards (*Cnemidophorus inornatus*), *Horm. Behav.*, 27, 269, 1993.

27. Pfaff, D.W., Schwartz-Giblin, S., McCarthy, M.M., and Kow, L.-M., Cellular and molecular mechanisms of female reproductive behaviors, in *The Physiology of Reproduction*, Vol. 2., Neill, J.D. and Knobil, E., Eds., Raven Press, New York, 1994, 107.

28. Witt, D., Young, L., and Crews, D., Progesterone and sexual behavior in males, *Psychoneuroendocrinology,* 19, 553, 1994.

29. Witt, D., Young, L., and Crews, D., Progesterone modulation of androgen-dependent sexual behavior in male rats, *Physiol. Behav.,* 57, 307, 1995.

30. Witt, D.M., Reigada, L.C., and Wengroff, B.E., Intrahypothalmic progesterone regulates androgen-dependent sexual behavior in male rats, *Soc. Neurosci. Abstr.,* 23, 1357, 1997.

31. Phelps, S.M., Lydon, J.P., O'Malley, B.W., and Crews, D., Progesterone receptor potentiates induction of male-typical sexual behavior by testosterone in male transgenic mice, *Horm. Behav.,* 34, 294, 1998.

32. Wagner, C.K., Nakayama, A.Y., and DeVries, G.J., Potential role of maternal progesterone in the sexual differentiation of the brain, *Endocrinology,* 139, 3658, 1998.

33. Moore, M.C., Hughes, D.K., and Knapp, R., Hormonal control and evolution of alternative male phenotypes: generalization of models for sexual differentiation, *Amer. Zool.,* 38, 133, 1998.

34. Goodwin, J. and Crews, D., Progesterone receptor mRNA expression in the hypothalamus of whiptail lizards: Regional and species differences, *J. Neurobiol.,* 39, 287, 1999.

35. Crews, D., Wade, J., and Wilczynski, W., Sexually dimorphic areas in the brain of whiptail lizards, *Br. Behav. Evol.,* 36, 262, 1990.

36. Wade, J. and Crews, D., The relationship between reproductive state and "sexually" dimorphic brain areas in sexually reproducing and parthenogenetic whiptail lizards, *J. Comp. Neurol.,* 309, 507, 1991.

37. Wade, J. and Crews, D., Sexual dimorphisms in the soma size of neurons in the brain of whiptail lizards (*Cnemidophorus* species), *Brain Res.,* 594, 311, 1992.

38. Rand, M. S. and Crews, D., The bisexual brain: sex differences and sex behaviour differences in sexual and parthenogenetic lizards, *Brain Res.,* 665, 163, 1994.

39. Godwin, J. and Crews, D., Sex differences in estrogen and progesterone receptor messenger ribonucleic acid regulation in the brain of little striped whiptail lizards, *Neuroendocrinol.,* 62, 293, 1995.

40. Wennstrom, K. and Crews, D., Effect of long-term castration and long-term androgen treatment on sexually dimorphic estrogen-inducible progesterone receptor mRNA levels in the ventromedial hypothalamus of whiptail lizards, *Horm. Behav.,* 34, 11, 1998.

41. Wade, J., Huang, J.-M., and Crews, D., Hormonal control of sex differences in the brain, behavior and accessory sex structures of whiptail lizards (*Cnemidophorus* species), *J. Neuroendocr.,* 5, 81, 1993.

42. Wibbels, T. and Crews, D., Putative aromatase inhibitor induces male sex determination in a female unisexual lizard and in a turtle with TSD, *J. Endocrinol.,* 141, 295, 1994.

43. Wennstrom, K. and Crews, D., Making males from females: the effects of aromatase inhibitors on a parthenogenetic species of whiptail lizard, *Gen. Comp. Endocrinol.,* 99, 316, 1995.

44. Wennstrom, K. and Crews, D., Volumetric analysis of sexually dimorphic limbic nuclei in normal and sex-reversed whiptail lizards, *Brain Res.,* 838, 104, 1998.

45. Arnold, A.P., Genetically triggered sexual differentiation of brain and behavior, *Horm. Behav.,* 30, 495, 1996.

46. Wade, J. and Arnold, A.P., Functional testicular tissue does not masculinize development of the zebra finch song system, *Proc. Natl. Acad. Sci. U.S.A.*, 93, 5264, 1996.

47. Wee, B.E.F. and Clemens, L.G., Characteristics of the spinal nucleus of the bulbocavernosus are influenced by genotype in the house mouse, *Brain Res.*, 424, 305, 1987.

48. Tiersch, T.R., Mitchell, M.J., and Wachtel, S.S., Studies on the phylogenetic conservation of the *SRY* gene, *Hum. Genet.*, 87, 571, 1991.

49. Tiersch, T.R., Simco, B.A., Davis, K.B., and Wachtel, S.S., Molecular genetics of sex determination in channel catfish: studies on *SRY, ZFY, Bkm* and human telomeric repeats, *Biol. Reprod.*, 47, 185, 1992.

9

Hormonal and Nonhormonal Mechanisms of Sexual Differentiation of the Zebra Finch Brain: Embracing the Null Hypothesis

Arthur P. Arnold

CONTENTS

I. Classic Concepts of Brain Sexual Differentiation: The Hormonal Hypothesis

For the last 40 years the dominant theory of sexual differentiation of the vertebrate brain has been a theory of hormonal action. The textbook version of this theory can be briefly stated as follows. In birds and mammals (and probably other vertebrate species with heteromorphic sex chromosomes), all sex differences in development arise because of some difference in the complement of genes on the sex chromosomes. In mammals, the Y chromosome has a testis-determining gene that induces the primitive gonadal ridge to develop into a testis. In the absence of this gene, an ovary develops. The testis-determining gene has now been proved to be *Sry*,[1,2] although it is clear that numerous other genes interact with *Sry* in the genetic cascade that leads to a testis or ovary (e.g., References 3 and 4). Once the gonads become dimorphic, any subsequent sexual differentiation in nongonadal tissues is thought to result from the differential action of gonadal steroid hormones. During fetal and neonatal life of mammals, testosterone secreted by the testes acts (by itself or after metabolism to estradiol) on the genital tubercle, brain, and other tissues to induce a masculine pattern of development. When testosterone levels are low as in females, a feminine pattern of phenotypic development ensues. The concept of sex-steroidal induction of sexually dimorphic somatic phenotype stems from the classic studies of investigators such as Lillie and Jost,[5,6] and was first successfully demonstrated to apply to behavior and neural development by Phoenix et al.[7]

The Phoenix et al.[7] paper had a galvanizing effect on the field of behavioral neuroendocrinology. They tested the general hypothesis that gonadal secretions early in life determine whether an animal acts in a masculine or feminine manner in adulthood. More specifically, their hypothesis was that testosterone causes a masculine pattern of behavioral development. In general, there are two direct tests of this hypothesis. One is to increase the level of testosterone in fetal or neonatal animals that have low testosterone levels (e.g., females), and then measure masculine or feminine aspects of behavior when these animals have reached adulthood. The second method is to reduce testosterone levels in animals (e.g., males) that normally have high levels, with the expectation that the manipulation would lead to reduced masculine development. Phoenix et al.[7] used the first method and exposed fetal female guinea pigs to testosterone by injecting the pregnant mothers. After the exposed females (and control females) reached adulthood, the investigators conducted standardized tests of copulatory behavior to measure the propensity of the androgenized females to act like a male or female. The result was that females exposed to testosterone showed more masculine behaviors and fewer feminine behaviors than control females. The conclusion was that testosterone had a permanent masculinizing effect (increased behaviors typical of males) and a permanent defeminizing effect (decreased behaviors typical of females) on behavioral development. Subsequent studies also demonstrated that castration of neonatal males induces behavioral demasculinization and feminization. Although Phoenix et al.[7] measured behavior, they astutely extrapolated their findings to the brain, and suggested that testosterone acts on the brain to induce masculine patterns of neural development. Soon after the Phoenix et al.[7] study, other investigators quickly expanded the list of phenotypes (dependent variables) that were sexually differentiated because of the action of testosterone and its metabolites.[8-11] For example, sex differences in neural control of gonadotrophin secretion was shown to be differentiated under androgenic control.

In the 1970s came the discovery that the brain is sexually dimorphic in its morphology and that the cellular organization of the brain was sexually differentiated by the action of gonadal steroid hormones.[12] In 1971 Raisman and Field[13] described a sex difference in the number of a specific synapse type in the preoptic region of the rat. They found that females given testosterone early in development developed a masculine number of synapses, and males castrated at birth developed a feminine number.[14] By 1976 large sex differences at the light microscopic level were discovered in songbirds.[15] Gorski et al.[16] found the volume of the sexually dimorphic region of the preoptic area (SDN-POA) could be made masculine in females if they were given testosterone early in life, or could be made feminine in males if they were castrated neonatally. These and other studies proved unequivocally that testosterone and/or its metabolites are necessary and sufficient to induce sex differences in mammalian neural development.[17-19]

In addition to these direct tests of the hypothesis that gonadal secretions induce brain sexual differentiation, a host of other studies lent indirect support. For example, male rats were found to have higher plasma levels of testosterone than females just before and after birth.[20,21] The brain regions that are sexually differentiated by steroids were found to express appropriate receptors for androgens and/or estrogens during the period of development when steroids act to influence sexual differentiation.[22] Moreover, drugs that block these receptors reduced or prevented masculinization of the brain. In some brain regions, conversion of testosterone to estrogen was found to be essential for brain. These regions were found to express aromatase, the enzyme required for the metabolism of androgen to estrogen, and drugs that inhibit aromatase were found to reduce or prevent masculinization.[23] Estrogen and androgen were found to have cellular effects (preventing cell death,[24] promoting synapse formation,[25,26] stimulating neurite outgrowth,[27] and preventing synapse elimination[28] that explained the mechanisms of steroid-induced changes in neural development that underlie sexual differentiation. All of these findings indirectly corroborated the hormonal hypothesis.

Although the phrase "sexual differentiation" of mammals normally evokes the concept of differences in *development* of the brain or other tissues that lead to *permanent* sex differences in morphology and function, it is important to realize that sex differences in adult function of the brain are also the result of more transient but nevertheless highly potent actions of sex steroid hormones in adulthood.[29] In mammals, adult males have higher circulating levels of androgens than do females, and females experience higher circulating sex differences in estrogen. These steroids act differently on the adult brain to cause changes in cell and synaptic morphology and function that are part of the series of factors leading to sex differences in neural function.[30,31]

II. Sexual Differentiation of the Neural Circuit for Song in Zebra Finches

Research on sex differences in the songbird brain has progressed in the context of the mammalian studies that are briefly described above. Indeed, as we shall see, many of the experimental manipulations used in songbirds are quite similar to those mentioned above in experiments on mammals.

Male zebra finches sing a quiet courtship song to attract the female. This behavior is activated by androgens in adulthood, as evidenced by the findings that castration reduces singing and testosterone treatment of castrates increases singing.[32,33] Females vocalize but never produce courtship song, even if given androgens in adulthood.[34] The sex difference in behavior results from large sex differences in the telencephalic neural circuit for song control.[15] Males have more neurons devoted to song, and these neurons are larger and occupy five to six times more volume in the brain than homologous areas in the female. Virtually all measures of neuronal morphology show a male-biased sex difference that pervades the entire neural circuit.[35]

The hormonal hypothesis developed in studies of mammals led to the idea that gonadal steroid hormonal secretions might also be responsible for sexual differentiation of the zebra finch song system. Thus, when this question was first addressed by Gurney and Konishi,[36] they administered androgen or estradiol to hatchling male and female zebra finches, in an effort to sex-reverse neural and behavioral development. At that time, the prevailing wisdom was that in birds estrogen is a feminizing hormone during development, in much the same way that testosterone is a masculinizing hormone in mammals.[37] In quail, for example, treatment of male embryos with estrogen blocked development of masculine behaviors, and inhibition of estrogen synthesis or action in female embryos caused development of masculine behavioral patterns.[38-40] Thus, Gurney and Konishi initially expected that estrogen treatment of male hatchlings would prevent masculine development of the song circuit. The surprising result was that estrogen had no such effect in males, but rather it significantly masculinized neural development of females.[36,41,42] Females treated with estradiol at hatching have song circuit regions that are more masculine as measured in several ways (e.g., neuron size and number, volumes of brain regions, length of dendrites; reviewed in Reference 35). Although this important result has been confirmed in numerous laboratories, no investigator has been able to sex-reverse completely the phenotype of the song circuit with any regime of steroid hormones. By using common measures of brain masculinity such as volume of song control regions or neuron size or number, estrogen (with or without androgen) treatment induces the song circuit to be half-masculine. For example, the volume of brain regions in females treated with estradiol at hatching is about half-way between the values for males and females.[43] Estrogenized females do sing a courtship song, unlike control females. Injection of nonaromatizable androgens or drugs that influence androgen receptors have relatively little effect on sexual differentiation of telencephalic brain regions,[42-44] so the most potent steroids are estrogens.

The effects of estrogen led to the hypothesis that male song system development is the result of endogenous estrogen action early in development. By analogy to mammalian sexual differentiation, the idea was that during early development males might secrete more estrogen than females. Alternatively, males might secrete more testosterone, which in turn would lead via aromatization to a higher level of estrogen in brain. Support for this hypothesis comes from direct and indirect experimental tests. The one direct test is the experiment described in the preceding paragraph, which shows that estrogen masculinizes females. Indirect support has come from three experimental findings. First, Hutchison et al.[45] found that hatchling males have higher plasma levels of estrogen than females. However, this result is in some doubt because two other studies failed to replicate it.[46,47] Second, estrogen receptors are found in and near HVC (high vocal center), one of the brain regions importantly involved in song.[48-51] Third, aromatase is expressed in high abundance in or near several regions within the telencephalic song circuit during developmental periods when estrogen can masculinize females.[51] These experiments establish the ability of estrogen to masculinize, and establish the sites of estrogen synthesis and action near HVC and other parts of the song circuit during an estrogen-sensitive developmental period.

In the course of these studies, several puzzling findings emerged. One was that the doses of estrogen used to masculinize female hatchlings is well above the physiological level of estrogen found in plasma of hatchling males.[46] This result suggests that a masculinizing dose of estrogen is not normally present in male plasma, at least after hatching. If masculine development requires estrogen, then the estrogen acting in brain might not be derived from the plasma. Instead, the high expression of aromatase near neural song regions could mean that local levels of estrogen in brain near the song system are higher than in plasma, and could represent the source of estrogen involved in sexual differentiation. In that case, the failure to observe a male-biased sex difference in plasma estrogen in early development[46,47] could simply reflect the fact that the estrogen required for masculinization is not synthesized in the testes. However, if the estrogen synthesized in brain comes from testicular androgen, then one would expect that young male finches would have higher levels of androgen than females. This prediction is not supported by the data because all three studies of plasma levels of sex steroids failed to find higher plasma levels of aromatizable androgen in males.[52-54] The male brain might still be the site of greater estrogen action, even if the secretion of sex steroids is not sexually dimorphic, if, for example, males have higher levels of estrogen receptors or aromatase. Although several studies have looked for sex differences in aromatase or estrogen receptors early in development, none has been found to date.[47,51,55-59] Thus, despite numerous experimental attempts, there are no data that support the idea that the male song circuit is exposed to higher levels of estrogen than the female circuit.

Particularly important is the finding that no investigator has substantially blocked masculine neural development in males with any endocrine or antiendocrine manipulation. The first experiments of this sort used several estrogen receptor blockers (tamoxifen, CI628, LY117018), which were administered to hatchling males for about the first month after hatching. None of these experiments prevented masculine neural development.[60,61] The significance of this finding is somewhat in doubt, however, because it is not entirely clear how effective these drugs were as antiestrogens in brain. Although tamoxifen proved to be a clean antiestrogen in zebra finch oviduct, it was ineffective in blocking the masculinization produced in hatchling females by estrogen.[62,63] Therefore there was no independent proof that tamoxifen was having its desired effect in males.

More convincing were experiments using fadrozole, a potent inhibitor of estrogen synthesis. Fadrozole inhibited aromatase activity by 75 to 100% *in vivo* and *in vitro*, measured in both young and adult zebra finches.[64] Thus, there was no doubt about the effectiveness of this drug. However, when fadrozole or the similar aromatase inhibitor vorozole was given to males for several weeks after hatching, the brain developed to be as masculine that of control males.[65-67] Males treated before hatching with fadrozole, on Days 3, 5, or 8 of embryonic life, also have a masculine neural circuit and song.[68-72] These studies suggest that high levels of estrogen synthesis during development are not necessary for masculine neural development. Two studies suggest that inhibition of steroid synthesis has some demasculinizing effect. Merten and Stocker-Buschina[73] found some demasculinization in two song nuclei caused by posthatching treatment with fadrozole, although the fadrozole-treated males sang and appeared to have neurons that were closer in phenotype to male that to female neurons. Posthatching treatment with the steroid synthesis inhibitor MK434 also is reported to have modest demasculinizing effects on portions of the song system,[74] suggesting a role for steroids in masculine neural development. However, no study using steroid synthesis inhibitors or steroid receptor blockers has substantially prevented masculine song system development in males, despite numerous attempts with drugs administered both before and after hatching.

The result that effective antiestrogenic drugs such as fadrozole, administered before or after hatching, cause relatively little or (most often) no demasculinization of males strongly conflicts with the hypothesis that masculine neural development is triggered by estrogen in males. Viewed from the perspective of this failure, the potent masculinizing effects of estrogen in females can be suspected to be a pharmacological effect. That is, until manipulations of estrogen action in males are found to produce substantial demasculinization, the idea that endogenous estrogen normally triggers masculine development is unsupported. The demonstration of estrogen-induced masculinization of females strongly implies that the masculinizing mechanisms are sensitive to estrogen, but there is little support for a critical role for estrogen in males.

However, most of the experiments that provide no support for a critical role for estrogen have involved negative results. Most of these experiments are an attempt to manipulate estrogen levels or estrogen action. When we fail to find an effect of the manipulation, we can only conclude that the manipulation was not effective under the conditions of the experiment. These experiments cannot prove the null hypothesis (that estrogen has no effect). Nevertheless, the increasing number of these studies gives the impression that estrogen synthesis can be reduced without any sizable effect on masculine development. This result conflicts with the idea that estrogen is important. As we begin to embrace the null hypothesis, however, we must remind ourselves that no tests of estrogen action can prove that estrogen has no effect.

Another series of experiments casts additional doubt on the idea that masculine neural development is induced by testicular secretions. Because differentiation of the avian ovary requires the synthesis of estrogen, fadrozole has the ability to block ovarian differentiation and promote development of testes in genetic females.[75] Female zebra finch embryos given fadrozole before or during gonadal differentiation (on Days 3, 5, or 8 of embryonic development) typically develop a testis on the right side and an ovotestis on the left.[68-71,76] The ovotestis characteristically has a central core of seminiferous tubules surrounded by some ovarian follicles. The extent of ovarian development on the left is variable, and in two animals the gonadal reversal was complete so that they possessed virtually no ovarian tissue. These gonadally reversed genetic females have a feminine song system. This result is striking because the clear prediction of the hormonal hypothesis is that animals with testes and lacking ovaries should have a masculine brain. A critical question is, of course, whether the testicular tissue found in fadrozole-treated females secretes the normal profile of hormones that come from testes in genetic males. Because the important secretory events are likely to occur in small embryos or hatchlings, it is difficult or impossible with current methods to measure specific hormones at the critical developmental stages. However, indirect evidence indicates that the testicular tissue is endocrinologically active. Adult females with testes in some cases possess plasma levels of androgens that are in the male range.[71,77] Moreover, the androgen-dependent syrinx (vocal organ) in these females is larger than in control females in adulthood.[70,77] Both the right testis and left ovotestis contain all of the stages of sperm development including mature sperm. To the extent that sperm development requires androgen, the presence of sperm suggests that androgen was synthesized and secreted. Although these gonadally reversed genetic females have impressive masculinization of gonadal phenotype, the organization of the germ cell layers in the seminiferous tubules is not completely masculine. The concentric organization of cells representing the various stages of sperm development is somewhat disrupted especially on the left side, so that one can readily discriminate the testes in genetic females from those of control males.

Because the presence of large amounts of endocrinologically active testicular tissue in these genetic females fails to induce substantial masculine development of the song system, we have concluded that testicular tissue seems not to be sufficient by itself to induce masculinization of the brain. Furthermore, because some individual females have a feminine phenotype in the song system but virtual lack of any ovarian tissue, we have reached a second conclusion, that the presence of ovarian tissue is not required for feminine patterns of neural development. Of these two conclusions, the latter is stronger. Because it is impossible to prove that the secretions of the testicular tissue in genetic females were identical to that of male testes, there is room for some reservation about the first conclusion. Despite this question, one is left with the impression that, if one could produce comparable sex reversal of the mammalian gonad, one would certainly have induced significant if not complete masculinization of the genitalia and brain.

III. Embracing the Null Hypothesis

The fundamental paradox of sexual differentiation of the zebra finch song system is that estrogen causes masculinization of genetic females, but that most manipulations of estrogen levels in genetic males fail to prevent masculinization. Moreover, genetic females with testes nevertheless have a feminine song system. The repeated failure to find large effects of endocrine manipulations on song system development of males leads one to suspect that the null hypothesis may be correct, that estrogens or androgens are not by themselves the single agent that induces sexual differentiation of the brain. How, then, can one prove the null hypothesis? Any experiment that manipulates hormone levels or hormone action can never prove the hypothesis that the hormone has no effect. The only viable route is to develop a new hypothesis that identifies an alternative causal agent for sexual differentiation. A direct experimental test of the new hypothesis would require manipulation of the causal agent to assess its effects on sexual differentiation.

What are the candidates for alternative causal agents? There are two broad classes. One class are hormonal signals other than gonadal steroids, which might act on the brain to induce or block brain masculinization or feminization. For example, inhibin is found at higher levels in plasma and gonads of male chicken embryos, compared with females.[78] If this protein can cross the blood–brain barrier and act on the brain in the embryos, it is a hormonal candidate for the inducing agent.

The second class are nonhormonal signals, which I have previously called "direct genetic" signals.[79-81] These represent genes that are expressed in the brain and act within neural cells to promote masculine or feminine neural development. The gonads themselves are a prime example of a tissue that is sexually differentiated by the action of genes that are expressed within the tissue itself. *Sry* is expressed in the fetal male mouse gonadal ridge just before the onset of morphological sexual differentiation.[1] Because an *Sry* transgene introduced into female mice causes testicular development, and deletion of this gene from XY genotype results in ovarian differentiation, it is clear that *Sry* induces testicular differentiation. Applying this model to neural development of zebra finches, one possibility is that sex chromosomal genes expressed in brain might bias neural development in a sex-specific manner. For example, female finches possess the W sex chromosome that males lack (in birds female sex chromosomes are ZW and males are ZZ), so that if any W genes were expressed in brain, these might promote a feminine pattern of neural development, or block a male pattern. Furthermore, if Z chromosome genes are expressed in brain, males might have a higher dose of Z genes than females, so that sex differences in gene dosage might promote sexual differentiation.

IV. Evidence Against the Hormonal Hypothesis in Other Systems

Strong evidence from other systems suggests that sexual differentiation of somatic tissues is not always triggered by sex steroid hormones. The first strong evidence came from work on the tamar wallaby.[82] Male wallabies develop the anlage for masculine genital structures (scrotum) prior to the differentiation of the gonads into testes or ovaries. The anlage for feminine structures such as mammary tissue also begins to differentiate before sex differences occur in the levels of gonadal steroids. Moreover, administration of gonadal steroids does not alter the early sexual differentiation of these tissues.[83] Clearly, sexual differentiation in this instance cannot have been induced by gonadal steroid hormones.

A second example comes from studies of cultures of rat hypothalamius and midbrain. Beyer et al. made dissociated cell cultures of male and female tissue from these brain regions. By 10 days *in vitro*, sex differences developed in the number of neurons that were immunoreactive for tyrosine hydroxylase or prolactin.[84-87] Because both male and female tissue were treated identically, the sex differences *in vitro* must have resulted from a cell-autonomous program of development. Because the tissues were harvested before the gonads are thought to produce sexually dimorphic levels of testosterone, it is unlikely that the cells had been sexually differentiated prior to harvest by the action of steroids. This conclusion is strengthened considerably by the finding that the masculine or feminine phenotype of the cultures is not disrupted by manipulations of the sex steroid endocrine enviroment of the pregnant dams prior to harvesting the cells.[85] Again, the differentiation of these neurons seems not to have been induced by gonadal steroids.

In both of the above cases, the inducer of sexual differentiation has not been identified, but the evidence strongly suggests that it is not a gonadal steroid.

V. Incompatibility of the Hormonal and "Direct Genetic" Hypotheses?

Because a great deal of experimental evidence indicates that many sexual dimorphisms in mammalian brain and behavior are induced by testosterone or its metabolites, can one not argue that other inducers including direct genetic inducers are unlikely to play a role? Does not evidence in favor of testosterone-induced brain masculinization prove that direct genetic factors are *not* important in those systems? For example, one sexual dimorphism that has been extensively studied is in rat lumbar spinal cord, which contains neurons in the spinal nucleus of the bulbocavernosus (SNB) which innervate striated muscles attached to the penis. This nucleus contains about three times the number of motoneurons as are present in the female SNB, and the neurons are twice as large in males. The sex difference in number of neurons can be completely reversed by prolonged treatment of females with testosterone during development,[24] and males lacking functional androgen receptor have a completely feminine number of neurons.[88] Using Occam's razor, one might conclude that one need invoke no factors other than androgens to explain the sex differentiation of SNB motoneuronal number. Thus, direct genetic factors would seem to be excluded. Of course, this argument is flawed. One cannot use an experiment that tests the role of testosterone as a strong test of other factors. For example, the experiments that demonstrated complete sex-reversal of SNB motoneuronal number generally used a high dose of testosterone, probably higher than normally occurs in males. It is therefore conceivable that lower physiological levels might not be as effective by themselves, and that they require the action of some other factor (direct genetic?) to cause complete masculinization. For example, a direct genetic factor in males might synergize with androgens to produce full masculinization of the SNB in males. Alternatively, a direct genetic factor in females might tend to inhibit the action of androgens. (One is reminded that there is extensive overlap in the levels of plasma testosterone measured in male and female rat fetuses around the time of prenatal sexual differentiation,[20] which raises the question of why some females are not masculinized as much as males. One potential answer is that they may not respond as well as genetic males.) To investigate the role of factors other than gonadal hormones, one must manipulate those factors while keeping testosterone levels constant.

Some experiments have manipulated the composition of genes on the Y chromosome in order to examine their effects on behavior. For example, aggressive behavior in mice differs across strains. Maxson and co-workers[89-91] have produced inbred congenic strains that differ in their Y chromosome but otherwise have the same autosomal background. Males of these congenic strains show different levels of aggressive behavior, a result that indicates that some Y genes influence aggression. These Y genes are candidates for direct genetic influences on aggressive behavior, which is a sexually dimorphic trait. However, the between-strain differences are also potentially attributable to Y effects on the level of androgens or the sensitivity of neural circuits to gonadal hormones. In the end, one can only determine if a specific action of a gene has a direct, nonhormonal effect on neural development by identifying the gene and establishing its molecular and cellular mechanism of action. If the gene is expressed in brain in appropriate places and at appropriate times of development, and manipulation of the gene product alters the course of sexual differentiation, one can build a case for a nonhormonal mechanism of action.

VI. Evolutionary Considerations

Sex differences in behavior and physiology are found in most vertebrate and many invertebrate groups. These sex differences allow the two sexes to occupy different ecological niches, and to play complementary roles in reproductive and social behaviors that impact on reproduction. The selective pressures that favored evolution of sexually dimorphic morphology and function must originally have stemmed from the evolution of two types of gametes. Because eggs and sperm came to be produced by different cellular processes, the gamete-producing cells and those that control them evolved into sexually dimorphic structures. One can speculate that once the vertebrate gonad evolved the use of steroid hormones for intercellular communication within the gonad, other tissues of the body evolved a sensitivity to those hormones as a mechanism of communication between the gonads and those tissues. For example, the brain expression of receptors for steroid hormones may have been favored so that the gonad could signal to the brain that gametes were being produced, which was advantageous because the brain could then coordinate behaviors and functions that are necessary for the exchange of gametes. For example, estrogenic action on hypothalamic cells serves the function of activating female reproductive behavior precisely at the time that gametes have become available in the estrus cycle. The evolution of sexual dimorphisms in brain structure and function presumably occurred when the selection pressure on males and females was different.

The developmental mechanisms that lead to sexual differentiation of the brain are the result of evolutionary forces that favored sex differences in neural function. Although sex differences in the brain are widespread among vertebrates, there are large differences among species in the brain regions that are sexually dimorphic, or in the quality of dimorphism found in an individual brain region. Even among rodents, for example, different areas of the hypothalamus are sexually dimorphic, and one particular measure of hypothalamic structure may be sexually dimorphic in only a subset of species. This heterogeneity of dimorphisms leads to the conclusion that sex-specific patterns of neural development have come and gone during evolution. Thus, sexually dimorphic patterns of development have evolved from a previously monomorphic state, and monomorphic developmental patterns have evolved from a previously sexually dimorphic state.

A second observation suggests, however, that sexually dimorphic development of any specific brain region may evolve or disappear relatively rapidly in evolution. For example, in the closely related species of the wren genus *Thryothorus*, some species possess a large sexual dimorphism in the neural circuit that controls vocal behavior, whereas other species are nearly monomorphic.[92,93] Because the genome of these species differ from each other only at a relatively small number of loci, there must be a small number of genes which can turn dimorphic neural development on or off. For example, genes that control the sensitivity to steroid hormones might either permit or prevent sexually dimorphic development in response to gonadal hormones.

When selection pressures exist that favor evolution of sexual differences in a specific circuit in the brain (for example, when male courtship singing is advantageous but singing in females is disadvantageous), any viable mutation that leads to a sex difference in development of the relevant neural circuits will be favored. In essence, the development of that specific neural circuit must come under the influence of a sexually dimorphic molecular signal that is already present in the animal. If the animal is already secreting sexually dimorphic levels of gonadal steroids, then evolution of the mechanisms within the neural circuit to respond to those hormones will be favored. In mammals, for example, development of a neural circuit might evolve a sensitivity to androgen, which would make development dimorphic and lead to sex differences in the function subserved by that circuit. Alternatively, the circuit could evolve a sensitivity to genes that are dimorphically expressed within the circuit itself, which would lead to sexual dimorphisms via direct genetic control of neural development. One might imagine that through the long millennia of vertebrate evolution, both kinds of dimorphic signals might have acquired control of sex-specific patterns of neural development. An interesting question is whether neural circuits are more likely to evolve sensitivity to hormones or direct genetic factors. On the one hand, gonadal hormones potentially have access to all cells in the body, so that this kind of signal is present ubiquitously and thus may be more available than gene products that are not expressed as widely. However, some Y chromosomal genes are known that are widely expressed in males only,[94] so that some Y chromosomal genes might be as available as testosterone within the brain. Evolution of sensitivity to some of these male-specific signals may therefore be as likely as sensitivity to hormonal signals.

VII. Conclusion

A rigorous test of the direct genetic hypothesis requires that we identify genes that are expressed in a sexually dimorphic pattern early in development in brain, and determine whether or not these genes act to promote sex-specific patterns of neural development. Genes on the sex chromosomes are the most interesting because these could be expressed in only one sex, or could be expressed in different doses in the two sexes. Some studies already provide evidence that Y chromosomal genes may expressed in brain.[95-97] Further studies are under way and should bear interesting fruit.

References

1. Goodfellow, P.N. and Lovell-Badge, R., SRY and sex determination in mammals, *Annu. Rev. Genet.*, 27, 71–92, 1993.
2. McElreavey, K., Barbaux, S., Ion, A., and Fellous, M., The genetic basis of murine and human sex determination: a review, *Heredity*, 75, 599–611, 1995.
3. Swain, A., Narvaez, V., Burgoyne, P.S., Camerino, G., and Lovell-Badge, R., Dax1 antagonizes Sry action in mammalian sex determination, *Nature*, 391, 761–767, 1998.
4. Vilain E. and McCabe E.R., Mammalian sex determination: from gonads to brain, *Mol. Genet. Metab.*, 65, 74–84, 1998.
5. Burns, R.K., Role of hormones in the differentiation of sex, in *Sex and Internal Secretions*, W.C. Young, Ed., Williams and Wilkins, Baltimore, 1961, 76–160.
6. Jost, A., Vigier, B., Prepin, J., and Perchellet, J.P., Studies on sex differentiation in mammals, *Recent Prog. Horm. Res.*, 29, 1–41, 1973.
7. Phoenix, C.H., Goy, R.W., Gerall, A.A., and Young, W.C., Organizing action of prenatally administered testosterone propionate on the tissues mediating mating behavior in the female guinea pig, *Endocrinology*, 65, 369–382, 1959.
8. Barraclough, C.A. and Gorski, R.A., Studies on mating behaviour in the androgen-sterilized female rat in relation to the hypothalamic regulation of sexual behaviour, *J. Endocrinol.*, 25, 175–182, 1962.
9. Harris, G.W. and Levine, S., Sexual differentiation of the brain and its experimental control, *J. Physiol.*, 181, 379–400, 1965.
10. Whalen, R.E. and Edwards, D.A., Hormonal determinants of the development of masculine and feminine behavior in male and female rats, *Anat. Rec.*, 157, 173–180, 1967.

11. Arai, Y. and Masuhiro, Y., Long-lasting effects of prepubertal administration of androgen on male hypothalamic–pituitary–gonadal system, *Endocrinol. Jpn.*, 15, 375–378, 1968.

12. Arnold, A.P. and Gorski, R.A., Gonadal steroid induction of structural sex differences in the CNS, *Annu. Rev. Neurosci.*, 7, 413–442, 1984.

13. Raisman, G. and Field, P.M., Sexual dimorphism in the preoptic area of the rat, *Science*, 173, 731–733, 1971.

14. Raisman, G. and Field, P.M., Sexual dimorphism in the neuropil of the preoptic area of the rat and its dependence on neonatal androgen, *Brain Res.*, 54, 1–29, 1973.

15. Nottebohm, F. and Arnold, A.P., Sexual dimorphism in vocal control areas of the song bird brain, *Science*, 194, 211–213, 1976.

16. Gorski, R.A., Gordon, J.H., Shryne, J.E., and Southam, A.M., Evidence for a morphological sex difference within the medial preoptic area of the rat brain, *Brain Res.*, 148, 333–346, 1978.

17. Arai, Y. and Matsumoto, A., Synapse formation of the hypothalamic arcuate nucleus during post-natal development in the female rat and its modification by neonatal estrogen treatment, *Psychoneuroendocrinology*, 3, 31–45, 1978.

18. Arai, Y. and Nishizuka, M., Morphological correlates of neuronal plasticity to gonadal steroids: sexual differentiation of the preoptic area, in " The Development of Sex Differences and Similarities in Behavior," M. Haug, R.E. Whalen, C. Aron, and K.L. Olsen, Eds., NATO ASI Series, *Behav. Soc. Sci.*, Series D, Vol. 73, Kluwer Academic, Boston, 1992, 311–323.

19. Suzuki, Y., Ishii, H., and Arai, Y., Prenatal exposure to androgen increases neuron number in the hypogastric ganglion, *Dev. Brain Res.*, 10, 151–154, 1983.

20. Weisz, J. and Ward, I.L., Plasma testosterone and progesterone titers of pregnant rats, their male and female fetuses and neonatal offspring, *Endocrinology*, 106, 306–316, 1980.

21. Picon, R., Testosterone secretion of foetal rat testes *in vitro*, *J. Endocrinol.*, 71, 231–238, 1976.

22. McEwen, B.S., Zigmond, R.E., and Gerlach, J.L., Sites of steroid binding and action in the brain, *Struct. Funct. Nerv. Tissue*, 5, 205–290, 1972.

23. McEwen, B.S., Lieberburg, I., Chaptal, C., and Krey, L.C., Aromatization: important for sexual differentiation of the neonatal rat brain, *Horm. Behav.*, 9, 249–263, 1977.

24. Nordeen, E.J., Nordeen, K.W., Sengelaub, D.R., and Arnold, A.P., Androgens prevent normally occuring cell death in a sexually dimorphic spinal nucleus, *Science*, 229 , 671–673, 1985.

25. Matsumoto, A. and Arai, Y., Effects of estrogen on early postnatal development of synaptic formation in the hypothalamic arcuate nucleus of female rats, *Neurosci. Lett.*, 2, 79–82, 1976.

26. Matsumoto, A. and Arai, Y., Effects of androgen on sexual differentiation of synaptic organization in the hypothalamic arcuate nucleus: an ontogenetic study, *Neuroendocrinology*, 33, 166–169, 1981.

27. Toran-Allerand, C.D., Sex steroids and the development of the newborn mouse hypothalamus and preoptic area *in vitro*: implications for sexual differentiation, *Brain Res.*, 106, 407–412, 1976.

28. Jordan, C.L., Letinsky, M.S., and Arnold, A.P., The role of gonadal hormones in neuromuscular synapse elimination in rats. II. Multiple innervation persists in the adult levator ani muscle after juvenile androgen treatment, *J. Neurosci.*, 9, 239–247, 1989.

29. Matsumoto, A., Arai, Y., Urano, A., and Hyodo, S., Molecular basis of neuronal plasticity to gonadal steroids, *Funct. Neurol.*, 10, 59–76, 1995.

30. Matsumoto, A., Hormonally induced neuronal plasticity in the adult motoneurons, *Brain Res. Bull.*, 44, 539–547, 1997.

31. Matsumoto, A., Hormonally induced synaptic plasticity in the adult neuroendocrine brain, *Zool. Sci.*, 9, 679–695, 1992.

32. Pröve, E., Der Einfluss von Kastration und Testosteronsubsitution auf das Sexualverhalten männlicher Zebrafinken (*Taeniopygia guttata castanotis* Gould), *J. Ornithol.*, 115, 338–347, 1974.

33. Arnold, A.P., The effects of castration and androgen replacement on song, courtship, and aggression in zebra finches (*Poephila guttata*), *J. Exp. Zool.*, 191, 309–326, 1975.

34. Arnold, A.P., Behavioral Effects of Androgen in Male Zebra Finches (*Poephila guttata*) and a Search for Its Sites of Action. Ph.D. thesis, Rockefeller University, New York, 1974.

35. Arnold, A.P., Developmental plasticity in neural circuits controlling birdsong: sexual differentiation and the neural basis of learning, *J. Neurobiol.*, 23, 1506–1528, 1992.

36. Gurney, M.E. and Konishi, M., Hormone-induced sexual differentiation of brain and behavior in zebra finches, *Science*, 208, 1380–1382, 1980.

37. Arnold, A.P. and Mathews, G.A., Sexual differentiation of brain and behavior in birds, in *Handbook of Sexology*, Vol. 6, *Pharmacology of Sexual Function*, J.M.A. Sitsen, Ed., Elsevier Science Publishers, Amsterdam, 1988, 122–144.

38. Adkins-Regan, E., Sexual differentiation in birds, *TINS*, 10, 517–522, 1987.

39. Balthazart, J. and Ball, G.F., Sexual differentiation of brain and behavior in birds, *Trends Endocrinol. Metab.*, 6, 21–29, 1995.

40. Balthazart, J., Tlemcani, O., and Ball, G.F., Do sex differences in the brain explain sex differences in the hormonal induction of reproductive behavior? What 25 years of research on the Japanese quail tells us, *Horm. Behav.*, 30, 627–661, 1996.

41. Gurney, M.E., Hormonal control of cell form and number in the zebra finch song system, *J. Neurosci.*, 1, 658–673, 1981.

42. Gurney, M.E., Behavioral correlates of sexual differentiation in the zebra finch song system, *Brain Res.*, 231, 153–172, 1982.

43. Jacobs, E.C., Grisham, W., and and Arnold, A.P., Lack of a synergistic effect between estradiol and dihydrotestosterone in the masculinization of the zebra finch song system, *J. Neurobiol.*, 27, 513–519, 1995.

44. Schlinger, B.A. and Arnold, A.P., Androgen effects on the development of the zebra finch song system, *Brain Res.*, 561, 99–105, 1991.

45. Hutchison, J.B., Wingfield, J.C., and Hutchison, R.E., Sex differences in plasma concentrations of steroids during the sensitive period for brain differentiation in the zebra finch, *J. Endocrinol.*, 103, 363–369, 1984.

46. Adkins-Regan, E., Abdelnabi, M., Mobarak, M., and Ottinger, M.A., Sex steroid levels in developing and adult male and female zebra finches, *Gen. Comp. Endocrinol.*, 78 , 93–109, 1990.

47. Schlinger, B.A. and Arnold, A.P., Plasma sex steroids and tissue aromatization in hatchling zebra finches: implications for the sexual differentiation of singing behavior, *Endocrinology*, 130, 289–299, 1992.

48. Nordeen, K.W., Nordeen, E.J., and Arnold, A.P., Estrogen accumulation in zebra finch song control nuclei: implications for sexual differentiation and adult activation of song behavior, *J. Neurobiol.*, 18, 569–582, 1987.

49. Gahr, M., Flugge, G., and Güttinger, H.-R., Immunocytochemical localization of estrogen-binding neurons in the songbird brain, *Brain Res.*, 402, 173–177, 1987.

50. Gahr, M. and Konishi, M., Developmental changes in estrogen-sensitive neurons in the forebrain of the zebra finch, *Proc. Nat. Acad. Sci. U.S.A.*, 85, 7380–7383, 1988.

51. Jacobs, E.C., Arnold, A.P., and Campagnoni, A.T., Developmental regulation of the distribution of aromatase and estrogen receptor mRNA expressing cells in the zebra finch brain, *J. Comp. Neurol.*, 1999.

52. Hutchison, J.B., Wingfield, J.C., and Hutchison, R.E., Sex differences in plasma concentrations of steroids during the sensitive period for brain differentiation in the zebra finch, *J. Endocrinol.*, 103, 363–369, 1984.

53. Adkins-Regan, E., Abdelnabi, M., Mobarak, M., and Ottinger, M.A., Sex steroid levels in developing and adult male and female zebra finches, *Gen. Comp. Endocrinol.*, 78 , 93–109, 1990.

54. Schlinger, B.A. and Arnold, A.P., Plasma sex steroids and tissue aromatization in hatchling zebra finches: implications for the sexual differentiation of singing behavior, *Endocrinology*, 130, 289–299, 1992.

55. Vockel, A., Prove, E., and Balthazart, J., Sex- and age-related differences in the activity of testosterone-metabolizing enzymes in microdissected nuclei of the zebra finch brain, *Brain Res.*, 511, 291–302, 1990.

56. Gahr, M., Developmental changes in the distribution of oestrogen receptor mRNA expressing cells in the forebrain of female, male and masculinized female zebra finches, *NeuroReport*, 7, 2469–2473, 1996.

57. Schlinger, B.A. and Arnold, A.P., Brain is the major site of estrogen synthesis in the male zebra finch, *Proc. Nat. Acad. Sci. U.S.A.*, 88, 4191–4194, 1991.

58. Wade, J., Schlinger, B.A., and Arnold, A.P., Aromatase and 5b-reductase activity in cultures of developing zebra finch brain: and investigation of sex and regional differences, *J. Neurobiol.*, 27, 240–251, 1995.

59. Jacobs, E.R., Arnold, A.P., and Campagnoni, A.T., Zebra finch estrogen receptor cDNA: Cloning and mRNA expression, *J. Steroid Biochem. Mol. Biol.*, 59, 135–145, 1996.

60. Mathews, G.A. and Arnold, A.P., Antiestrogens fail to prevent masculine ontogeny of the zebra finch song system, *Gen. Comp. Endocrinol.*, 80, 48–58, 1990.

61. Mathews, G.A., Brenowitz, E.A., and Arnold, A.P., Paradoxical hypermasculinization of the zebra finch song system by an antiestrogen, *Horm. Behav.*, 22, 540–551, 1988.

62. Mathews, G.A. and Arnold, A.P., Tamoxifen fails to block estradiol accumulation, yet is weakly accumulated by the juvenile zebra finch anterior hypothalamus: an autoradiographic study, *J. Neurobiol.*, 22, 970–975, 1991.

63. Mathews, G.A. and Arnold, A.P., Tamoxifen's effects on the zebra finch song system are estrogenic, not antiestrogenic, *J. Neurobiol.*, 22, 957–969, 1991.

64. Wade, J., Schlinger, B.A., Hodges, L.L., and Arnold, A.P., Fadrozole, a potent and specific inhibitor of aromatase in the zebra finch brain, *Gen. Comp. Endocrinol.*, 94, 53–61, 1994.

65. Wade, J. and Arnold, A.P., Post-hatching inhibition of aromatase activity does not alter sexual differentiation of the zebra finch song system, *Brain Res.*, 639, 347–350, 1994.

66. Balthazart, J., Absil, P., Fiasse, V., and Ball, G.F., Effects of the aromatase inhibitor R76713 on sexual differentiation of brain and behavior in zebra finches, *Behavior*, 131, 225–260, 1994.

67. Foidart, A. and Balthazart, J., Sexual differentiation of brain and behavior in quail and zebra finches: studies with a new aromatase inhibitor, R76713, *J. Steroid Biochem. Mol. Biol.*, 53, 267–275, 1995.

68. Springer, M.L. and Wade, J., The effects of testicular tissue and prehatching inhibition of estrogen synthesis on the development of courtship and copulatory behavior in zebra finches, *Horm. Behav.*, 32, 46–59, 1997.

69. Wade, J., Springer, M.L., Wingfield, J.C., and Arnold, A.P., Neither testicular androgens nor embryonic aromatase activity alters morphology of the neural song system in zebra finches, *Biol. Reprod.*, 55, 1126–1132, 1996.

70. Wade, J. and Arnold, A.P., Functional testicular tissue does not masculinize development of the zebra finch song system. *Proc. Nat. Acad. Sci. U.S.A.*, 93, 5264–5268, 1996.

71. Wade, J., Gong, A., and Arnold, A.P., Effects of embryonic estrogen on differentiation of the gonads and secondary sexual characteristics of male zebra finches, *J. Exp. Zool.*, 278, 405–411, 1997.

72. Gong, A. and Arnold, A.P., Pre-hatching inhibition of aromatase activity masculinizes syringeal and gonadal tissue but not the song system in zebra finch females, *Soc. Neurosci. Abstr.*, 22, 756, 1996.

73. Merten, M.D.P. and Stocker-Buschina, S., Fadrozole induces delayed effects on neurons in the zebra finch song system, *Brain Res.*, 671, 317–320, 1995.

74. Grisham, W., Tam, A., Greco, C.M., Schlinger, B.A., and Arnold, A.P., A putative 5a-reductase inhibitor demasculinizes portions of the zebra finch song system, *Brain Res.*, 750, 122–128, 1997.

75. Elbrecht, A. and Smith, R.G., Aromatase enzyme activity and sex determination in chickens, *Science*, 255, 467–470, 1992.

76. Gong, A. and Arnold, A.P., Pre-hatching inhibition of aromatase activity masculinizes syringeal and gonadal tissue but not the song system in zebra finch females, *Soc. Neurosci. Abstr.*, 22, 756, 1996.

77. Gong, A., Freking, F.W., Wingfield, J.C., Schlinger, B.A., and Arnold, A.P., Effects of fadrozole on phenotype of gonads, syrinx, and neural song system in zebra finches, *Gen. Comp. Endocrinol.*, not yet published.

78. Rombauts, L., Vanmontfort, D., Verhoeven, G., and Decuypere, E., Immunoreactive inhibin in plasma, amniotic fluid, and gonadal tissue of male and female chick embryos, *Biol. Reprod.*, 46, 1211–1216, 1992.

79. Arnold, A.P., Wade, J., Grisham, W., Jacobs, E.C., and Campagnoni, A.T., Sexual differentiation of the brain in songbirds, *Dev. Neurosci.*, 18, 124–136, 1996.

80. Arnold, A.P., Genetically triggered sexual differentiation of brain and behavior, *Horm. Behav.*, 30, 495–505, 1996.

81. Arnold, A.P., Sexual differentiation of the zebra finch song system: positive evidence, negative evidence, null hypotheses, and a paradigm shift, *J. Neurobiol.*, 33, 572–584, 1997.

82. Renfree, M.B. and Short, R.V., Sex determination in marsupials: evidence for a marsupial-eutherian dichotomy, *Philos. Trans. R. Soc. London B: Biol. Sci.*, 322, 41–53, 1988.

83. Shaw, G., Renfree, M.B., Short, R., and O, W.-S., Experimental manipulation of sexual differentiation in wallaby pouch young treated with exogenous steroids, *Development*, 104, 689–701, 1988.

84. Beyer, C., Pilgrim, C., and Reisert, I., Dopamine content and metabolism in mesencephalic and diencephalic cell cultures: sex differences and effects of sex steroids, *J. Neurosci.*, 11, 1325–1333, 1991.

85. Beyer, C., Eusterschulte, B., Pilgrim, C., and Reisert, I., Sex steroids do not alter sex differences in tyrosine hydroxylase activity of dopaminergic neurons in vitro, *Cell Tissue Res.*, 270, 547–552, 1992.

86. Beyer, C., Kolbinger, W., Froehlich, U., Pilgrim, C., and Reisert, I., Sex differences of hypothalamic prolactin cells develop independently of the presence of sex steroids, *Brain Res.*, 593, 253–256, 1992.

87. Pilgrim, C. and Reisert, I., Differences between male and female brains — developmental mechanisms and implications, *Horm. Metab. Res.*, 24, 353–359, 1992.

88. Breedlove, S.M. and Arnold, A.P., Hormone accumulation in a sexually dimorphic motor nucleus of the rat spinal cord, *Science*, 210, 564–566, 1980.

89. Maxson, S.C., Didier- Erickson, A., and Ogawa, S., The Y chromosome, social signals, and offense in mice, *Behav. Neural Biol.*, 52, 251–259, 1989.

90. Maxson, S.C., Searching for candidate genes with effects on an agonistic behavior, offense, in mice, *Behav. Gene.*, 26, 471–476, 1996.

91. Maxson, S.C., Sex differences in genetic mechanisms for mammalian brain and behavior, *Biomed. Rev.*, 7, 85–90, 1997.

92. Brenowitz, E.A., Arnold, A.P., and Levin, R.N., Neural correlates of female song in tropical duetting birds, *Brain Res.*, 343, 104–112, 1985.

93. Brenowitz, E.A. and Arnold, A.P., Interspecific comparisons of the size of neural song control regions and song complexity in duetting birds: evolutionary implications, *J. Neurosci.*, 6, 2875–2879, 1986.

94. Jencius, M.J. and Page, D.C., A proposed path by which genes common to mammalian X and Y chromosomes evolve to become X inactivated, *Nature*, 394, 776–780, 1998.

95. Clepet, C., Schafer, A., Sinclair, A., Palmer, M., Lovell-Badge, R., and Goodfellow, P., The human *SRY* transcript, *Hum. Mol. Genet.*, 2, 2007–2012, 1993.

96. Harry, J., Koopman, P., Brennan, F., Graves, J., and Renfree, M.B., Widespread expression of the testis- determining gene *SRY* in a marsupial, *Nat. Genet.*, 11, 347–349, 1995.

97. Lahr, G., Maxson, S.C., Mayer, A., Just, W., Pilgrim, C., and Reisert, I., Transcription of the Y chromosomal gene, *Sry*, in adult mouse brain, *Mol. Brain Res.*, 33, 179–182, 1995.

10

Sexual Differentiation of a Neuromuscular System

Scott E. Christensen, S. Marc Breedlove, and Cynthia L. Jordan

CONTENTS

I. Introduction

Nearly 20 years ago a sexually dimorphic neuromuscular system was described that offered a relatively simple behavioral model for exploring steroidal influences upon the development and adult function of the nervous system. The system consists of striated perineal muscles called the bulbocavernosus (BC) and levator ani (LA) and their innervating motoneurons in the spinal nucleus of the bulbocavernosus (SNB). In adult male rats the BC and LA attach exclusively to the base of the penis — the BC wraps around the bulb of the penis, while the LA attaches to the bulb of the penis on one side and encircles the rectum before attaching to the contralateral bulb. The SNB motoneurons innervating these muscles are in an unusual dorsomedial position within the ventral horn of the fifth and sixth lumbar segments. The target muscles BC and LA are rhythmically active during penile erections and ejaculation, and appear to be crucial for the proper formation of copulatory plugs.[1,2] The male rat must provide adequate tactile stimulation of the female's vagina and cervix for pregnancy to complete to term,[3,4] and the BC/LA muscles also mediate these penile reflexes that provide this stimulation *in copula*. So these muscles and their motoneurons play a crucial role in male rat reproductive behavior.

In adult female rats the BC is entirely absent while the LA is a very slender muscle encircling the rectum. Likewise, the adult female rat has significantly fewer and smaller SNB motoneurons than do males. Most of the SNB motoneurons in females, and a minority of them in males, innervate the external anal sphincter.[5] Students of sexual differentiation will immediately surmise, correctly, that testicular androgen, especially testosterone (T), is responsible for the sexual dimorphism of the adult SNB system. Students of natural selection will not be surprised to learn that androgens instigate this sexual dimorphism by altering a bewildering variety of cellular processes during fetal, neonatal, pubertal, and adult periods. In this chapter we will try to organize what we know about where androgen is acting and what androgen is doing to direct the masculine development of the SNB system. For clarity, we organize the discussion chronologically around the ontogeny of the SNB system. We hope that delineating how androgen acts upon a simple neural system will guide efforts to understand the action of androgens in more complex neural systems.

II. Prenatal Processes

Cihak et al.[6] provided the first definitive study of BC/LA development in rats. They report that both male and female rats form a recognizable BC/LA, but that the muscles in female rats begin undergoing involution just before birth. Agreeing with other researchers that the rat LA is closely related to the BC, they point out that the two muscles emerge from the same anlagen. This same muscle anlagen splits earlier in development to provide the external anal sphincter (EAS). Later findings that the LA, BC, and EAS motoneurons all reside within a single spinal cord nucleus, the SNB, conforms with the idea that the three muscles are developmentally related.

Meanwhile, the spinal cord precursors that will eventually produce SNB motoneurons are all preparing for mitosis on the 12th day of embryonic development (E12). This and subsequent notations will regard the day of sperm-positive vaginal smears as E1. By this reckoning, most rat dams deliver their pups on E23, which we will also denote as the first day of life (P1). Thus injection of radiolabeled thymidine on E12 will label the majority of SNB cells in adulthood. Injection of radiolabeled thymidine on E13 labels only a minority of SNB motoneurons in adulthood, and injections on E14 label fewer still. Injections on any day from E15 to P3 label no SNB motoneurons at all.[7] In regard to these birthdates, the SNB motoneurons do not seem to differ from other motoneurons of the lumbar and sacral spinal cord. These data suggest that androgen has no effect upon the neurogenesis of SNB motoneurons, because fetal male rats do not produce detectable levels of circulating androgen until E15 or later.[8,9]

On E18 both males and females have about one third as many motoneurons in the SNB region as do adult males.[10] On E20, 2 days later, the number of SNB motoneurons is much greater in both sexes, but males display significantly more than do females. On E22, the day before birth, the number of SNB motoneurons has increased in males but remained constant in females. From the day of birth on, the number of motoneurons seen in the SNB region declines in both sexes, but more precipitately in females than in males, resulting in the adult sex difference by P10. Androgen treatment of developing female rats greatly attenuates the loss of SNB motoneurons.

The increase in the number of SNB motoneurons *before* birth is probably due to a secondary migration of motoneurons from the more conventional ventrolateral position to the dorsomedial position of adult SNB cells. It cannot be due to additional neurogenesis, as the increase in number between E18 and E20 is well after mitosis is finished on E14. Alternatively, it is possible that the motoneurons are in position early but differentiate into recognizable motoneurons from E18 to E20. However, injection of horseradish peroxidase (HRP) into the target muscles of embryonic rats suggests otherwise. When HRP is injected on E18, retrogradely labeled motoneurons are found almost exclusively in the ventrolateral position. Progressively later injections label progressively fewer ventrolateral and progressively more dorsomedial motoneurons, suggesting that the cells are migrating during this period.[11] The spindlelike morphology of a few labeled motoneurons in an intermediate position also resembles that of migrating motoneurons. Presumably it is only motoneurons innervating the BC/LA and external anal sphincter that undergo this secondary migration to a dorsomedial position. If so, then it seems likely that the target muscles instruct the motoneurons to begin the migration. Curiously, prenatal treatment of female fetuses with one androgen, dihydrotestosterone propionate (DHTP), inhibits this migration, resulting in an anomalous, ventrolateral position of motoneurons innervating the BC.[12] Treatment with another androgen, testosterone propionate (TP), does not have this effect. As we'll see later, the target muscles appear to express androgen receptors sometime before birth while motoneurons will not become androgen sensitive until the second week of life. Therefore, the androgenic influence upon this secondary migration is likely acting upon the target muscles, and any differential effect of T and DHT may be due to differential fates of the two steroids in the fetal muscles.

III. Neonatal Processes

The decline in SNB motoneurons *after* birth is due to cell death (apoptosis), as evidenced by the appearance of pyknotic profiles associated by other workers with degenerating motoneurons.[13] The profiles are found in the SNB region only during those developmental periods during which motoneuron numbers are declining; more are seen in females (who lose many SNB cells) than in males (who lose few). The same androgen treatment that prevents the loss of SNB cells in developing females also reduces the number of pyknotic profiles.[10] Moreover, androgen treatment given after the presumed period of cell death does not increase the number of SNB motoneurons seen in adulthood.[14-16] About the time the motoneurons die, their target muscles undergo a similar involution. Thus, the critical nature of the critical period for androgenic increase in the number of SNB motoneurons is clear: once the cells are dead, they can no longer be rescued by androgen. As we'll see, this does not mean that androgen has no further influence upon the SNB system, only that the number of SNB motoneurons is set early in life.

Androgen receptors rather than estrogen receptors are crucial for SNB masculinization, because rats with a defective gene for the androgen receptor (*tfm*), develop a feminine SNB system.[17] Many experiments indicate that the androgenic sparing of the SNB system is a primary effect upon the target muscles and that the sparing of SNB motoneurons themselves is a secondary consequence of the preservation of their targets. For example, if the afferents from the brain to the spinal cord are severed, androgen still spares both SNB motoneurons and their targets, indicating that brain afferents are not necessary for androgen to spare the system.[18] Likewise, androgen can spare the BC/LA muscles from involution even if the motoneurons are removed entirely in newborn female rats,[19] indicating that the motoneurons are not necessary for androgenic sparing of the muscles. Furthermore, small doses of antiandrogen are more effective at blocking the action of systemic androgen if they are provided to the muscles rather than systemically,[20] suggesting that androgen acts directly on the target muscles to spare them. Moreover, androgen spares SNB motoneurons at a time in development before they develop their own androgen receptors.[21] Finally, when female rats that are mosaic for androgen sensitivity are exposed to perinatal androgen, the androgen-insensitive SNB cells are just as likely to be spared as their neighboring SNB cells with functional androgen receptors.[22] This latter experiment proves conclusively that androgen does not act upon the motoneurons themselves to spare them and, by elimination, strengthens the argument that androgen acts upon the muscles to spare both them and their motoneurons. However, muscle is a rather heterogeneous tissue, composed of muscle fibers, fibroblasts, satellite cells, and glia, and we have no data at present to indicate which of these cell(s) are responding to the perinatal androgen.

Nancy Forger[23,24] has presented data to suggest that androgen spares the SNB system by regulating either the production of a neurotrophic factor or the sensitivity of the system to some such factor. Perinatal delivery of ciliary neurotrophic factor (CNTF) to the BC/LA complex spares both the muscles and their motoneurons from cell death, if only for the duration of the treatment.[23] This experiment raised the possibility that androgen spares the system by inducing the release of CNTF from some tissues in the region of the target muscles. However, the finding that knockout mice lacking the gene for CNTF display a normal SNB system at birth shows that CNTF is not necessary for the sparing of the system and suggests that CNTF is not normally involved in that process. But perhaps androgen affects a CNTF-*like* factor, since male mice lacking the gene for the CNTF receptor α (CNTFRα) have significantly fewer SNB cells at birth, equivalent to female controls.[24] Indeed, the finding that androgen treatment regulates the expression of CNTFRα in BC/LA muscles is consistent with the idea that androgen makes the muscles sensitive to CNTF-like substances and normally induces the release of a CNTF-like factor from some nearby tissues. Presumably, androgen does not act upon the SNB motoneurons themselves to produce any neurotrophic factor, because SNB motoneurons with defective copies of the androgen receptor gene are just as likely to be spared as androgen-insensitive SNB cells.[22]

IV. Pubertal Processes

As we have seen, the SNB system undergoes sexual differentiation during the perinatal period of development. Simply described, the system develops prenatally in both sexes independently of androgen, but its survival depends critically on the presence of androgen around birth. Although sexual differentiation of the SNB system is complete by P10 (i.e., the system in female rats is essentially gone), the SNB system in males at P10 is in many ways still quite immature. Moreover, androgen still has a role in regulating subsequent developmental stages, namely, neuromuscular synapse elimination and the growth and retraction of SNB motoneuronal dendrites.

A. Neuromuscular Synapse Elimination

HRP studies make it clear that axons from SNB motoneurons find their targets well before birth.[11] Moreover, electrical stimulation of the pudendal nerve as early as 3 days before birth evokes BC/LA muscles to contract,[25] suggesting that the connections formed between SNB motoneurons and their target muscles prior to cell death are also functional. However, despite the early onset of functional innervation, such connections remain remarkably immature well after birth. This was demonstrated in a series of studies that used the LA to evaluate the postnatal development of SNB efferent connections.[26-28] Perhaps the most obvious sign of immaturity is the fact that a high level of multiple innervation typical of newborn muscles persists in the LA through the first two weeks after birth: at P14, over 90% of LA muscle fibers are innervated by more than one motoneuron rather than being predominantly singly innervated as is the case for other rat muscles by this age.[26] Moreover, many LA synapses at P14 still retain the appearance of navigating growth cones[26] and have lower synaptic strengths,[28] more typical of neuromuscular synapses at birth.[29] During the next two weeks, neuromuscular innervation of the LA matures rapidly. By P28, the number of synaptic contacts per LA muscle fiber is reduced to one on most fibers, and LA synapses lose their growth cone appearance and also gain significantly in strength.

The loss of multiple innervation in developing skeletal muscles is commonly referred to as neuromuscular synapse elimination. As described for other rat muscles, this process in the LA involves the selective retraction of axonal branches without any net change in the number of innervating motoneurons or target muscle fibers.[26,28] Thus, SNB motoneurons come to innervate far fewer LA muscle fibers than they do initially, and through synapse elimination individual SNB motoneurons gain (nearly) exclusive control over a group of muscle fibers. However, unlike other rat muscles that undergo synapse elimination during the first two weeks after birth, synapse elimination in the LA occurs primarily during the third and fourth postnatal weeks: between P14 and P28, the proportion of muscle fibers multiply innervated drops from >90% to about 30%.[26] Thus, the degree of synapse elimination in the LA is similar to that seen in other skeletal muscles but is unique in that the time course is delayed.

The greater androgen sensitivity of the LA relative to other skeletal muscles suggested that androgen somehow controls this delay in synapse elimination. This hypothesis was tested by manipulating systemic androgen levels during the period of synapse elimination.[27] Male rats at P7 were castrated to decrease the already low level of circulating androgens,[30] and given oil vehicle, while other P7 castrates were given exogenous androgen to increase androgen titers to the range postpubertal rats experience. The results of this experiment indicate that both endogenous and exogenous androgens delay synapse elimination in the LA. Specifically, a week after androgen deprivation (at P14), LA muscles from oil-treated castrates exhibit significantly less multiple innervation (73%) than normal, suggesting that endogenous androgen, despite its low level, normally delays this process.[27] On the other hand, androgen treatment during synapse elimination prevents much of the normal decline in multiple innervation. By P28, about 70% of LA muscle fibers in androgen-treated muscles are still multiply innervated compared to the normal 30% multiple innervation at this age.[27,28] In short, androgen treatment maintains synapses that would otherwise be eliminated.

If androgen treatment halts synapse elimination, then one might expect that other indexes of neuromuscular development would be halted as well.

To the extent that this has been examined, this does not appear to be the case. For example, as multiple inputs are eliminated, motor nerve terminals develop their adult morphology, losing their growth cone appearance and acquiring a highly branched but circumscribed terminal arbor by P28. Androgen treatment beginning on P7 does not perturb this aspect of development. Neuromuscular contacts at P28 have an adult appearance and no longer look like growth cones despite being formed by multiple axons.[27] Likewise, increases in synaptic strength develop normally in such androgen-treated muscles.[28] This same juvenile androgen treatment also has no effect on the growth of dendrites belonging to SNB motoneurons, although the growth of SNB cell bodies[31] and LA muscle fibers is accelerated.[27] Thus, it appears that exogenous androgen given between P7 and P28 regulates the loss of efferent synapses from SNB motoneurons without altering the growth and elaboration of SNB motoneuronal dendrites. Moreover, while *endogenous* androgens during this same prepubertal period only transiently delay synapse elimination, prepubertal androgens exert rather profound effects on the growth SNB dendrites,[31] as we'll see later.

The effect of androgen treatment on synapse elimination is permanent.[32] The same high level of multiple innervation seen at P28 is also seen more than a year after the end of juvenile (P7 to P35) androgen treatment.[32] Since such muscles were from animals castrated at P7, exposure to androgen after P35 was minimal. It appears that once androgen prevents synapses from being eliminated during the normal period of synapse elimination, such synapses are then maintained indefinitely. Thus, like the effect of androgen on developmental cell death which determines the adult number of SNB motoneurons, androgen also determines the adult number of SNB efferent synapses by regulating developmental synapse elimination. Recent evidence suggests that some adult sex differences in brain connectivity may also be engendered through a hormonal regulation of synapse elimination.[33]

The androgenic regulation of synapse elimination has raised many of the same questions that have already been discussed for the androgenic control of cell death. The obvious questions concern where and how androgen acts to control the fate of synapses. Although neither of these questions is unequivocally answered to date, recent work provides hints to both. The effect of androgen treatment on synapse elimination was initially demonstrated using TP, leaving open the question of whether this hormone exerts its effect on synapse elimination as an androgen, an estrogen, or both. Answering this question might point to or exclude possible sites of hormone action, depending on whether androgen or estrogen receptors are involved. This question was answered in a study in which male rats castrated at P7 were given one of five possible treatments during synapse elimination: estradiol (as estradiol benzoate, EB), DHTP, a combination of EB and DHTP, TP, or oil vehicle. Results of this experiment show that TP influences synapse elimination by acting as an androgen and not as an estrogen: DHTP, either alone or in combination with EB, mimicked the effect of TP on synapse elimination, whereas EB had little or no effect.[34] These results also implicate androgen receptors (ARs) as having a role in the hormonal regulation of synapse elimination and have led to histological studies of AR expression as a means of localizing potential sites of androgen action.

Results based on steroid autoradiography and AR immunocytochemistry (ICC) suggest that SNB motoneurons develop ARs and therefore the ability to respond directly to androgen by P10,[35,21] well within the critical period for the androgenic regulation of synapse elimination.[36] Moreover, results based on AR ICC[21] and biochemical binding assays[37] indicate that ARs are present in the LA by birth and continue to be expressed throughout the synapse elimination period. The presence of ARs in both SNB motoneurons and their target, the LA muscle, during the synapse elimination period indicate that both are potential sites of action.

The effect of androgen on synapse elimination may be mediated via neurotrophic factors as has been suggested for the effect of androgen on SNB cell death. CNTF also mimics the effects of androgen on synapse elimination in the LA, maintaining multiple innervation that would ordinarily be lost.[38] For example, LA muscles at P28 exposed daily to exogenous CNTF for 2 weeks contain a higher proportion of multiply innervated fibers (70 to 80%) than vehicle-treated controls (25 to 30%). Unlike androgen, however, CNTF and other neurotrophic factors influence synapse elimination in parts of the nervous system that are not obviously sexually dimorphic or sensitive to gonadal steroids, including other skeletal muscles[39-42] and thalamic inputs to the visual cortex in the brain.[43] Thus, while androgen may have selective action on only some systems by virtue of ARs, other molecular agents such as neurotrophic factors may be relevant to synapse elimination in general. Given that synapse elimination is one cellular mechanism by which gonadal hormones control the adult number of synapses and that neurotrophic factors appear generally involved in developmental synapse elimination, an interaction between sex steroids and neurotrophic factors is likely involved in regulating synaptic number and mediating sexual differentiation of the brain.

B. Dendritic Growth and Retraction

At the same time neuromuscular synapses retract, the cell bodies and dendrites of SNB motoneurons grow substantially. Based on retrograde labeling of SNB motoneurons with cholera toxin conjugated to HRP (CT-HRP), the length of SNB dendrites increases by about fivefold between P7 and P28, achieving a length that is almost double their adult length.[31] The most significant increases in somal size also occur during this same period; on average, SNB somata more than double in size, nearly achieving their adult size by P28. Thus, by the end of the first postnatal month, SNB dendrites have achieved a length considerably longer than in adulthood, while SNB somata are nearly at their adult size. During the next 3 weeks (P28 to P49), the dendrites of SNB motoneurons retract to their adult length, while SNB somata grow until they achieve their adult size by P49. In summary, while the adult size of both SNB dendrites and somata has fully emerged by P49, the pattern of development for these two morphological features is different. Development of SNB dendrites involve two distinct phases, exuberant growth followed by retraction, while development of somata involves a steady increase in size.

Through a series of elegant experiments, Dale Sengelaub and his colleagues[31,44,45,47-50] have identified and characterized several critical factors involved in the emergence of mature dendritic lengths and somal sizes for SNB motoneurons. Unexpectedly, endogenous androgens were found to play a crucial role in the exuberant growth of SNB dendrites. Because the level of endogenous androgens throughout the first postnatal month is quite low (plasma T $\approx$ 0.5 ng/ml) compared with adult levels (plasma T $\approx$ 3ng/ml[30]) one might expect prepubertal androgens to have negligible effects on the morphology of SNB motoneurons. In some ways this expectation is borne out since prepubertal androgens seem to make no contribution to the growth of SNB somata[31] and have only minor effects on the pattern of efferent connections from SNB motoneurons in the LA.[27] However, during the prepubertal period (up to P28), SNB dendrites grow exuberantly in the presence of endogenous androgens and fail to grow without gonadal androgens. Males castrated at P7 and deprived of their own gonadal androgens for 3 weeks show, at P28, dendritic lengths markedly below the normal values seen at this age; indeed, dendritic lengths are equivalent to those typically seen at P7.[31] On the other hand, replacing androgen by treating such castrated males with TP allows SNB dendrites to grow normally during this same period (P7 to P28). While the growth of SNB motoneuron cell bodies continues unimpaired in oil-treated castrates through P28, androgen treatment of castrates during this same period accentuates the growth of SNB motoneuronal somata, presumably because the resulting level of circulating androgens exceeds what is normally experienced at this time.

While development of the SNB system is generally viewed as an androgen- and not an estrogen-dependent process (i.e., androgens and not estrogens regulate cell death and synapse elimination to control the adult number of SNB motoneurons and efferent synapses), both androgenic and estrogenic hormones appear to be involved in the prepubertal growth of SNB dendrites. Treatment with either the nonaromatizable androgen, DHTP, or the aromatized product of T, estrogen (as EB), from P7 to P28 results in intermediate dendritic lengths at P28,[44] while normal lengths are achieved only if DHTP and EB are given together during this period.[45] These results suggest that androgenic and estrogenic hormones each partially support the prepubertal growth of SNB dendrites, and both are required to support fully normal dendritic growth during this period. The effect of an aromatase inhibitor, fadrozole, on dendritic growth further suggests that estrogens are normally involved in the development of SNB motoneuronal dendrites. Treating gonadally intact males between P7 and P28 with fadrozole partially prevents the normal growth of SNB motoneuronal dendrites, without impairing either normal body weight gain or growth of SNB somata.[45] Because aromatase is necessary for the conversion of T into E, the selective effect of fadrozole on the prepubertal growth of SNB dendrites strongly argues that estrogens normally participate in the growth of SNB dendrites.

Where and how gonadal hormones are acting to regulate SNB dendritic and somal expansion during the prepubertal period is largely unanswered. While we know a great deal about the ontogeny of ARs in the SNB system,[21,35,37] we know considerably less about when and where estrogen receptors develop. In adulthood, androgen acts on the target muscles to regulate SNB dendritic length and on the SNB motoneurons to regulate somal size (discussed in detail below). Moreover, both SNB motoneurons and target muscles have ARs prepubertally. Taken together, these findings raise the possibility that androgens act on either one or both of these sites to promote the prepubertal growth of SNB dendrites and cell bodies. While current evidence argues against estrogen acting on either SNB cell bodies or dorsal root afferents,[46-49] brain afferents or the target muscles remain possible sites of action for the estrogenic regulation of dendritic growth.

After P28 the pattern of growth for SNB cell bodies and dendrites diverges: while each achieves its adult size by P49, cell bodies of SNB motoneurons grow and the dendrites of SNB motoneurons retract. Reminiscent of the effect of androgen treatment on the retraction of SNB motor axon terminals,[27] androgen treatment during the period of dendritic retraction similarly impedes this process.[31] While androgen treatment before P28 has no effect on the growth of SNB dendrites, despite being supraphysiological for that age, androgen treatment beginning on P28 appears to prevent the normal retraction of SNB dendrites. Between P28 and P49, the length of SNB dendrites normally declines by about 45%, but SNB dendrites in animals treated with androgen during this period are maintained at the same exuberant length, resulting in significantly longer dendrites than normal at P49.[31] In contrast to the permanent effect of androgen treatment on neuromuscular synapse elimination,[27,32] the effect of androgen treatment on SNB dendritic retraction appears transient, since dendritic lengths are normal by P70.[31] On the other hand, while endogenous androgens are required for dendrites to grow prepubertally, they have little or no impact on the retraction of SNB dendrites during puberty (second postnatal month), since animals castrated at P28 and given oil vehicle during the retraction period have normal dendritic lengths at P49. The fact that the level of endogenous androgens climbs steeply during this time suggests that the growth and retraction of SNB dendrites have markedly different thresholds for androgens: dendritic growth during the prepubertal period is exquisitely sensitive to androgens while dendritic retraction during puberty is less so. Moreover, since supraphysiological levels of androgen before P28 have no effect on dendritic growth, the normal low level of prepubertal androgens is apparently sufficient to achieve maximal androgen-stimulated growth.

Interestingly, while dendrites undergo dramatic changes in length between P7 and P49, the characteristic *distribution* of dendrites seen in adulthood is already evident by P7.[31] In the transverse plane, SNB dendrites throughout postnatal development have a radial, multipolar arbor with dendrites projecting predominantly to three areas: dorsally toward and beyond the central canal, contralaterally to the SNB on the other side of the spinal cord, and ipsilaterally to the motoneurons in the ventral horn on the same side. Despite the significant impact of gonadal hormones on the development of SNB dendritic lengths, gonadal hormones appear to have no role postnatally in determining or modifying the characteristic shape of the dendritic arbor for SNB motoneurons.

Another sexually dimorphic motor pool at the same lumbar level as the SNB is the dorsolateral nucleus (DLN). Although the SNB and DLN appear very similar in some respects (e.g., both innervate sexually dimorphic penile muscles and contain a sexually dimorphic number of motoneurons,[51,52] both have ARs,[14,35,21] and undergo androgen-regulated cell death[10,53]), the adult shape of the dendritic arbor and the manner in which dendrites develop to their adult length are strikingly different for these two motoneuronal groups. The adult shape of the dendritic arbor for SNB motoneurons is multipolar with bilateral projections, whereas the adult shape of the dendritic arbor for DLN motoneurons is bipolar and strictly ipsilateral.[54] Moreover, the dendrites of DLN motoneurons grow without a period of retraction to reach their adult length at P49.[50] In short, while adult dendritic length is evident by P49 for both SNB and DLN motoneurons, they are achieved through different routes: dendritic expansion followed by retraction for the SNB and steady dendritic growth for the DLN. Like the SNB, however, the prepubertal growth of DLN dendrites is sensitive to androgens.[50]

Comparing the adult shape and development of SNB and DLN dendrites has shed some light on the mechanisms that might determine dendritic morphology. For example, while dendritic length is highly sensitive to postnatal androgens, the shape of the dendritic arbor apparently is not.[31,50] An interesting set of experiments suggests that cues intrinsic to the spinal cord may determine the distinctive shape of SNB and DLN dendritic arbors. As mentioned earlier, females exposed to DHTP during the few days before birth had the BC and LA muscles as adults, but are innervated by motoneurons in the DLN.[12] More extensive characterization of such DHTP-treated females revealed that about half of the BC motoneurons were appropriately located in the SNB and the other half were ectopically located in the DLN.[55] Given that DLN and SNB dendritic arbors are distinctly different, one could then ask whether BC motoneurons in the DLN would have an arbor characteristic of DLN or SNB motoneurons. The answer is that BC motoneurons in the unusual DLN position have an arbor identical to normal DLN motoneurons, bipolar and ipsilateral, whereas BC motoneurons in the SNB of such DHTP-treated animals have a multipolar arbor characteristic of SNB motoneurons in normal males.[55] However DHTP accomplishes this mismatch between position in the spinal cord and target, one clear message of these results is that the distinctive shapes of dendritic arbors adopted by different motoneuronal groups is probably not determined by the target muscles but rather is determined by local cues in the spinal cord.

The SNB is positioned medially, whereas the DLN is positioned laterally. Thus, the dendritic fields of SNB motoneurons from the two sides of the spinal cord overlap substantially,[50] whereas DLN dendritic fields from the two sides do not. Recent evidence suggests that dendro–dendritic interactions may be one factor driving the retraction of SNB dendrites and may underlie the different patterns of dendritic growth seen in the SNB and DLN. If developing motoneurons are deprived of their targets, they tend to die. This fact was taken advantage of to alter the amount of dendro–dendritic interactions between the right and left SNB. On the day of birth, the right BC and LA muscles only were excised in gonadally intact male rats pups and their spinal cords were assessed at various times thereafter.[50] As expected, this manipulation resulted in a loss of SNB motoneurons on the right but not on the left side. At P28, SNB dendrites on the left side where the target remained intact exhibited the same exuberant length as in normals, suggesting that the loss of contralateral SNB dendrites has no effect on the prepubertal growth of SNB dendrites. However, at P49, when dendritic length in normal males had decreased to adult values, the length for SNB dendrites remained exuberant in animals lacking a contralateral SNB, comparable to what was present at P28 and significantly more than normal values at P49. Such exuberant lengths were maintained into adulthood (P70). Interestingly, SNB cell bodies grew at a normal pace even though their dendrites failed to undergo retraction. In sum, several observations suggest that dendro–dendritic interactions may be an important factor driving the retraction of SNB dendrites in development: (1) SNB dendrites from the left and right side normally overlap extensively in development and also undergo developmental retraction. (2) Conversely, DLN dendrites from the two sides neither overlap nor retract in development. (3) Experimentally reducing the overlap of SNB dendrites from the two sides appears to prevent the normal developmental regression of SNB dendrites. If such dendro–dendritic interactions are important, it suggests that developing SNB dendrites from the two sides are engaged in some sort of competition, possibly competing for synaptic inputs or the neurotrophic factors that such afferents might supply. It is noteworthy that androgens in adulthood increase both the length of SNB dendrites[56] as well as the number of afferent inputs[57,58] and that neurotrophic factors mimic many of the effects of steroid hormones on neurons.[23,38] Thus, androgens may transiently prevent dendritic retraction in development by increasing the number of synaptic inputs to SNB motoneurons and/or the neurotrophic factors such afferents may supply.

V. Androgenic Influences Upon the Adult SNB System

Although androgen permanently organizes the SNB in a sexually dimorphic fashion early in development, androgen also exerts transient, activational changes on adult SNB morphology. In 1981, Breedlove and Arnold[17] showed that males castrated 28 days previously had significantly smaller SNB motoneuronal somata and nuclei than did gonadally intact males. TP treatment of male castrates prevented these shrinkages. A similar effect of androgen was found in females; that is, castrated TP-treated adult females had significantly larger somata and nuclei than females treated with vehicle alone. Likewise, *tfm* males, which have high levels of circulating androgens but lack a functional androgen receptor, had smaller SNB motoneurons relative to their wild-type brothers. The effect of androgen on SNB cell size was substantial: in TP-treated females, the nuclei and somata of SNB motoneurons were more than 20% larger than their vehicle-treated controls. Later studies found this effect is specific to androgens: TP and the nonaromatizable androgen DHTP induce somatic and nuclear changes whereas, estrogen, another metabolite of T, does not.[59] Together, these results demonstrate that the morphology of SNB motoneurons in adulthood is dependent on the presence of androgen and its receptor. These findings pointed to the SNB as an easily accessible model for investigating the mechanisms and sites of action involved in steroid-induced neural plasticity. Subsequent research has borne out this idea; considerable work on the SNB has increased our understanding of the effect of androgen on this system and, more generally, continues to shed light on the complexities of steroid action within the mature nervous system.

The effects of androgen on motoneuronal morphology is also evidenced in the DLN. Motoneurons in the DLN, which innervate the ischiocavernosus muscle (IC) and external urethral sphincter (EUS), respond to androgens with an increase in soma and nucleus size;[54] motoneurons of the retrodorsolateral nucleus (RDLN), which control sexually monomorphic limb muscles, show little or no such changes.[60] Indeed, within the SNB and DLN, only those motoneurons which innervate sexually dimorphic muscles respond to androgen. Specifically, Collins et al.[61] showed that HRP-labeled SNB motoneurons innervating the BC and LA muscles, but not the external anal sphincter (EAS) muscle, increase in size after androgen exposure. Likewise, within the DLN, only the motoneurons innervating the IC muscles respond morphologically to androgen; the remaining motoneurons that innervate the EUS do not. Because the BC and IC are required for successful intromission and copulation,[2] androgens appear to selectively regulate the morphology of motoneurons that innervate muscles essential for normal male mating behavior.

Kurz et al.[56] showed that circulating androgens also have a profound effect on the adult dendritic arborization of SNB motoneurons. Dendritic lengths of CT-HRP-labeled SNB motoneurons of males castrated in adulthood 6 weeks beforehand were reduced by 56% compared with intact males. A separate group of male rats which were castrated and, 6 weeks later, given a 28-day T replacement regimen, showed an SNB dendritic length similar to intact controls. As expected, somata of CT-HRP-labeled SNB cells were also reduced in castrates as compared with T-treated castrates. Thus, androgens substantially affect both cell size and the degree of dendritic arborization in the adult SNB and these striking changes can wax and wane as androgen levels vary.

The activational effect of androgen on SNB morphology is not limited to experimental hormone manipulations: Forger and Breedlove[62] showed that plasticity of SNB motoneurons can be correlated with natural fluctuations in circulating androgens. Outside their breeding season, the reproductive system of the male white-footed mouse regresses. Photoperiods are the most predictable signal of seasonal change and can be used to induce these alterations in reproductive function in the laboratory. Animals kept in short days exhibit gonadal involution and a reduction in seminal vesicle and BC/LA muscle weight in comparison with animals kept in long days. Forger and Breedlove[62] found that males kept in short days also have significantly smaller SNB somata, nuclei, and dendrites than do males kept in long days. In keeping with the effect found in rats by Kurz et al.,[56] SNB motoneurons in white-footed mice that were castrated and given T replacement for 4 weeks had longer dendrites (22% more) than castrates that received no T replacement. Thus, in this species, the size of SNB dendritic arbor, somata, and nuclei is under androgenic control and is contingent upon naturally fluctuating T levels.

SNB motoneurons show consistent, marked changes in somatic and dendritic morphology in the presence of androgen; it is not surprising, then, that androgenic effects are detected at the ultrastructural level as well. Leedy et al.[57] showed that the proportion of SNB neuronal membrane apposing synaptic terminals is significantly increased in male castrates given T for 6 weeks relative to untreated castrates. Indeed, a similar difference in synaptic inputs was found with only 48 h of T treatment — an effect that occurred prior to any change in soma size. Matsumoto et al.[58] later confirmed the effects of T on SNB synaptic input with a more in-depth analysis of the specific ultrastructural changes involved. Similar to the effect found by Leedy et al.[57] they showed that the number of synaptic contacts made onto the somatic and proximal dendritic membrane of SNB motoneurons was increased in T-treated castrates relative to the control-treated castrates. In addition, the size of synaptic profiles, the occurrence of direct contact between SNB motoneurons and the number of double synapses were also increased by T administration.

As the surface area of the neuronal membrane expands, input resistance is decreased which, in turn, increases the threshold for excitation. Thus, the substantial androgen-induced increase in SNB cell size likely affects neuronal activity. Moreover, an increase of synapses by androgens likely reflects the addition of new functional inputs that modulate neuronal activity within the SNB. The short period of time required to cause changes in synaptic number raises two interesting possibilities. First, given that penile reflexes of long-term castrates can be restored within 6 h of T replacement,[63] the rapid change in functional inputs in the presence of T suggests that such changes may underlie this return of erectile capability. Second, because an increase in synaptic inputs precedes somal changes, it is possible that an increase in functional inputs may drive the increase in soma size.

Electrotonic coupling and its cellular substrate — gap junctions — are typically not detected between adult mammalian motoneurons.[65] A preliminary report by Collins and Erichsen,[66] however, suggested that adult SNB motoneurons are exceptional in that they are electrically coupled. It was also shown, on the basis of electron microscopy, that gap junctional plaques are present between adjacent SNB motoneurons and that their number and size are reduced in the absence of adult androgens.[58] It was later demonstrated by *in situ* hybridization that androgen also regulates the expression of connexin 32 (Cx32), a gap junction gene expressed by SNB motoneurons, suggesting that this underlies the difference in gap junction number. Castration results in a significant reduction in Cx32 mRNA expression in SNB motoneuronal soma and proximal dendrites; T replacement restores these values.[67] Indeed, T treatment for 2 days was sufficient to restore the amount of Cx32 mRNA to that seen in intact males. The modulation of this gene by androgen appears selective to motor pools that are sexually dimorphic: RDLN motoneurons produce Cx32 yet T treatment does not affect its expression. While Cx32 appears to be a required molecular component for gap junctions between some cells,[68] Cx32 also appears to be involved in pore formation within cells.[69,70] Thus androgen-regulated Cx32 may not be the exclusive substrate for associated changes in gap junctions.

The androgenic control of gap junctions and dendritic arborization are also likely to have consequences on male sexual behavior. The SNB lies near the midline of the spinal cord and the motoneurons on each side appear to be connected via a robust dendritic bundle; these connections presumably underlie bilateral activation of SNB motoneurons which in turn drive synchronous activation of the left and right BC/LA muscles.[71] Similarly, gap junctions enable rapid and synchronous activation of SNB motoneurons. Thus, both the dendritic profile and the presence of gap junctions between SNB motoneurons suggest coordinated activity of left and right SNB motoneurons, to ensure that both sides of the penile musculature act in concert during copulation. The regulation of SNB motoneuronal gap junctions and dendrites, then, are presumably two more means by which androgens regulate the function of this neuromuscular system.

There are several reports that androgens also regulate the expression of structural proteins within the SNB; such alterations are likely the molecular correlates of somatic and dendritic changes. The cytoskeletal protein, β-tubulin, is a major component of microtubules, found in small neurons residing in motor pools. Expression of β-tubulin mRNA in the SNB is reduced after castration and restored with T treatment, whereas androgens have no effect on the level of β-tubulin gene expression in RDLN motoneurons.[72] A similar trend was also found for β-actin expression, another cytoskeletal protein associated with neuronal membranes and postsynaptic densities.[73] Although the molecular events determining the production and function of these proteins are unknown, the androgen sensitivity of their genes for the SNB but not the RDLN suggest that β-tubulin and β-actin constitute two plausible molecular mechanisms involved in the maintenance of adult male SNB morphology. More candidates will likely be discovered.

Thus far we have discussed changes in SNB morphology induced by *systemic* androgens; however, crucial to understanding SNB plasticity is the determination of where in the system androgens act to induce these changes. For any given structural or molecular change, there are at least four possible sites where androgens could be operating: (1) spinal, or (2) supraspinal afferents which synapse onto SNB motoneurons, (3) the target muscles which the motoneurons innervate, and (4) the SNB motoneurons themselves.

Rand and Breedlove[74] found that androgens increase the size of the BC/LA muscles by acting on the muscles locally. Small capsules containing T or the antiandrogen, hydroxyflutamide (hFl), were sutured on each side of the BC muscle of male castrates. Thus, for each rat, the BC muscle on one side received T while that on the other side received hFl. This approach established local differences in T exposure. The T-exposed side was significantly larger than the hFl-exposed contralateral side. Because the differences in T exposure were lost at more distant tissues (i.e., the SNB and seminal vesicles) androgens could only have been operating locally on the muscles to increase their size.

Rand and Breedlove[75] used this same technique to determine whether androgens acting on target muscles control SNB motoneuronal morphology. BC muscles of castrates were treated as above, and after 30 days one side of the BC muscle was injected with CT-HRP to visualize the innervating motoneurons. They found that dendritic length was substantially increased only in SNB motoneurons innervating the T-treated side. The difference in arborization was comparable with the effect of systemic T found by Kurz et al.[56] — a particularly striking finding considering that these animals received little or no systemic androgen. Surprisingly, SNB soma size was identical for motoneurons innervating the T-treated and hFl-treated muscles. To further elucidate target control of dendritic arborization, the investigators divided the dendritic projections of labeled SNB motoneurons into three fields representing the major fields occupied by SNB dendrites: ipsilateral to the DLN, contralateral to the other SNB, and dorsal along the central canal. They found that androgen treatment in the muscle affected dendrites in only two fields, the contralateral and dorsal fields. In contrast, dendritic length within the ipsilateral field was similar across muscle treatments. However, systemic androgen induces a significant increase in arborization within all three fields.[56] Thus, it appears that, although dendritic length is largely controlled via androgen acting on the target muscle, ipsilateral projections are controlled via androgen acting elsewhere, perhaps on the motoneurons themselves. This study provided strong evidence that the dendritic arbor of SNB motoneurons is regulated by their targets. While this regulatory mechanism has yet to be identified, a likely scenario is that, in the presence of androgens, the BC muscle releases some retrograde signal(s) that selectively stimulates a large part of the dendritic arbor of SNB motoneurons to grow.

The study by Rand and Breedlove[75] suggested that androgens do not act on the muscle to induce somatic changes. An earlier study by Araki et al.,[76] however, suggested the opposite. To assess the role of target muscles in SNB somata size, Araki cut the pudendal nerve (which contains SNB axons) of gonadally intact males and reattached it to either the androgen-insensitive soleus muscle or the BC/LA muscle. They found that SNB soma size was reduced in motoneurons reinnervated to the soleus muscle relative to SNB motoneurons reinnervating the BC/LA muscle — a predicted difference if SNB cell size is indeed controlled by target muscles.

These conflicting findings were largely reconciled in a recent study using early androgenized female rats mosaic for the testicular feminization mutation (*tfm*; 77). The adult SNB of these mosaic animals possess an equal number of *tfm* and wild-type motoneurons. Thus, two populations of SNB motoneurons are present within the same animal: half express functional androgen receptors while the remaining half do not.[22] Importantly, the mosaic BC/LA muscle fibers show a wild-type pattern of AR immunoreactivity (ir), indicating that the muscle is likely uniformly androgen-sensitive as in wild-type males.[22] To assess whether androgen also acts on the muscle to regulate SNB somata size, mosaic *tfm* females were androgenized in development (to promote SNB motoneuron and target survival in females) and later given either T capsules or blanks as adults. By distinguishing *tfm* SNB motoneurons from wild-type motoneurons on the basis of AR-ir (i.e., *tfm* motoneurons exhibit little or no nuclear AR-ir whereas wildtype motoneurons exhibit distinct nuclear AR-ir[78]) it was shown that only wild-type motoneurons respond to adult androgen with an increase in soma size. These results demonstrate that androgen acts directly on the motoneurons to increase their size and that this increase in soma size is dependent on the presence of ARs residing in the nucleus. The results of Araki et al.,[76] suggest that the SNB cells must be innervating the BC/LA muscles to demonstrate this response. Perhaps factors specifically from those muscles increase AR expression in the motoneurons to enable them to respond in a cell-autonomous fashion.

There is evidence that AR gene expression may be regulated by neurotrophic factors from the target muscles. Axotomy of adult SNB motoneurons in gonadally intact males significantly reduces AR-ir within these cells,[79,80] an effect which is reversed by reinnervation of the BC/LA muscles.[79] Because reinnervation restores AR-ir, some signal from the target muscles may control the expression of the AR gene in SNB motoneurons. However, given the means by which this was demonstrated, axonal injury and regeneration per se remained a reasonable explanation for the loss and recovery of

AR-ir in SNB motoneurons. Al-Shamma and Arnold[81] addressed this possibility by administering the axonal transport blocker, vinblastine, to intact SNB axons: in vinblastine-treated SNB motoneurons, AR-ir was reduced to a levels seen in axotomized SNB motoneurons. This result provides persuasive evidence that a retrograde signal from the target muscle may contribute significantly to the control of AR expression in the SNB.

In an effort to identify this signal, Al-Shamma and Arnold[81] assessed several neurotrophic factors known to affect motoneuronal survival, including brain-derived neurotrophic factor (BDNF), ciliary neurotrophic factor (CNTF), neurotrophin-4, and glial cell line–derived neurotrophic factor. Of these, only BDNF, applied to the severed ends of SNB axons, prevented the downregulation of AR-ir in axotomized SNB motoneurons. Thus, BDNF may be the retrograde signal by which AR expression is regulated by the target muscle. Whether BDNF normally regulates AR expression *in vivo* remains to be seen.

From the findings thus described it appears that ARs present in the nucleus confers androgen sensitivity to SNB motoneurons; however, two reports indicate that, while they may be necessary, they may not be sufficient. Lubischer and Arnold[80] showed that SNB motoneurons axotomized at P14 and allowed to reinnervate with their target did not respond to T in adulthood with an increase in soma size. Yet if axotomy was performed only a week later (P21), SNB motoneurons later showed a normal response to adult androgens. However, differences in AR expression do not appear to underlie these differences in androgen sensitivity; specifically, SNB axotomized at P14 and P21 and allowed to reinnervate their targets exhibit the same degree of AR-ir.[80] This suggests that, in addition to ARs, a signal provided by the target muscle during a critical period in development is required for establishing androgen sensitivity in SNB motoneurons. Jordan[82] also suggests that ARs alone are insufficient to confer androgen sensitivity. As noted previously, the EAS and EUS motoneurons within the SNB and DLN, respectively, do not respond to androgens with an increase in size.[61] Again, it seemed likely that the androgen insensitivity of these cells was attributable to a relative lack in AR expression. Contrary to this expectation, however, Jordan[82] found that AR-ir in EUS motoneurons did not differ from IC motoneurons within the DLN. For both subpopulations of the DLN, over 90% of the motoneurons showed robust AR-staining in their nuclei, although soma size for one subpopulation is androgen sensitive and not for the other. This result suggests that a factor, in addition to the steroid-bound AR, is required for androgenic regulation of cell size. For example, an accessory protein to the AR might normally bind directly to DNA or the AR complex itself to facilitate AR transcriptional processes related to motoneuronal plasticity. Given that a putative accessory protein is involved, then its expression is likely induced via a signal from the target muscle during development and expressed regionally by only specific motoneuronal subgroups in adulthood. Much work is still required in the identification of cofactors specific to the AR. However, as they are discovered, the accessibility of the SNB system may facilitate the investigation of their function *in vivo*.

Recently, Breedlove[83] demonstrated that significant changes in SNB soma size could be attributed solely to sexual activity, independent of androgenic action. This study took advantage of the fact that male copulatory behavior can be maintained by circulating androgens much less than normal physiological levels.[63] Adult male rats given T capsules producing a lower-than-normal level of circulating androgens were allowed to copulate with receptive females. After 28 days, these animals had significantly smaller SNB somata and nuclei than their noncopulating controls despite equivalent T treatment. It is perhaps surprising that SNB cell size in copulators was *reduced* relative to their controls. One possible explanation is that the effects of androgens and sexual activity on SNB morphology normally counterbalance one another; only when androgenic influence is submaximal can the effect of sexual activity be detected. Whatever the explanation, this finding suggests that morphological changes in the adult SNB, although clearly affected by androgens, may also respond to hormonally independent, activity-driven mechanisms. The response of the SNB to such signals is not restricted to adulthood: in early development, both anogenital stimulation and circulating androgen are necessary to establish a male-typical number of SNB motoneurons.[84] In a more general sense, these findings demonstrate that, as sexually dimorphic neural structures drive behavior, so too can behavior cause observable changes in neuronal morphology.

That activity per se can affect SNB cell size makes likely the notion that androgenic influences on SNB morphology may in part involve activity-driven mechanisms. Because the N-methyl-d-aspartate (NMDA) version of the glutamate receptor is involved in activity-dependent neural plasticity in the hippocampus, and that genes encoding for the NMDA receptor are also expressed in motoneurons of the rat lumbar spinal cord,[85] this receptor could participate in androgen-induced plasticity of SNB motoneurons. In a recent study,[86] MK-801, a selective NMDA receptor antagonist, was administered to androgen-treated adult male castrates. Blocking NMDA receptors interfered with the influence of androgen on SNB motoneuronal somata. Specifically, animals treated with MK-801 and T had significantly smaller SNB somata than did T-treated, saline animals. MK-801 had no effect on SNB somata in animals that did not receive T, indicating that the MK-801 effect is dependent on the presence of T. This result suggests that NMDA receptors and androgens normally interact to control SNB morphology. There are several possible means by which this interaction could occur. First, androgens may increase NMDA receptor expression within the SNB; this would result in greater NMDA receptor activation which may, in turn, induce signals mediating somal growth. An interaction of androgen and the NMDA receptor could also occur through a modulation of neuronal activity. That is, androgens may increase the activity of SNB motoneurons independently of NMDA receptors, which then facilitates activation of NMDA receptors (by virtue of their voltage-gated properties). Future studies could assess NMDA receptor expression and their distribution along the membrane of SNB motoneurons, in the presence or absence of androgen. Furthermore, assuming that NMDA receptor activation drives SNB plasticity in an LTP-like fashion, interruption of LTP signaling pathways delineated in the hippocampus may identify more precisely the molecular mechanisms involved in the androgen-induced plasticity of the adult SNB.

Finally, aging takes its toll on this neuromuscular system. Matsumoto[87] found that the number and size of synaptic contacts onto SNB motoneuronal somata are reduced in aged rats, which probably relates to the slightly reduced levels of circulating androgen seen at this stage.

Numerous structural, ultrastructural, and molecular changes in the SNB have been shown to vary with androgen exposure. Work on the adult SNB has shown that androgens operate on different sites to induce specific neural plastic responses. The dissociation between the sites of action involved in somatic and dendritic growth is best illustrated with the *tfm* mosaic analysis: whereas dendritic growth seems largely regulated by muscle targets, an increase in soma size appears to be a cell-autonomous response of SNB motoneurons to androgens. The mosaic analysis developed for this study provides a useful tool in assessing where androgen acts to influence other measures of SNB plasticity.

Androgenic influences on adult SNB morphology and the role of the AR in this system are not yet fully understood. The role of steroid-sensitive spinal and supraspinal afferents in inducing changes in the SNB remain largely unknown. The retrograde signal provided by BC/LA muscles to maintain the dendritic arbor, while presumably a neurotrophic factor, has yet to be identified. Moreover, target-derived neurotrophic factors probably play an important role in the expression of ARs in SNB motoneurons. Also, a putative accessory protein may work in conjunction with the AR complex to induce morphological changes, and its presence appears to be dependent on target innervation during a critical developmental period. Finally, recent work has introduced the role of activity-driven mechanisms in adult SNB morphology. Sexual activity can influence motoneuron morphology independently of androgen. Likewise, blocking the NMDA receptor interferes with the influence of androgen on SNB morphology; the unique activity-dependent properties of this receptor likely underlie this effect. Plasticity within the adult SNB takes several forms and appears to involve a myriad of cellular mechanisms which operate at a number of different sites. This complexity is likely the rule, rather than the exception, in other steroid-sensitive neural systems. Unraveling such complexity will be a formidable task. Given its tractability, the SNB will continue to serve as a useful model system in addressing the many questions regarding the role of sex steroids in development of the nervous system and adult plasticity that underlies hormone-sensitive behaviors.

References

1. Sachs, B.D., Role of striated penile muscles in penile reflexes, copulation, and induction of pregnancy in the rat, *J. Reprod. Ferti.*, 66, 433, 1982.
2. Hart, B.L., and Melese-d'Hospital, Y., Penile mechanisms and the role of the striated penile muscles in penile reflexes, *Physiol. Behav.*, 31, 807, 1983.
3. Chester, R.V., and Zucker, I., Influence of male copulatory behavior on sperm transport, pregnancy and pseudopregnancy in female rats, *Physiol. Behav.*, 5, 35, 1970.
4. Adler, N.T., Effects of the male's copulatory behavior on successful pregnancy of the female rat, *J. Comp. Physiol. Psychol.*, 69, 613, 1969.
5. McKenna, K.E. and Nadelhaft, I., The organization of the pudendal nerve in the male and female rat, *J. Comp. Neuro.*, 248, 532, 1986.
6. Cihak, R., Gutmann, E., and Hanzlikova, V., Involution and hormone-induced persistence of the muscle sphincter (levator) ani in female rats, *J. Anat.*, 106, 93, 1970.
7. Breedlove, S.M., Jordan, C.L., and Arnold, A.P., Neurogenesis of motoneurons in the sexually dimorphic spinal nucleus of the bulbocavernosus in rats, *Dev. Brain Res.*, 9, 39, 1983.
8. Feldman, S.C. and Bloch, E., Developmental pattern of testosterone synthesis by fetal rat testes in response to luteinizing hormone, *Endocrinology*, 102, 999, 1978.
9. Picon, R., Testosterone secretion by foetal rat testes *in vitro, J. Endocrinol.*, 71, 231, 1976.
10. Nordeen, E.J., Nordeen, K.W., Sengelaub, D.R., and Arnold, A.P., Androgens prevent normally occurring cell death in a sexually dimorphic spinal nucleus, *Science*, 229, 671, 1985.
11. Sengelaub, D.R. and Arnold, A.P., Development and loss of early projections in a sexually dimorphic rat spinal nucleus, *J. Neurosci.*, 6, 1613, 1986.
12. Breedlove, S.M., Hormonal control of the anatomical specificity of motoneuron to muscle innervation in rats, *Science*, 227, 1357, 1985.
13. Hamburger, V., Cell death in the development of the lateral motor column in the chick embryo, *J. Comp. Neurol.*, 160, 535, 1975.
14. Breedlove, S.M. and Arnold, A.P., Hormone accumulation in a sexually dimorphic motor nucleus in the rat spinal cord, *Science*, 210, 564, 1980.
15. Breedlove, S.M. and Arnold, A.P., Hormonal control of a developing neuromuscular system: I. Complete demasculinization of the spinal nucleus of the bulbocavernosus in male rats using the anti-androgen flutamide, *J. Neurosci.*, 3, 417, 1983.
16. Breedlove, S.M. and Arnold, A.P., Hormonal control of a developing neuromuscular system: II. Sensitive periods for the androgen induced masculinization of the rat spinal nucleus of the bulbocavernosus, *J. Neurosci.*, 3, 424, 1983.

17. Breedlove, S.M. and Arnold, A.P., Sexually dimorphic motor nucleus in the rat lumbar spinal cord: response to adult hormone manipulation, absence in androgen-insensitive rats, *Brain Res.*, 225, 297, 1981.

18. Fishman, R.B. and Breedlove, S.M., The androgenic induction of spinal sexual dimorphism is independent of supraspinal afferents, *Dev. Brain Res.*, 23, 255, 1985.

19. Fishman, R.B. and Breedlove, S.M., Neonatal androgen maintains sexually dimorphic perineal muscles in the absence of innervation, *Muscle Nerve*, 11, 553, 1988.

20. Fishman, R.B. and Breedlove, S.M., Local perineal implants of anti-androgen block masculinization of the spinal nucleus of the bulbocavernosus, *Dev. Brain Res.*, 70, 283, 1992.

21. Jordan, C.L., Padgett, B.A., Hershey, J., Prins, G., and Arnold, A.P., Ontogeny of androgen receptor immunoreactivity in lumbar motoneurons and in the sexually dimorphic levator ani muscle of male rats, *J. Comp. Neurol.*, 379, 88, 1997.

22. Freeman, L.M., Watson, N.V., and Breedlove, S.M., Androgen spares androgen-insensitive motoneurons from apoptosis in the spinal nucleus of the bulbocavernosus in rats, *Horm. Behav.*, 30, 424, 1996.

23. Forger, N.G., Roberts, S.L., Wong, V., and Breedlove, S.M., Ciliary neurotrophic factor maintains motoneurons and their target muscles in developing rats, *J. Neurosci.*, 13, 4720, 1993.

24. Forger, N.G., Howell, M., Bengston, L., Mackenzie, L., Dechiara T., and Yancopoulos G., Sexual dimorphism in the spinal cord is absent in mice lacking the ciliary neurotrophic factor receptor, *J. Neurosci.*, 17, 9605, 1997.

25. Rand, M.N. and Breedlove, S.M., Ontogeny of functional innervation of bulbocavernosus muscles in male and female rats, *Dev. Brain Res.*, 33, 150, 1987.

26. Jordan, C.L., Letinsky, M.S., and Arnold, A.P., Synapse elimination occurs late in the hormone-sensitive levator ani muscle of the rat, *J. Neurobiol.*, 19, 335, 1988.

27. Jordan, C.L., Letinsky, M.S., and Arnold, A.P., The role of gonadal hormones in neuromuscular synapse elimination in rats. i. Androgen delays the loss of multiple innervation in the levator ani muscle, *J. Neurosci.*, 9, 229, 1989.

28. Jordan, C.L., Pawson, P.A., Arnold, A.P., and Grinnell, A.D., Hormonal regulation of motor unit size and synaptic strength during synapse elimination in the rat levator ani muscle, *J. Neurosci.*, 12, 4447, 1992.

29. Fladby, T. and Jansen, J.K.S., Postnatal loss of synaptic terminals in the partially denervated mouse soleus muscle, *Acta Physiol. Scand.*, 129, 73, 1987.

30. Corpechot, C., Baulieu, E.E., and Robel, P., Testosterone, dihydrotestosterone and androstanediols in plasma, testes and prostates of rats during development, *Acta Endocrinol.*, 96, 127, 1981.

31. Goldstein, L.A., Kurz, E.M., and Sengelaub, D.R., Androgen regulation of dendritic growth and retraction in the development of a sexually dimorphic spinal nucleus, *J. Neurosci.*, 10, 935, 1990.

32. Lubischer, J.L., Jordan, C.L., and Arnold, A.P., Transient and permanent effects of androgen during synapse elimination in the levator ani muscle of the rat, *J. Neurobiol.*, 23, 1, 1992.

33. Mong, J.A., Glaser, E., and McCarthy, M.M., Gonadal steroids promote glial differentiation and alter neuronal morphology in the developing hypothalamus in a regionally specific manner, *J. Neurosci.*, 19, 1464, 1999.

34. Jordan, C.L., Watamura, S., and Arnold, A.P., Androgenic, not estrogenic, steroids alter neuromuscular synapse elimination in the rat levator ani, *Developmental Brain Research*, 84, 215, 1995.

35. Jordan, C.L., Breedlove, S.M., and Arnold, A.P., Ontogeny of steroid accumulation in spinal lumbar motoneurons of the rat: implications for androgen's site of action during synapse elimination, *J. Comp. Neurol.*, 313, 441, 1991.

36. Jordan, C.L., Letinsky, M.S., and Arnold, A.P., Critical period for the androgenic block of neuromuscular synapse elimination, *J. Neurobiol.*, 21, 760, 1990.

37. Fishman, R.B., Chism, L., Firestone, G.L., and Breedlove, S.M., Evidence for androgen receptors in sexually dimorphic perineal muscles of neonatal male rats: absence of androgen accumulation by the perineal motoneurons, *J. Neurobiol.*, 21, 694, 1990.

38. Jordan, C.L., Ciliary neurotrophic factor may act in target musculature to regulate developmental synapse elimination, *Dev. Neurosci.*, 18, 185, 1996.

39. English, A.W. and Schwartz, G., Both basic fibroblast growth factor and ciliary neurotrophic factor promote the retention of polyneuronal innervation of developing skeletal muscle fibers, *Dev. Biol.*, 169, 57, 1995.

40. Kwon, Y.W. and Gurney, M.E., Brain-derived neurotrophic factor transiently stabilizes silent synapses on developing neuromuscular junctions, *J. Neurobiol.*, 29, 503, 1996.

41. Kwon, Y.W., Abbondanzo, S.J., Stewart, C.L., and Gurney, M.E., Leukemia inhibitory factor influences the timing of programmed synapse withdrawal from neonatal muscles, *J. Neurobiol.*, 28, 35, 1995.

42. Jordan, C.L., Morphological effects of ciliary neurotrophic factor treatment during neuromuscular synapse elimination, *J. Neurobiol.*, 31, 29, 1996.

43. Cabelli, R.J., Andreas, H., and Shatz, C., Inhibition of ocular dominance column formation by infusion of NT-4/5 or BDNF. *Science*, 267, 1662, 1995.

44. Goldstein, L.A. and Sengelaub, D.R., Differential effects of dihydrotestosterone and estrogen on the development of motoneuron morphology in a sexually dimorphic rat spinal nucleus, *J. Neurobiol.*, 25, 878, 1994.

45. Burke, K.A., Kuwajima, M., and Sengelaub, D.R., Aromatase inhibition reduces dendritic growth in a sexually dimorphic rat spinal nucleus, *J. Neurobiol.*, 38, 301, 1999.

46. Simerly, R.B., Chang, C., Muramatsu, M., and Swanson, L.W., Distribution of androgen and estrogen receptor mRNA-containing cells in the rat brain: an in situ hybridization study, *J. Comp. Neurol.*, 294, 76, 1990.

47. Taylor, S. R., Widows, M.R., and Sengelaub, D.R., Estrogenic influences and possible site of action in development of motoneuron morphology in a sexually dimorphic rat spinal nucleus, *Soc. Neurosci. Abstr.*, 21, 40, 1995.

48. Hays, T.C., Goldstein, L.A., Mills, A.C., and Sengelaub, D.R., Motoneurons development after deafferentation: II. Dorsal rhizotomy does not block estrogen-supported growth in the dorsolateral nucleus, *Dev. Brain Res.*, 91, 20, 1996.

49. Goldstein, L.A., Mills, A.C., and Sengelaub, D.R., Motoneuron development after deafferentation: I. Dorsal rhizotomy does not alter the growth in the spinal nucleus of the bulbocavernosus (SNB), *Dev. Brain Res.*, 91, 11, 1996.

50. Goldstein, L.A. and Sengelaub, D.R., Motoneuron morphology in the dorsolateral nucleus of the rat spinal cord, normal development and androgenic regulation, *J. Comp. Neurol.*, 338, 588, 1993.

51. Schroder, H.D., Organization of the motoneurons innervating the pelvic muscles of the male rat, *J. Comp. Neurol.*, 192, 567, 1980.

52. Jordan, C.L., Breedlove, S.M., and Arnold, A.P., Sexual dimorphism in the dorsolateral motor nucleus of the rat lumbar spinal cord and its response to neonatal androgen, *Brain Res.*, 249, 309, 1982.

53. Sengelaub, D.R. and Arnold, A.P., Hormonal control of neuron number in sexually dimorphic spinal nuclei of the rat: I. Testosterone-regulated death in the dorsolateral nucleus, *J. Comp. Neurol.*, 280, 622, 1989.

54. Kurz, E.M., Brewer, R.G., and Sengelaub, D.R., Hormonally mediated plasticity of motoneuron morphology in the adult rat spinal cord, a cholera-toxin-HRP study, *J. Neurobiol.*, 22, 976, 1991.

55. Kurz, E.M., Bowers, C.A., and Sengelaub, D.R., Morphology of rat spinal motoneurons with normal and hormonally altered specificity, *J. Comp. Neurol.*, 292, 638, 1990.

56. Kurz, E.M., Sengelaub, D.R., and Arnold, A.P., Androgens regulate the dendritic length of mammalian motoneurons in adulthood, *Science*, 232, 395, 1986.

57. Leedy, M.G., Beattie, M.S., and Bresnahan, J.C., Testosterone-induced plasticity of synaptic inputs to adult mammalian motoneurons, *Brain Res.*, 424, 386, 1987.

58. Matsumoto, A., Arnold, A.P., Zampighi, G.A., and Micevych, P.E., Androgenic regulation of gap junctions between motoneurons in the rat spinal cord, *J. Neurosci.*, 8, 4177, 1988.

59. Forger, N.G., Fishman, R.B., and Breedlove, S.M., Differential effects of testosterone metabolites upon the size of sexually dimorphic motoneurons in adulthood, *Horm. Behav.*, 26, 204, 1992.

60. Leslie, M., Forger N.G., and Breedlove S.M., Sexual dimorphism and androgen effects on spinal motoneurons innervating the rat flexor digitorum brevis, *Brain Res.*, 561, 269, 1991.

61. Collins, W.F. III, Seymour, A.W., and Klugewicz, S.W., Differential effect of castration on the somal size of pudendal motoneurons in the adult male rat, *Brain Res.*, 577, 326, 1992.

62. Forger, N.G. and Breedlove, S.M., Seasonal variation in mammalian striated muscle mass and motoneuron morphology, *J. Neurobiol.*, 18, 155, 1987.

63. Hart, B.L., Testosterone regulation of sexual reflexes in spinal male rats, *Science*, 155, 1282, 1967.

64. Walton, K.D. and Navarrette, R., Postnatal changes in motoneurone electrotonic coupling studied in the *in vitro* rat lumbar spinal cord, *J. Physiol.*, 433, 283, 1991.

65. Mazza, E., Nunez-Abades, P.A., Spielmann, J.M., and Cameron, W.E., Anatomical and electrotonic coupling in developing genioglossal motoneurons of the rat, *Brain Res.*, 598, 127, 1992.

66. Collins, W.F. III and Erichsen, J.T., Direct excitatory interactions between rat penile motoneurons, *Soc. Neurosci. Abstr.*, 14 181, 1988.

67. Matsumoto, A., Arai, Y., Urano, A., and Hyodo, S., Effect of androgen on the expression of gap junction and beta-actin mRNAs in adult rat motoneurons, *Neurosci. Res.* 14, 133, 1992.

68. Bennett, M.V.L., Barrio, L.C., Bargiello, T.A., Spray, D.C., Hertzberg, E., and Saez, J.C., Gap junctions: new tools, new answers, new questions, *Neuron*, 6, 305, 1991.

69. Yamamoto, T., Hertzberg, E.L., and Nagy, J.I., Subsurface cisterns in alpha-motoneurons of the rat and cat: immunohistochemical detection with antibodies against connexin 32, *Synapse*, 8, 119, 1991.

70. Scherer, S.S., Deschenes, S.M., Xu, Y., Grinspan, J.B., Fischbeck, K H., and Paul, D.L., Connexin32 is a myelin-related protein in the PNS and the CNS, *J. Neurosci.*, 15, 8281, 1995.

71. Rose, R.D. and Collins, W.F. III, Crossing dendrites may be a substrate for synchronized activation of penile motoneurons, *Brain Res.*, 337, 373, 1985.

72. Matsumoto, A., Arai, Y., and Hyodo, S., Androgenic regulation of expression of beta-tubulin messenger RNA in motoneurons of the spinal nucleus of the bulbocavernosus, *J. Neuroendocrinol.*, 5, 357, 1993.

73. Matsumoto, A., Arai, Y., Urano, A., and Hyodo, S., Androgen regulates gene expression of cytoskeletal proteins in adults rat motoneurons, *Horm. Behav.*, 28, 357, 1994.

74. Rand, M.N. and Breedlove, S.M., Androgen locally regulates rat bulbocavernosus and levator ani size, *J. Neurobiol.*, 23, 17, 1992.

75. Rand, M.N. and Breedlove, S.M., Androgen alters the dendritic arbors of SNB motoneurons by acting upon their target muscles, *J. Neurosci.*, 15, 4408, 1995.

76. Araki, I., Harada, Y., and Kuno, M., Target-dependent hormonal control of neuron size in the rat spinal nucleus of the bulbocavernosus, *J. Neurosci.*, 11, 3025, 1991.

77. Watson, N.V., Freeman, L.M., and Breedlove, S.M., Neuronal size in the spinal nucleus of the bulbocavernosus (SNB): Direct modulation by androgen in animals with mosaic genetic androgen insensitivity, *Soc. Neurosci. Abstr.*, 22, 697, 1996.

78. Freeman, L.M., Padgett, B.A., Prins, G.S., and Breedlove, S.M., Distribution of androgen receptor immunoreactivity in the spinal cord of wild-type, androgen-insensitive and gonadectomized male rats, *J. Neurobiol.*, 27, 51, 1995.

79. Al-Shamma, H.A. and Arnold, A.P., Importance of target innervation in recovery from axotomy-induced loss of androgen receptor in rat perineal motoneurons, *J. Neurobiol.*, 28, 341, 1995.

80. Lubischer, J.L. and Arnold, A.P., Evidence for target regulation of the development of androgen sensitivity in rat spinal motoneurons, *Dev. Neurosci.*, 17, 106, 1995.

81. Al-Shamma, H.A. and Arnold, A.P., Brain-derived neurotrophic factor regulates expression of androgen receptors in perineal motoneurons, *Proc. Natl. Acad. Sci. U.S.A.*, 94, 1521, 1997.

82. Jordan, C.L., Androgen receptor (AR) immunoreactivity in rat pudendal motoneurons: Implications for accessory proteins, *Horm. Behav.*, 32, 1, 1997.

83. Breedlove, S.M., Sex on the Brain, *Nature*, 389, 801, 1997.

84. Moore, C.L., Dou, H., and Juraska, J.M., Maternal stimulation affects the number of motor neurons in a sexually dimorphic nucleus of the lumbar spinal cord, *Brain Res.*, 572, 52, 1992.

85. Toelle, T.R., Berthele, A., Laurie, D.J., Seeburg, P.H., and Zieglgansberger, W., Cellular and subcellular distribution of NMDAR1 splice variant mRNA in the rat lumbar spinal cord, *Eur. J. Neurosc.*, 7, 1235, 1995.

86. Christensen, S.E. and Breedlove, S.M., MK-801 blocks the activational effect of testosterone on SNB morphology, *Soc. Neurosci. Abstr.*, 24, 199, 1998.

87. Matsumoto, A., Synaptic changes in the perineal motoneurons of aged rats, *J. Comp. Neurol.*, 400, 103, 1998.

11

Development of Sexually Dimorphic Forebrain Pathways

Richard B. Simerly

CONTENTS

I. Introduction

The functional properties of the brain depend on the organization of complex neural circuits that form during development under the influence of both genetically determined developmental programs and environmental factors. The major mechanism for the storage of information affecting behavior is the formation of new connections between functionally related sets of neurons, a process that continues in adulthood.[1] Although many neural circuits are formed during the growth of the brain in response to neural activity,[2] some circuits are formed by the elimination of unused or inappropriate connections that develop during an initial proliferation of more widespread projections.[3] Thus, the ability of an animal to elaborate an appropriate response to a particular stimulus depends on cellular mechanisms responsible for the

0-8493-1165-9/00/$0.00+$.50
© 2000 by CRC Press LLC

architecture of functional neural circuits, as well as on the plasticity of those circuits throughout life. The cellular and molecular mechanisms that control the development of neural connections are among the most actively researched areas in neuroscience. The results of work carried out largely during the last few decades has led to the general view that developing axons are guided to their targets by molecular guidance cues, followed by a refinement of connections by neural activity and other environmental factors, which sculpt the detailed pattern of connections that persist into adulthood in order to match the architecture of neural systems with their functional requirements.[4] Neural pathways necessary for many instinctive behaviors and physiological functions are largely formed at birth and are presumably due mainly to activity-independent developmental mechanisms.[5] For example, the neural circuits that underlie behavioral imprinting must be present at birth, but are permanently influenced by experience during a restricted postnatal critical period. Similarly, the neural pathways that control gonadotropin secretion, and therefore ovulation, are formed well before they are activated during puberty in response to levels of circulating ovarian steroid hormones. Moreover, these pathways develop differently in males and females under the influence of sex steroid hormones during the first week of life. Treatment of female, but not male, rats with estradiol after the 18th postnatal day results in a surge in gonadotropin secretion.[6] Thus, the development of the major connections between groups of neurons that transduce the positive feedback actions of estrogen on gonadotropin secretion are likely controlled by inherent sex-specific, hormonally directed developmental programs, although steroid hormones continue to influence the fine structure of synaptic connections in neuroendocrine pathways throughout adult life.[7-9]

Sex steroid hormones exert profound effects on mammalian forebrain development including regulation of the volume of sexually dimorphic nuclei, the density of certain classes of synapses, and the number of neurotransmitter specific neurons.[10-15] In addition to regulating the number of neurons in sexually dimorphic nuclei, sex steroids appear to regulate the development of sexually dimorphic patterns of connectivity in forebrain regions thought to play important roles in mediating reproductive function. A seminal observation was that estradiol causes a dramatic proliferation of neurites extending from explant cultures of mouse preoptic tissue.[16] One of the first morphological sex differences identified in the mammalian forebrain was a greater number of certain types of synapses onto neurons in the dorsal part of the medial preoptic area;[17] however, the source of these inputs has not been identified. Similarly, hormonally determined sex differences in synaptic density were identified in the medial amygdala and several hypothalamic nuclei (see Reference 12 for review), but technical limitations prevented determination of which cells provide these sexually dimorphic inputs. Immunohistochemical methods have demonstrated sex differences in neurotransmitter-specific innervation patterns to the lateral septum from neurons in the bed nuclei of the stria terminalis (BST) and medial amygdala,[18,19] or to the medial preoptic nucleus by serotonergic neurons in the brain stem.[20,21] Sexually dimorphic innervation patterns of vasopressin[22,23] and tyrosine hydroxylase[24] containing fibers are present in the anteroventral periventricular nucleus (AVPV), and the preoptic part of the periventricular nucleus contains a sexually dimorphic plexus of enkephalin immunoreactive fibers.[25] Although each of these sex differences has been shown to be dependent on the neonatal hormone environment, evidence is lacking for either a direct action of hormones on developing axons or for hormonally regulated target dependence of sexually dimorphic connections in these pathways.

II. Sexually Dimorphic Forebrain Pathways

Many forebrain regions that contain high densities of hormone-sensitive neurons share strong connections with each other, and it has been suggested that together they comprise a limbic–hypothalamic neural network that coordinates the influence of circulating gonadal steroids on distinct aspects of reproductive function.[26,27] Arai[28,29] and his colleagues found sex differences in synaptic morphology in many of these same regions and was the first to recognize that together they form the backbone of a dimorphic forebrain circuit involved in sexually differentiated behaviors and physiological responses. Anatomical evidence collected largely during the last two decades has defined a system of interconnected sexually dimorphic nuclei in the mammalian forebrain that mediates neural control of reproduction.[30] In general, these nuclei are larger in male animals and provide more intense inputs to target regions in males. The medial preoptic nucleus (MPN) contains the cell group first identified by Gorski and co-workers[31] as the sexually dimorphic nucleus of the preoptic area and lies at the center of this forebrain circuit.[20] In addition to containing one of the highest densities of hormone-sensitive neurons in the forebrain,[32,33] the MPN has strong bidirectional connections with the other sexually dimorphic nuclei in the hypothalamus and limbic region.[30,34] Thus, sex steroid hormones may not only exert developmental effects directly on sexually dimorphic cell groups, but may also coordinate their development transsynaptically.

Reproductive neural pathways can be viewed as essentially consisting of sensory and motor components, with intrahypothalamic integrative circuits interposed between the sensory and the motor parts of the circuit.[35,36] Within this theoretical framework the motor neurons of ovulation are those that contain gonadotropin-releasing hormone (GnRH). The activity of these neurons is influenced by visceral sensory information ascending from the brain stem via adrenergic pathways, as well as by other sensory modalities such as olfaction. Of particular importance to reproduction are pheromonal cues from the vomeronasal organ relayed centrally by the accessory olfactory bulb.[37] This olfactory information is transmitted to the hypothalamus primarily by two sexually dimorphic nuclei, the medial nucleus of the amygdala and principal nucleus of the bed nuclei of the stria terminalis (BSTp; see Reference 38 for nomenclature). Other primary sensory pathways, such as those transmitting visual, auditory, and somatosensory information, appear to reach the sexually dimorphic hypothalamic nuclei by way of the ventral subiculum and ventral part of the lateral septum.[39] Reproduction in rodents is highly dependent on circadian influences thought to be relayed to the hypothalamus by the suprachiasmatic nucleus. This nucleus sends its major projection to the subparaventricular zone of the hypothalamus. Through the widespread projections of neurons in the subparaventricular zone this sensory information diverges to influence a wide variety of behaviors and physiological responses (see Reference 35). One notable exception to this pattern is the AVPV, which appears to receive a direct and sexually dimorphic input from the suprachiasmatic nucleus.[40,41]

III. Transduction of Positive Feedback: The AVPV

A substantial body of evidence indicates that the sexually differentiated pattern of gonadotropin secretion is due to sex differences in neural systems regulating the release of GnRH from the hypothalamus,[42,43] yet GnRH neurons themselves do not appear to express receptors for estrogen.[30,44] The arcuate nucleus of the hypothalamus appears to play an important role in regulating gonadotropin secretion,[45,46] and morphological and neurochemical sex differences have been reported for this important nucleus. However, the preoptic region has long been appreciated as an essential site for the integration of endocrine and sensory influences on gonadotropin secretion.[47] The AVPV of the preoptic region is a sexually dimorphic nucleus that appears to be an essential part of the neural circuits that control gonadotropin secretion.[48,49] The AVPV lies in the rostral tip of the periventricular zone of the hypothalamus and is unusual among sexually dimorphic nuclei in that it is larger in female animals.[50] The AVPV does not contain GnRH neurons,[51,52] but ablation of the AVPV reliably blocks spontaneous ovulation, induces constant estrus, and abolishes the ability of progesterone to induce a preovulatory surge of luteinizing hormone (LH) or prolactin in female rats.[53,55] Moreover, the AVPV contains high densities of neurons that express receptors for estradiol (E_2) and progesterone,[30,56,57] and implants of E_2[58] or antiestrogens[59] into the region of the AVPV cause significant changes in LH secretion. Thus, the AVPV appears to be an essential component of neural pathways that convey positive feedback of ovarian steroids on the phasic secretion of LH that precedes ovulation. Consistent with its proposed functional role, we recently observed terminals derived from AVPV neurons in close apposition to a subpopulation of GnRH neurons in the region of the vascular organ of the lamina terminalis (OVLT).[60] The distribution of this subpopulation of GnRH neurons appears to overlap with that of neurons that express FOS immunoreactivity during proestrus,[61] as well as with the distribution of

GnRH neurons proposed to be involved in the initiation of the preovulatory LH surge.[62] Moreover, hormone treatments that cause a surge in gonadotropin secretion result in a high density of FOS immunoreactive neurons in the AVPV that project to the region containing this same population of GnRH neurons.[63] AVPV neurons also appear to send massive projections to tuberoinfundibular (TIDA) neurons in the arcuate nucleus of the hypothalamus[60] that regulate secretion of prolactin,[64] and this projection is much stronger in females than it is in males.[65] These observations are consistent with the general view that the AVPV plays a particularly important role in the transduction of the sexually dimorphic positive feedback effects of ovarian steroid hormones on secretion of LH and prolactin.

The organization of inputs to the AVPV suggests that it is influenced by a variety of sensory modalities.[66] The strongest inputs come from other parts of the periventricular zone of the hypothalamus, such as the arcuate nucleus, or from telencephalic regions that innervate the periventricular zone. The BSTp is such as region sending particularly strong projections to parts of the hypothalamus that contain high densities of hormone-sensitive neurons, and to regions that are sexually dimorphic. For example, in male rats the BSTp provides a massive projection to both the MPN and AVPV.[67] The BSTp is larger in adult male rats[68] and contains an abundance of neurons that express high levels of receptors for estrogen (ER) and androgen receptors (AR).[33] Moreover, expression of ARs and ERs in the BSTp and the AVPV appears to be sexually dimorphic in adult rats. AR expression in the BSTp tends to be higher in male rats, while levels of ER expression are generally higher in females.[33,69-71] The majority of neurons in the BSTp appears to express ER and AR mRNA during the postnatal period.[72] Because the BSTp contains high levels of the enzyme aromatase,[73] which converts testosterone to estradiol,[74,75] either receptor may mediate the effects of testosterone on the development of BSTp neurons. In contrast to the BSTp, which is larger in males, the AVPV is larger[50] and appears to contain more neurons in female rats.[76] Furthermore, administration of either testosterone or estradiol during the first 5 days of life decreases (defeminizes) the volume of the AVPV in genetic females.[77-79] Thus, the connection between the BSTp and the AVPV represents a hormone-sensitive interface between the telencephalon and diencephalon that links two sexually dimorphic nuclei having divergent developmental histories: a hormonally induced reduction of neurons in the AVPV of males, concomitant with an increase in the number of neurons in the BSTp.[68,76]

IV. Development of the Projection from the BSTp to the AVPV

In order to compare the pattern of projections from the BSTp in male rats with that of females, *Phaseolus vulgaris* leucoagglutinin (PHA-L) was injected iontophoretically into the BSTp and the distribution of labeled fibers evaluated immunohistochemically.[80] In males, the BSTp appears to send its strongest projections to regions that contain high densities of neurons that express receptors for sex steroid hormones. From the BSTp fibers pass rostrally to innervate the medial preoptic nucleus, the AVPV, and the ventral part of the lateral septal nucleus; other fibers course through the stria terminalis to provide a strong return projection to the medial nucleus of the amygdala. In addition, a descending pathway passes through the periventricular zone of the hypothalamus and forms dense terminal fields in the periventricular and arcuate hypothalamic nuclei. Perhaps the strongest projection passes caudally along the medial forebrain bundle and forms a dense terminal field in the ventral premammillary nucleus. The projection pattern of the BSTp in female rats is largely the same as that observed in males,[80] with the notable exception that the BSTp provides only a weak input to the AVPV[81] (Figure 11.1). Postnatal treatment of female rats with testosterone caused a dramatic sex reversal in the density of BSTp inputs to the AVPV, yet had a less-pronounced effect on other terminal fields, such as that in the MPN, suggesting that sex steroid hormones exert a target-specific influence on the development of BSTp projections.[80] The demonstration of morphological or neurochemical sexual dimorphisms in connectivity has usually depended on rigorous quantification, often at the ultrastructural level. The projection from the BSTp to the AVPV is unique among hodological sex differences with respect to its magnitude, suggesting that it may be a particularly useful experimental model system for studying the development of sexually dimorphic connections in the mammalian forebrain.

Despite numerous examples of sexually dimorphic patterns of development we at present know surprising little about the cellular mechanisms directing formation of sex-specific neural circuits. For conceptual purposes the development of the pathway from the BSTp to the AVPV can be divided into two distinct phases. First, axons of neurons in the BSTp must find their way out of the BSTp and into the medial preoptic area. Second, these axons must find their targets (e.g., the MPN and AVPV) before forming a dense plexus of terminals in each nucleus. The work of Toran-Allerand and colleagues[82,83] suggests that estrogen can indeed promote neurite extension from hypothalamic explants. Thus, sex steroids may promote proliferation of axonal growth from the BSTp differently in males and females during early postnatal life. Alternatively, extension of axons from the BSTp during the postnatal period may be a developmental program established through prenatal hormone exposure, or may be an intrinsic property of these neurons. In other neural systems, such as the retinotectal pathway, axons innervate broad regions of presumptive targets, which are then sculpted back to more specific terminal fields by a combination of neural and molecular activities.[3,4,84] Recently, we used DiI axonal labeling to examine the development of the sexually dimorphic projection from the BSTp to the AVPV.[81] The results of these studies indicate that the projection from the BSTp to the AVPV is established between postnatal Days 9 and 10 (P9, P10) in male rats and appears to be maintained during the juvenile period. Although labeled fibers extended

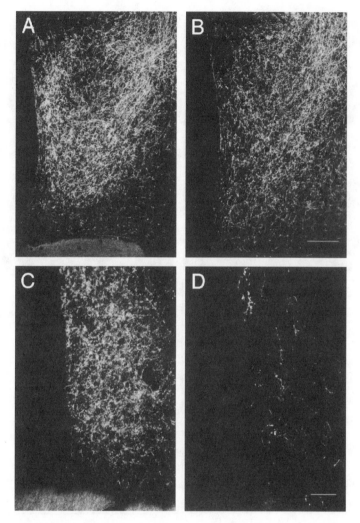

FIGURE 11.1

Darkfield photomicrographs of frontal sections through the preoptic region which compare the distribution and density of PHA-L-labeled fibers in the AVPV (C and D) and MPNm (A and B) of adult male (A and C) and female (B and D) rats. Scale bars represent 20 μm in A and B and 50 μm in C and D. (From Hutton, L.A. et al., *J. Neurosci.*, 18(8), 3003–3013, 1998. With permission.)

from the BSTp toward the preoptic region in both male and female neonates, a strong connection with the AVPV was not apparent in female rats at any of the ages studied, and by P10 the density of labeled axons in the AVPV of males was approximately 20-fold greater than that of females (Figure 11.2). These results are consistent with a previous report of greater synaptic density in the AVPV of males relative to that of females.[79] Although our observation that axons extended from the BSTp in both males and females would seem to argue against hormone induction of neurite outgrowth, it was not possible to

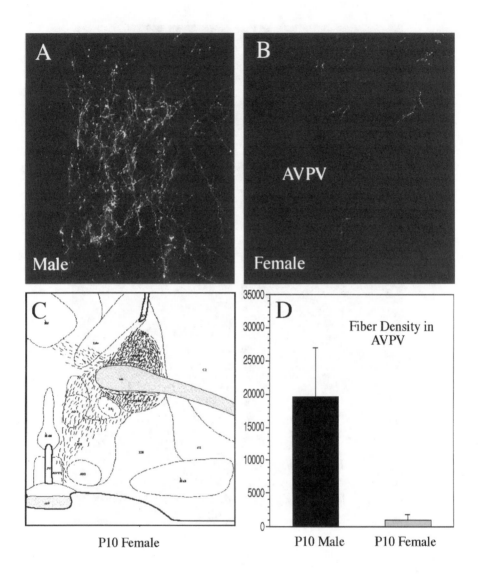

FIGURE 11.2
Sex difference in development of projections from the BSTp to the AVPV. (A, B) Confocal images of DiI-labeled fibers in the AVPV of male (A) and female (B) P10 rats. (C, D) The distribution of labeled fibers in females is illustrated (C), and the relative density of labeled fibers in the AVPV of male and female rats is represented graphically (D).

compare accurately the density of axons extending from the BSTp in our experiments. Nevertheless, it was clear that axonal extension from the BSTp is substantial in both sexes; however, the different hormone environments in which this process takes place may influence the magnitude of outgrowth. As developing axons pass from the BSTp through the medial preoptic area, they appear to be attracted toward the medial part of the MPN and the AVPV. This observation is consistent with a directed axonal guidance mechanism similar to that which occurs in the development of thalamocortical projections to layer IV of the visual cortex where axons from the lateral geniculate nucleus terminate in specific regions without sending collaterals to inappropriate targets.[85] Similarly, it is unlikely that the BSTp innervates the AVPV in both sexes and then modifies the density of this input in females since our results indicate that few axons ever reach the AVPV. The interpretation that the projection from the BSTp to the AVPV is primarily due to directed growth is also supported by the observation that labeled axons from the BSTp did not appear to be highly branched as they traversed the medial preoptic area until they reached their targets in the MPN and AVPV. It is also notable that, although the projection from the BSTp to the MPN was also weaker in females, it remained much more substantial than that to the AVPV. Together, these findings suggest that a sex- and region-specific activity influences the development of the projection from the BSTp to the AVPV producing a sexually dimorphic architecture in pathways that convey olfactory information to the hypothalamus.

Although the molecular events underlying the sex-specific innervation pattern of the AVPV by BSTp neurons remain unclear, sex steroid hormones may regulate the expression of a chemoattractive factor which acts to attract axons derived from neurons in the BSTp to the AVPV in male rats. Alternatively, expression of chemorepulsive factors by AVPV neurons may be regulated in a sex-specific manner under the control of testosterone during the perinatal period. Although a few such chemorepulsive factors have been identified,[86] none has been reported to be regulated by sex steroid hormones. Differences in the degree of axonal branching and arborization could contribute to the sexually dimorphic innervation of the AVPV. Testosterone has been shown to increase dendritic branching in the preoptic area,[13,87,88] and a similar mechanism involving axonal arborization could also contribute to the profound difference we find in the pattern of terminations in the AVPV between males and females. There may also be hormone-inducible factors that promote the branching of axons locally within the AVPV and MPNm of males. Peptide growth factors, such as brain-derived neurotrophic factor (BDNF), are expressed in the optic tectum and have been shown to promote branching of axons from retinal ganglion cells.[89] BDNF is expressed in the AVPV of male rats (R. Simerly and M. Kirigiti, unpublished observation), which may promote the branching of BSTp axons that have reached their target much in the way it promotes dendritic spine formation in hippocampal cultures.[90] Thus, there are several plausible mechanisms that may contribute to the development of the sexually dimorphic projection from the BSTp to the AVPV. Hormonal and other factors may stimulate outgrowth of axons from the BSTp, or it may preferentially regulate the survival of BSTp neurons that project to the AVPV. The enhanced density of BSTp terminals observed in the AVPV of males may also be due to increased axonal branching induced by a sexually dimorphic expression of membrane associated factors by AVPV neurons. The possible expression of chemoattractive or chemorepulsive factors by AVPV neurons should also be explored and it is likely that several of these possible mechanisms act together to control the development of the dimorphic projection from the BSTp to the AVPV.

Expression of sex steroid hormone receptors by discrete populations of neurons during developmental critical periods is thought to be the primary molecular event underlying the hormonal control of neuronal development. Although ER and AR expression in the bed nuclei of the stria terminalis and medial preoptic area of neonatal animals has been assessed with both biochemical and histochemical methods (see References 69, 71, and 91 for reviews), with few exceptions, expression in the AVPV and BSTp has not been addressed specifically. Sex differences in levels of ER binding and mRNA have been reported for the medial preoptic area,[69,91] and the AVPV of adult female rats contains more neurons that bind exogenous estradiol[92] and express ER immunoreactivity.[93] AR expression in the AVPV is relatively low in adult animals, but it is quite high in the BSTp.[33] Moreover, levels of AR mRNA in the BSTp are significantly higher in neonatal males, relative to that of females.[72] We used *in situ* hybridization to confirm expression patterns of ER and AR mRNA in the AVPV and BSTp of neonatal rats. Consistent with earlier work we found high levels of ER mRNA in both the AVPV and BSTp with the highest levels in females (Figure 11.3). In contrast, AR mRNA is virtually absent in the AVPV of neonatal rats (Figure 11.4), but is quite abundant in the BSTp, as it is in adult rats. Although AR mRNA containing neurons are not strictly localized to this nucleus, they are more abundant and express

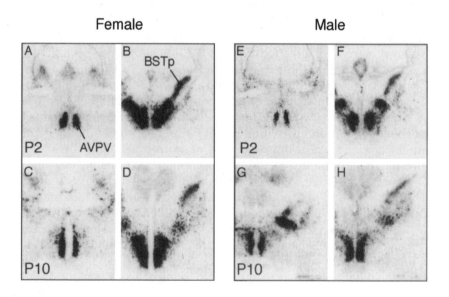

FIGURE 11.3

Ontogeny of ER gene expression in the AVPV and BSTp. Autoradiographic images that compare expression of ER mRNA in the AVPV and BSTp of female (left) and male (right) rats during the first 10 postnatal days (P0 to P10).

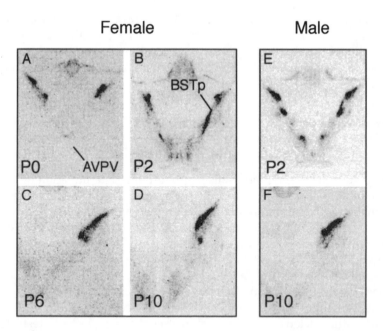

FIGURE 11.4

Ontogeny of AR gene expression in the AVPV and BSTp. Autoradiographic images that compare expression of ER mRNA in the AVPV and BSTp of female (left) and male (right) rats during the first 10 postnatal days (P0 to P10).

higher levels of message than other parts of the BST. Thus, we have confirmed that neurons in the BSTp express high levels of both ER and AR mRNA in male and female neonatal rats, but the AVPV displays an abundance of only ER mRNA, and the level of ER expression appears to be higher in female rats. Thus, either E_2 or androgen may influence development of the projection from the BSTp to the AVPV, but developmental actions of sex steroid hormones on the AVPV are likely to be mediated primarily by the ER.

One possible mechanism for regulation of the development of the projection from the BSTp to the AVPV by sex steroids is hormonal control of neuronal number in each nucleus. Because the BSTp is larger in males it is not surprising that it provides a more robust projection to its targets in males. Indeed, a detailed comparison between BSTp terminal fields in the hypothalamus of male and female rats confirms that most regions have a somewhat greater density of terminals in males relative to that found in homologous regions in females[80] (G. Gu and R. Simerly, unpublished observations). However, the 20-fold greater density of projections from the BSTp to the AVPV in male rats is much more than would be predicted simply on the basis of an approximately twofold greater number of neurons in the BSTp. Similarly, numbers of projection neurons are often influenced by the size of their targets,[94-96] but the AVPV contains fewer neurons in males relative to females, which would argue for a weaker input in males if target size were controlling the number of afferents from the BSTp. Thus, exposure to sex steroids during the perinatal period decreases the number of cells in the AVPV, while increasing the innervation of the remaining cells, resulting in a hormonally directed increase in the convergence of BSTp inputs onto AVPV neurons in males.

V. Sexual Differentiation of Dopamine Neurons in the AVPV

Potential developmental mechanisms underlying determination of the size of sexually dimorphic nuclei include hormonal influences on neurogenesis, neuronal growth and differentiation, neuronal migration and cell death (see References 10, 13, 14, 97 and 98 for reviews). Evidence for an effect of sex steroids on neurogenesis remains inconclusive, but clear examples of hormone-mediated cell death have been reported. Arai and colleagues[79,99] have addressed this issue in the AVPV where they determined that most neurons are born between embryonic Days 13 and 18, and a sex difference in neurogenesis was not observed. Prenatal hormone treatments did not affect the volume of the nucleus, but postnatal treatment of females with testosterone decreased, and castration of newborn males increased, the volume of the AVPV in adult animals,[77] but these changes are apparently not detectable until the peripubertal period.[79,100] A significant amount of the changes in the volume of the AVPV appear to be due to hormonal induction of cell death. Treatment of newborn female rats with testosterone decreased the number of BrdU-labeled neurons in the AVPV (born during a discrete prenatal period) that survive into adulthood, increased the incidence of pyknotic cells, and increased both the number of apoptotic neurons and the duration of apoptosis.[76,101] A similar reduction in cell number and increase in apoptosis in the AVPV can be induced by estrogen treatment[102] consistent with the predominance of the ER in AVPV neurons.

Clear identification of the cellular events mediating the development of sex differences in cell number are impeded by certain technical limitations. For example, volumetric measurements do not distinguish between changes in cell number, cell type, or alterations in the architecture of the neuropil. Similarly, analysis of pycnotic cells or DNA fragmentation is usually carried out by evaluating developmental "snapshots" that lack adequate temporal resolution to appreciate fully dynamic events such as cell death. Moreover, because hormones act in region- and cell-type-specific ways, it is often difficult to ascertain the effects of hormone treatments on specific neuronal populations accurately without stable cell markers. For instance, sex steroids increase the number of enkephalin neurons in the AVPV, yet decrease the number of dynorphin cells.[103] An additional confound is that differences in levels of steroids during the neonatal period may regulate gene expression, thereby altering levels of cellular markers and obscuring changes in the number of cells displaying a particular phenotype.

The AVPV contains a sexually dimorphic population of dopaminergic neurons, which are more abundant in females and differentiate under the influence of sex steroids during the postnatal period.[24,104,105] Recently, we used tyrosine hydroxylase (TH) as a marker for these cells to study their postnatal development. TH was chosen because cells labeled for TH undergo dramatic changes in number, and TH expression appears to be relatively resistant to acute changes in exposure to sex steroid hormones,[104,105] thereby minimizing activational effects that confound interpretation. Although we have known for some time that the development of the sex difference in the number of dopaminergic neurons in the AVPV was dependent on the neonatal steroid environment, we had not defined the time course of this differentiation. We recently used RT-PCR to compare levels of TH mRNA in microdissected samples of the AVPV from male and female neonatal rats killed on postnatal Days 2, 4, and 10 (P2, P4, and P10). Relative to overall levels of mRNA, TH mRNA is present in comparable amounts in male and female rats between birth and the second postnatal day (P2), but the ratio of TH mRNA to that of a control message decreased in males, while remaining stable in females, resulting in a marked sex difference by P10 (Figure 11.5A). Consistent results were obtained by using *in situ* hybridization and immunohistochemistry to compare the number of dopaminergic neurons in the AVPV of postnatal male and female rats. Cellular levels of TH mRNA appear to increase in both males and females during the first few days of life with a sex difference in the number of detectable cells first apparent on P4, and clearly established by P10. TH immunohistochemical staining follows a similar time course: similar numbers of TH immunoreactive neurons in the AVPV of male and female neonatal rats between P0 and P2 with a clear sex difference apparent by the middle of the first week of life (R. Simerly and M. Zee, unpublished observations). Thus, the sexual dimorphism in AVPV dopamine neurons appears to develop postnatally during the middle of the first postnatal week.

Because the AVPV receives strong inputs from other hormone-sensitive regions, such as the BSTp, the developmental actions of sex steroids may be mediated transsynaptically. That sex steroids cause the observed changes in the number of dopaminergic neurons by acting directly on the AVPV is consistent with preliminary findings using organotypic explant cultures derived from the region of the AVPV in male and female newborn rats. Explants from males maintained in defined conditions in the presence of 10^7 M testosterone for 2, 4, or 10 days *in vitro* were compared with explants from females cultured in the same medium lacking testosterone. The results were similar to those observed *in vivo*: numbers of TH mRNA containing cells were similar in explants from males (hormone treated) and females after 2 days *in vitro*, but became different by the fourth day *in vitro* (Figure 11.5B). In addition, sex steroids appear to act on the AVPV to determine the number of dopamine neurons that mature, independent of genetic sex. Explants of the AVPV prepared from newborn male and female rats, and maintained in media with or without testosterone (10^7 M), were used to demonstrate that testosterone treatment results in a masculine pattern of TH-immunoreactive neurons regardless of

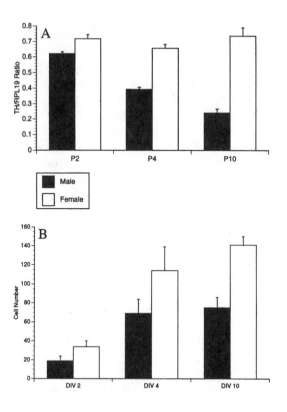

FIGURE 11.5

Ontogeny of TH gene expression in the AVPV. (A) RT PCR was used to compare the expression of TH mRNA in neonatal male and female rats. Animals were sacrificed on the second (P2), fourth (P4), and tenth (P10) postnatal day. Bars depict the mean ratios of optical densities for TH and RPL19 (±S.E.M.) at each age. Relative to RPL19, TH mRNA levels appear to be relatively constant in female rats, but decline during the first 10 days of life in males. (B) Semiquantitative analysis of TH mRNA expression in explants derived from the region of the AVPV in male and female rats maintained for 2, 4, or 10 days *in vitro* (DIV) as determined with *in situ* hybridization. Bars depict mean numbers of cells detected in explants for each experimental group. The numbers of TH mRNA-containing neurons were similar in explants from males and females after 2 DIV, but were significantly lower in males after 4 and 10 DIV.

whether the explants are derived from male or female rats (Figure 11.6). In both rats and mice the major difference in exposure to sex steroid hormones during postnatal life occurs during the first 24 h.[106] Therefore, in a recent series of experiments we exposed explants from newborn rats to either testosterone or E2 for 24 h, after which they were maintained for 9 days in media that lacked any sex steroids. The results of these studies indicate that exposure to either testosterone or E2 for 24 h appears to masculinize the expression of TH in the AVPV.[107] Moreover, the effects of steroid exposure appear to be permanent, but if the hormone treatments are delayed until after P6 they are ineffective, suggesting a lasting hormone-induced change in neuronal survival or transmitter phenotype that occurs during a restricted developmental window. That the

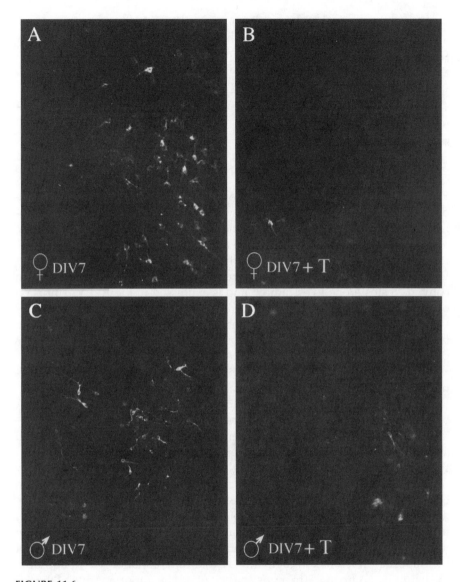

FIGURE 11.6
Hormonal control of TH gene expression *in vitro*. Immunohistochemistry was used to compare the number and distribution of TH-immunoreactive neurons in organotypic AVPV explants derived from female (A and B) or male (C and D) rats and maintained in culture for 7 days (DIV7) in the absence or presence of testosterone (+T; 10^{-7} *M*).

sexual differentiation of TH-immunoreactive neurons in the AVPV occurs *in vitro* as it does *in vivo* suggests that this model will prove useful for the study of hormonal regulation of neuronal development. Although it seems clear that sex steroid exposure enhances cell death in the AVPV,[79] it remains unclear whether sex steroids influence the number of dopamine neurons in the AVPV by promoting survival of these cells, or effect a lasting change in neurotransmitter phenotype. In order to resolve this issue it will be necessary to identify cellular markers for dopaminergic neurons that persist despite changes in TH gene expression, or that can be used to identify apoptotic dopaminergic neurons prior to loss of TH gene expression.

VI. Sexual Differentiation of Dopamine Neurons in Mice

Estrogen is as effective as testosterone in defeminizing the pattern of gonadotropin secretion, and pretreatment with antiestrogen blocks the developmental actions of testosterone on gonadotropin secretion.[42,82,108] Moreover, administration of estradiol during the first 5 days of life is as effective as testosterone in reducing (defeminizing) the volume of the AVPV.[79] These observations are generally interpreted to reflect the action of aromatase, which converts testosterone to estradiol[74] and is expressed in the AVPV.[73] Infusion of antisense oligonucleotides that correspond to the ER protects female rats from the defeminizing effects of neonatal testosterone treatment,[109] lending further support to the notion of an essential role of the ER in certain aspects of sexual differentiation.

One of the earliest demonstrations of the ability of sex steroids to defeminize gonadotropin secretion was carried out in mice,[110] however, the molecular mechanisms underlying these effects have been difficult to resolve. A variety of pharmacological approaches have been utilized, but the interpretation of the results of these experiments is often complicated by the mixed specificity of the compounds used, their access to the brain, or their differential metabolism in various brain regions (see References 111 and 112 for reviews). The coexpression of the ER and AR in certain regions[56,113] can also obscure the relative contributions of each receptor to sexual differentiation. Although definitive evidence for involvement of the α-form of the ER (ERα) in the regulation of neuronal development has proved difficult to obtain, previous pharmacological studies have provided much of what we know about the important role of E_2 in regulating sexual differentiation.[74,82] In addition, the naturally occurring Tfm mouse has clarified the important role of the AR in the development of certain neural systems.[114] Recently, a new animal model was developed in which the ERα was disrupted by homologous recombination, resulting in mice that appear to lack functional estrogen receptors.[115,116] These ERα knockout mice (ERKOα) are infertile, show altered reproductive behavior and neuroendocrinological responses, as well as changes in aggressive behavior;[116-119] (see Ogawa and Pfaff, Chapter 2 this volume). Immunostaining for ERα was virtually abolished in the AVPV of ERKOα mice, but a few immunoreactive cells were detected.[120] The reason for this residual staining remains unclear. One possibility is that recombinant events in some cells result in copies of ERα mRNA that get translated into protein. Another possible explantation is a low degree of cross-reactivity with other ER species such as the ERβ, which is also expressed in the AVPV.[121] Nevertheless, it is clear that detectable ERα immunoreactivity has been eliminated in nearly all neurons in the AVPV of the ERKOα mouse, which should have a dramatic effect on the development and function of the AVPV.

Both the BSTp and AVPV appear to be sexually dimorphic in mice and show similar patterns of hormone receptor gene expression as that described previously for rats[122] (Chen and Simerly, unpublished data). Moreover, the sexual dimorphism in the number of dopaminergic neurons in the AVPV is also conserved in C57BL6 mice. Therefore, we used ERKOα mice to test whether sexual differentiation of dopaminergic neurons in the AVPV is dependent on a functional ERα.[120] The number of TH-immunoreactive neurons in the AVPV was found to be three times that of wild-type males (Figure 11.7). In contrast, the AVPV contains the same number of TH-immunoreactive neurons in Tfm mice as in wild-type males indicating that sexual differentiation of this population of neurons is not dependent on an intact androgen receptor (Figure 11.8). The number of TH-immunoreactive neurons in the AVPV of female ERKOα mice remained higher than that of wild-type males, but TH staining appeared to be attenuated slightly, relative to that of wild-type females (Figure 11.9). Thus, the sexual differentiation of dopamine

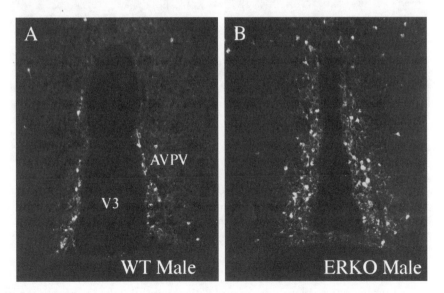

FIGURE 11.7
TH Expression in the AVPV of the male ERKOα mouse. Images of immunohistochemically stained sections through the AVPV to illustrate the density and distribution of neurons that express TH in the AVPV of wild type (+/+) male (A) or ERKOα male (B) mice. Wild-type females have significantly more TH-immunoreactive neurons in the AVPV relative to that of wild-type males. Disruption of the ERα in the male (B) prevents the normally occurring loss of TH cells

neurons in the AVPV appears to be dependent on the action of E_2 during the neonatal period that is mediated by the ERα, and independent of hormonal influences mediated by the androgen receptor (Figure 11.10). However, it should be noted that all neurons in the AVPV do not show the same sensitivity to the developmental actions of E_2. For example, proenkephalin neurons are sexually dimorphic in rats with more neurons present in the AVPV of males and treatment with testosterone during the neonatal period reverses this sex difference.[103] The AVPV of mice also contains a high density of proenkephalin neurons, but this population is not sexually dimorphic and is not affected by deletion of the ERα,[120] suggesting that the action of E_2 on the development of AVPV neurons shows considerable cell specificity, and that at least some AVPV cells undergo a normal pattern of development in the ERKOα mouse.

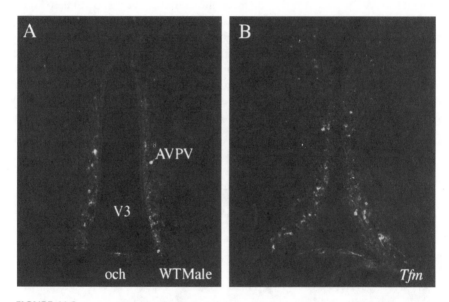

FIGURE 11.8
Development of dopaminergic neurons in Tfm male mice. Images illustrate the density and distribution of TH immunoreactive neurons in the AVPV of wild-type (C57BL/6J) male (A) and Tfm male (B) mice. No difference in the number of TH-stained neurons was observed, suggesting that development of the sex difference in dopaminergic neurons in the AVPV is independent of the androgen receptor.

Although receptor knockout animals represent the most direct way of testing the involvement of receptors in molecular signaling, the interpretation of such experiments is not always straightforward. Steroid hormones act on the brain throughout the life of the animal, and the ERα and AR function as transcriptional modulators that display a variety of protein–protein interactions that are only now being characterized. Thus, the effects of mutations in the ERα or AR genes will certainly interfere with receptor-mediated actions of E_2 and T on brain development, but may also influence other processes and

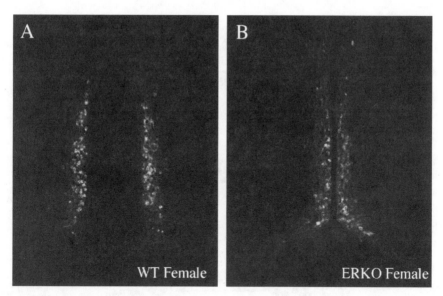

FIGURE 11.9

TH expression in the AVPV of the female ERKOα mouse. Immunofluorescence images of TH-immunoreactive neurons in the AVPV of wild-type (+/+) female (A), and ERKOα female mice (B). Disruption of the ERα in the female significantly reduces cellular staining, but the number of TH-immunoreactive neurons is similar to that of wild-type females.

signaling cascades. The absence of predicted developmental effects is equally difficult to interpret since the action of E_2 may be mediated by the ERα. Also troubling is the possibility that circulating levels of sex steroid hormones are dramatically different during the perinatal period in ERKOα mice. Adult female ERKOα mice have much higher levels of circulating E_2, relative to that of wild-type animals,[117,119] but it is not known if this is also the case early in development.

VII. Conclusion

The results of our analysis of TH development in ERKOα and Tfm mice would suggest that the ERα plays a major role in the development of the AVPV, but the high levels of AR expression in the BSTp in neonatal animals indicate that androgen may also influence the development of the projection from the BSTp to the AVPV. It would be especially interesting to determine whether different aspects of pathway formation are affected in Tfm and ERKOα mice. For example, growth of axons out of the BSTp may be perturbed only in Tfm animals, whereas proliferation of terminals in the AVPV may be dependent primarily on the ERα. Since the ERβ is also expressed in

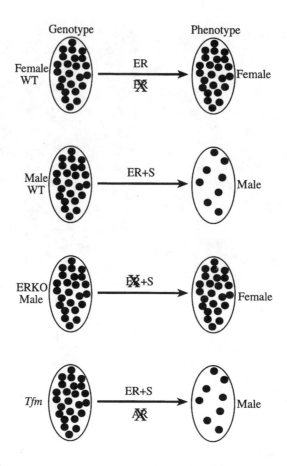

FIGURE 11.10

ERα dependent sexual differentiation. Schematic illustration to summarize the sexual differentiation of dopaminergic neurons in the AVPV of mice. The female pattern of dopaminergic neurons appears to be the default phenotype and is expressed in either the absence of high levels of sex steroids (Female WT), or in animals that lack a functional ERα (ERKOα male; ERKOα female). High levels of circulating sex steroid hormones during the perinatal period bring about the male pattern of dopaminergic neurons (Male WT). This loss of TH expression, or loss of neurons capable of expressing the dopaminergic phenotype, is dependent on a functional ERα, but is not noticeably influenced by the mutation of the androgen receptor (*Tfm*).

the AVPV it may contribute to the development of AVPV connections or transmitter phenotype as well. ERβ knockout mice are now available[123] so it is feasible to test the role of the ERβ directly. A potential complication is the possible involvement of receptor-independent effects of sex steroid hormones in regulating neuronal development (see References 126 and 127). Given these complexities we are indeed fortunate to have a rich literature detailing the actions of sex steroid hormones on neuronal function and development, which complements current experiments and will provide considerable help in interpreting their results. In future studies it is clear that the ability to utilize genetically manipulated animals to uncover the molecular mechanisms responsible for regulating hormonally directed development of neural pathways and cellular phenotype represents a major advantage of establishing a robust experimental model of sexual differentiation in mice. For example, certain aspects of the development of AVPV connections may prove to be ER dependent, implying an important role for aromatase, which can be tested directly by using aromatase knockout mice currently available.[124,125] Similarly, it is likely that molecules shown to have chemotropic activities in other neural systems may play a role in directing the development of AVPV connections, and a molecular genetic approach should prove useful in determining if hormones affect their function. Future studies may also combine the molecular advantages of genetically manipulated mice with the experimental access and control afforded by *in vitro* model systems to uncover signaling mechanisms that underlie hormonal control of synaptogenesis. Although a considerable amount of work remains before we have a clear understanding of the molecular events controlling the development of sexually dimorphic pathways in the mammalian forebrain, the current rate of progress promises improving insight into the genesis of sex-specific brain architecture.

References

1. Bailey, C.H. and Kandel, E.R., Structural changes accompanying memory storage, *Annu. Rev. Physiol.*, 55, 397–426, 1993.
2. Purves, D., *Neural Activity and the Growth of the Brain*, Cambridge University Press, Cambridge, U.K., 1994.
3. Cowan, W.M., Fawcett, J.W., O'Leary, D.D., and Stanfield, B.B., Regressive events in neurogenesis, *Science*, 225, 1258–1265, 1984.
4. Goodman, C.S. and Shatz, C.J., Developmental mechanisms that generate precise patterns of neuronal connectivity, *Cell*, Vol. 72/*Neuron*, Vol. 10(*Suppl.*), January, 77–98, 1993.
5. Catalano, S.M. and Shatz, C.J., Activity-dependent cortical target selection by thalamic axons, *Science*, 281, 559–562, 1998.
6. Ojeda, S.R. and Urbanski, H.F., Puberty in the rat, in *The Physiology of Reproduction*, Knobil, E. and Neill, J.D., Eds., Raven Press, New York, 1994, 363–409.
7. McEwen, B.S., Coirini, H., Westlind Danielsson, A., Frankfurt, M., Gould, E., Schumacher, M., and Woolley, C., Steroid hormones as mediators of neural plasticity, *J. Steroid. Biochem. Mol. Biol.*, 39, 223–232, 1991.
8. Matsumoto, A., Hormonally induced synaptic plasticity in the adult neuroendocrine brain, *Zool. Sci.*, 9, 679–695, 1992.
9. Woolley, C.S., Weiland, N.G., McEwen, B.S., and Schwartzkroin, P. A., Estradiol increases the sensitivity of hippocampal CA1 pyramidal cells to NMDA receptor-mediated synaptic input: correlation with dendritic spine density, *J. Neurosci.*, 17, 1848–1859, 1997.

10. Arnold, A.P. and Gorski, R.A., Gonadal steroid induction of structural sex differences in the central nervous system, *Ann. Rev. Neurosci.*, 7, 413–442, 1984.
11. De Vries, G.J., Sex differences in neurotransmitter systems, *J. Neuroendocrinol.*, 2, 1–13, 1990.
12. Matsumoto, A., Synaptogenic action of sex steroids in developing and adult neuroendocrine brain, *Psychoneuroendocrinology*, 16, 25–40, 1991.
13. Tobet, S.A. and Fox, T.O., Sex differences in neuronal morphology influenced hormonally throughout life, in *Handbook of Behavioral Neurobiology*, Vol. 11, *Sexual Differentiation*, Gerall, A.A., Moltz, H., and Ward, I.L., Eds., Plenum Press, New York, 1992, 41–83.
14. Kawata, M., Roles of steroid hormones and their receptors in structural organization in the nervous system, *Neurosci. Res.*, 24, 1–46, 1995.
15. Cooke, B., Hegstrom, C.D., Villeneuve, L.S., and Breedlove, S.M., Sexual differentiation of the vertebrate brain: principles and mechanisms, *Front. Neuroendocrinol.*, 19, 323–362, 1998.
16. Toran-Allerand, C.D., Sex steroids and the development of the newborn mouse hypothalamus and prepoptic area *in vitro*: implications for sexual differentiation, *Brain Res.*, 106, 407–412, 1976.
17. Raisman, G. and Field, P.M., Sexual dimorphism in the preoptic area of the rat, *Science*, 173, 731–733, 1971.
18. De Vries, G.J., Buijs, R.M., and Swaab, D.F., Ontogeny of the vasopressinergic neurons of the suprachiasmatic nucleus and their extrahypothalamic projections in the rat brain — present of a sex difference in the lateral septum, *Brain Res.*, 218, 67–78, 1981.
19. Caffé, A.R., Van Leeuwen, F.W., and Luiten, P.G.M., Vasopressin cells in the medial amygdala of the rat project to the lateral septum and ventral hippocampus, *J. Comp. Neurol.*, 261, 237–252, 1987.
20. Simerly, R.B., Swanson, L.W., and Gorski, R.A., Demonstration of a sexual dimorphism in the distribution of serotonin-immunoreactive fibers in the medial preoptic nucleus of the rat, *J. Comp. Neurol.*, 225, 151–166, 1984.
21. Simerly, R.B., Swanson, L.W., and Gorski, R.A., Cells of origin of a sexually dimorphic serotonergic input to the medial preoptic nucleus of the rat, *Brain Res.*, 324, 185–189, 1984.
22. Hoorneman, E.M.D. and Buijs, R.M., Vasopressin fiber pathways in the rat brain following suprachiasmatic nucleus lesioning, *Brain Res.*, 243, 235–241, 1982.
23. De Vries, G.J., Buijs, R.M., Van Leeuwen, F.W., Caffé, A.R., and Swaab, D.F., The vasopressinergic innervation of the brain in normal and castrated rats, *J. Comp. Neurol.*, 233, 236–254, 1985.
24. Simerly, R.B., Swanson, L.W., and Gorski, R.A., The distribution of monoaminergic cells and fibers in a periventricular nucleus involved in the control of gonadotropin release: Immunohistochemical evidence for a dopaminergic sexual dimorphism, *Brain Res.*, 330, 55–64, 1985.
25. Watson, R.E., Jr., Hoffmann, G.E., and Wiegand, S.J., Sexually dimorphic opioid distribution in the preoptic area: manipulation by gonadal steroids, *Brain Res.*, 398, 157–163, 1986.
26. Stumpf, W.E., The brain: an endocrine gland and hormone target, in *Anatomical Neuroendocrinology*, Stumpf, W.E. and Grant, L.D.S., Eds., Karger, Basel, 1975, 2–8.
27. Pfaff, D.W., *Estrogens and Brain Function*, Springer-Verlag, New York, 1980.

28. Arai, Y., Synaptic correlates of sexual differentiation, *Trends Neurosci.*, 4, 291–293, 1981.

29. Arai, Y., Synaptic sexual differentiation of the neuroendocrine brain, in *Endocrine Correlates of Reproduction*, Ochiai, K., Ed., Japan Science Society Press/Springer-Verlag, Berlin, 1984, 29–40.

30. Simerly, R.B., Hormonal regulation of limbic and hypothalamic pathways, in *Neurobiological Effects of Sex Steroid Hormones*, Micevych, P.E. and Hammer, R.P., Jr., Eds., Cambridge University Press, Cambridge, U.K., 1995, 85–114.

31. Gorski, R.A., Gordon, J.H., Shryne, J.E., and Southam, A.M., Evidence for a morphological sex difference within the medial preoptic area of the rat brain, *Brain Res.*, 148, 333–346, 1978.

32. Pfaff, D.W. and Keiner, M., Atlas of estradiol-concentrating cells in the central nervous system of the female rat, *J. Comp. Neurol.*, 151, 121–158, 1973.

33. Simerly, R.B., Chang, C., Muramatsu, M., and Swanson, L.W., Distribution of androgen and estrogen receptor mRNA-containing cells in the rat brain: an *in situ* hybridization study, *J. Comp. Neurol.*, 294, 76–95, 1990.

34. Simerly, R.B. and Swanson, L.W., The organization of neural inputs to the medial preoptic nucleus of the rat, *J. Comp. Neurol.*, 246, 312–342, 1986.

35. Swanson, L.W., The hypothalamus, in *Handbook of Chemical Neuroanatomy*, Vol. 5, *Integrated Systems of the CNS*, Part I, Björklund, A., Hökfelt, T., and Swanson, L.W., Eds., Elsevier Sciences Publishers, Amsterdam, 1987, 1–124.

36. Simerly, R.B., Anatomical substrates of hypothalamic integration, in *The Rat Nervous System*, Paxinos, G., Ed., Academic Press, San Francisco, 1995, 353–376.

37. Simerly, R.B., Hormonal control of neuropeptide gene expression in sexually dimorphic olfactory pathways, *Trends Neurosci.*, 13, 104–110, 1990.

38. Ju, G. and Swanson, L.W., Studies on the cellular architecture of the bed nuclei of the stria terminalis in the rat: I. Cytoarchitecture, *J. Comp. Neurol.*, 280, 587–602, 1989.

39. Risold, P.Y. and Swanson, L.W., Structural evidence for functional domains in the rat hippocampus, *Science*, 272, 1484–1486, 1996.

40. Watts, A.G., Swanson, L.W., and Sanchez-Watts, G., Efferent projections of the suprachiasmatic nucleus: I. Studies using anterograde transport of Phaseolus vulgaris leucoagglutinin in the rat, *J. Comp. Neurol.*, 258, 204–229, 1987.

41. Van der Beek, E.M., Horvath, T.L., Wiegant, V.M., and Buijs, R.M., Evidence for a direct neuronal pathway from the suprachiasmatic nucleus to the gonadotropin-releasing hormone system: combined tracing and light- and electron-microscopical immunocytochemical studies, *J. Comp. Neurol.*, 384, 569–579, 1997.

42. Barraclough, C.A., Sex differentiation of cyclic gonadotropin secretion, in *Advances in the Biosciences*, Vol. 25, Kaye, A.M. and Kaye, M., Eds., Pergamon Press, Oxford, 1979, 433-450.

43. Gerall, A.A. and Givon, L., Early androgen and age-related modifications in female rat reproduction, in *Handbook of Behavioral Neurobiology*, Gerall, A.A., Moltz, H., and Ward, I.L., Eds., Plenum Press, New York, 1992, 313–354.

44. Shivers, B.D., Harlan, R.E., Morrell, J.I., and Pfaff, D., Absence of oestradiol concentration in cell nuclei of LHRH-immunoreactive neurones, *Nature*, 304, 345–347, 1983.

45. Fink, G., Gonadotropin secretion and its control, in *The Physiology of Reproduction*, Knobil, E. and Neill, J., Eds., Raven Press, New York, 1988, 1349–1376.

46. Kalra, S.P. and Kalra, P.S., Neural regulation of luteinizing hormone secretions in the rat, *Endocrine Rev.*, 4, 311–351, 1983.
47. Gorski, R.A., Localization and sexual differentiation of the nervous structures which regulate ovulation, *J. Reprod. Fertil.*, Suppl. 1, 67–88, 1966.
48. Terasawa, E., Wiegand, S.J., and Bridson, W.E., A role for medial preoptic nucleus on afternoon of proestrus in female rats, *Am. J. Physiol.*, 238, E533–E529, 1980.
49. Herbison, A.E., Multimodal influence of estrogen upon gonadotropin-releasing hormone neurons, *Endocrine Rev.*, 19, 302–330, 1998.
50. Bleier, R., Byne, W., and Siggelkow, I., Cytoarchitectonic sexual dimorphisms of the medial preoptic and anterior hypothalamic areas in guinea pig, rat, hamster, and mouse, *J. Comp. Neurol.*, 212, 118–130, 1982.
51. Terasawa, E. and Davis, G.A., The LHRH neuronal system in female rats: relation to the medial preoptic nucleus, *Endocrinol. Jpn.*, 30, 405–417, 1983.
52. Simerly, R.B. and Swanson, L.W., The distribution of neurotransmitter-specific cells and fibers in the anteroventral periventricular nucleus: implications for the control of gonadotropin secretion in the rat, *Brain Res.*, 400, 11–34, 1987.
53. Wiegand, S.J. and Terasawa, E., Discrete lesions reveal functional heterogeneity of suprachiasmatic structures in regulation of gonadotropin secretion in the female rat, *Neuroendocrinology*, 34, 395–404, 1982.
54. Popolow, H.B., King, J.C., and Gerall, A.A., Rostral medial preoptic area lesions' influence on female estrous processes and LHRH distribution, *Physiol. Behav.*, 27, 855–861, 1981.
55. Ronnekleiv, O.K. and Kelly, M.J., Luteinizing Hormone-releasing hormone neuronal system during the estrous cycle of the female rat: effects of surgically induced persistent estrus, *Neuroendocrinology*, 43, 564–576, 1986.
56. Simerly, R.B., Distribution and regulation of steroid hormone receptor gene expression in the central nervous system, in *Advances in Neurology*, Vol. 59, Seil, F.J., Ed., Raven Press, New York, 1993, 207–226.
57. Blaustein, J.D. and Olster, D.H., Gonadal steroid hormone receptors and social behaviors, in *Advances in Comparative and Environmental Physiology*, Vol. 3, Balthazart, J., Ed., Springer-Verlag, Berlin, 1989, 31–104.
58. Akema, T., Tadakoro, Y., and Kimura, F., Regional specificity in the effect of estrogen implantation within the forebrain on the frequency of pulsatile luteinizing hormone secretion in the ovariectomized rat, *Neuroendocrinology*, 39, 517–523, 1984.
59. Peterson, S.L. and Barraclough, C.A., Suppression of spontaneous LH surges in estrogen-treated ovariectomized rats by microimplants of antiestrogens into the preoptic brain, *Brain Res.*, 484, 279–289, 1989.
60. Gu, G.B. and Simerly, R.B., Projections of the sexually dimorphic anteroventral periventricular nucleus in the female rat, *J. Comp. Neurol.*, 384, 142-164, 1997.
61. Lee, W.-S., Smith, M.S., and Hoffman, G.E., Luteinizing hormone-releasing hormone neurons express Fos protein during the proestrous surge of luteinizing hormone, *Proc. Natl. Acad. Sci. U.S.A.*, 87, 5163–5167, 1990.
62. King, J.C., Tai, D.W., Hanna, I.K., Pfeiffer, A., Haas, P., Ronsheim, P. M., Mitchell, S.C., Turcotte, J.C., and Blaustein, J.D., A subgroup of LHRH neurons in guinea pigs with progestin receptors is centrally positioned within the total population of LHRH neurons, *Neuroendocrinology*, 61, 265–275, 1995.

63. Le, W.W., Berghorn, K.A., Rassnick, S., and Hoffman, G.E., Periventricular preoptic area neurons coactivated with luteinizing hormone (LH)-releasing hormone (LHRH) neurons at the time of the LH surge are LHRH afferents, *Endocrinology,* 140, 510–519, 1999.

64. Neill, J.D. and Nagy, G.M., Prolactin secretion and its control, in *The Physiology of Reproduction,* Vol. 1, Knobil, E. and Neill, J.D., Eds., Raven Press, New York, 1994, 1833–1860.

65. Reinoso, B.S. and Simerly, R.B., Hormone-sensitive sexually dimorphic neurons in the anteroventral periventricular nucleus project to the arcuate nucleus of the hypothalamus, *Soc. Neurosci. Abstr.,* 17, 1229, 1991.

66. Simerly, R.B., Organization and regulation of sexually dimorphic neuroendocrine pathways, *Behav. Brain Res.,* 92, 194–203, 1997.

67. Simerly, R.B., Young, B.J., Capozza , M.A., and Swanson, L.W., Estrogen differentially regulates neuropeptide gene expression in a sexually dimorphic olfactory pathway, *Proc. Natl. Acad. Sci. U.S.A.,* 86, 4766–4770, 1989.

68. del Abril, A., Segovia, S., and Guillamón, A., The bed nucleus of the stria terminalis in the rat: regional sex differences controlled by gonadal steroids early after birth, *Dev. Brain Res.,* 32, 295–300, 1987.

69. DonCarlos, L.L. and McAbee, M., Estrogen receptor mRNA levels in the preoptic area of neonatal rats are responsive to hormone manipulation, *Brain Res. Dev.,* 84(2), 253–260, 1995.

70. Lisciotto, C.A. and Morrell, J.I., Sex differences in the distribution and projections of testosterone target neurons in the medial preoptic area and the bed nucleus of the stria terminalis of rats, *Horm. Behav.,* 28, 492–502, 1994.

71. Roselli, C.E., Sex differences in androgen receptors and aromatase activity in microdissected regions of the rat brain, *Endocrinology,* 128, 1310–1316, 1991.

72. McAbee, M. and DonCarlos, L.L., Hormonal regulation of androgen receptor (AR) mRNAin the neonatal rat forebrain, *Soc. Neurosci. Abstr.,* 22, 558, 1996.

73. Shinoda, K., Nagano, M., and Osawa, Y., Neuronal aromatase expression in preoptic, strial, and amygdaloid regions during late prenatal and early postnatal development in the rat, *J. Comp. Neurol.,* 343, 113–129, 1994.

74. MacLusky, N.J., Philip, A., Hurlburt, C., and Naftolin, F., Estrogen formation in the developing rat brain: sex differences in aromatase activity during early post-natal life, *Psychoneuroendocrinology,* 10, 355–361, 1985.

75. Lephart, E.D., A review of brain aromatase cytochrome P450, *Brain Res. Rev.,* 22, 1–26, 1996.

76. Sumida, H., Nishizuka, M., Kano, Y., and Arai, Y., Sex differences in the anteroventral periventricular nucleus of the preoptic area and in the related effects of androgen in prenatal rats, *Neurosci. Lett.,* 151, 41–44, 1993.

77. Ito, S., Murakami, S., Yamanouchi, K., and Arai, Y., Prenatal androgen exposure, preoptic area and reproductive functions in the female rat, *Brain Dev.,* 8, 463–468, 1986

78. Murakami, S. and Arai, Y., Neuronal death in the developing sexually dimorphic periventricular nucleus of the preoptic area in the female rat: Effect of neonatal androgen treatment, *Neurosci. Lett.,* 102, 185–190, 1989.

79. Arai, Y., Nishizuka, M., Murakami, S., Miyakawa, M., Machida, M., Takeuchi, H., and Sumida, H., Morphological correlates of neuronal plasticity to gonadal steroids: sexual differentiation of the preoptic area, in *The Development of Sex Differences and Similarities in Behavior,* Haug, M., Whalen, R.E., Aron, C., and Olsen, K.L., Eds., Kluwer Academic Publishers, Dordrecht, 1993, 311–323.

80. Gu, G.B. and Simerly, R.B., Target specific hormonal regulation of sexually dimorphic projections from the principal nucleus of the bed nuclei of the stria terminalis, *Soc. Neurosci. Abstr.*, 23, 341, 1997.

81. Hutton, L.A., Gu, G.B., and Simerly, R.B., Development of a sexually dimorphic projection from the bed nuclei of the stria terminalis to the anteroventral periventricular nucleus in the rat, *J. Neurosci.*, 18(8), 3003–3013, 1998.

82. Toran-Allerand, C.D., On the genesis of sexual differentiation of the central nervous system: morphogenetic consequences of steroidal exposure and possible role of a-fetoprotein, *Prog. Brain Res.*, 61, 63–98, 1984.

83. Toran-Allerand, C.D., Organotypic culture of the developing cerebral cortex and hypothalamus: relevance to sexual differentiation, *Psychoneuroendocrinology*, 16, 7–24, 1991.

84. O'Leary, D.D.M., Development of connectional diversity and specificity in the mammalian brain by the pruning of collateral projections, *Curr. Opin. Neurobiol.*, 2, 70–77, 1992.

85. Agmon, A., Yang, L.T., O'Dowd, D.K., and Jones, E.G., Organized growth of thalamocortical axons from the deep tier of terminations into layer IV of developing mouse barrel cortex, *J. Neurosci.*, 13, 5365–5382, 1993.

86. Goodman, C.S., Mechanisms and molecules that control growth cone guidance, *Annu. Rev. Neurosci.*, 19, 341–377, 1996.

87. Greenough, W.T., Carter, C.S., Steerman, C., and DeVoogd, T.J., Sex differences in dendritic patterns in hamster preoptic area, *Brain Res.*, 126, 63–72, 1977.

88. Hammer, R.P.J. and Jacobson, C.D., Sex difference in dendritic development of the sexually dimorphic nucleus of the preoptic area in the rat, *Int. J. Dev. Neuroscience*, 2, 77–85, 1984.

89. Sawai, H., Clarke, D.B., Kittlerova, P., Bray, G.M., and Aguayo, A.J., Brain-derived neurotrophic factor and neurotrophin-4/5 stimulate growth of axonal branches from regenerating retinal ganglion cells, *J. Neurosci.*, 16, 3887–3894, 1996.

90. Murphy, D.D., Cole, N.B., and Segal, M., Brain-derived neurotrophic factor mediates estradiol induced dendritic spine formation in hippocampal neurons, *Proc. Natl. Acad. Sci. U.S.A.*, 95, 11412–11417, 1998.

91. MacLusky, N.J. and Brown, T.J., Control of gonadal steroid receptor levels in the developing brain, in *Neural Control of Reproductive Function*, Lakoski, J.M., Perez-Polo, J. R., and Rassin, D.K., Eds., Alan R. Liss, New York, 1989, 45–59.

92. Bloch, G.J., Kurth , S.M., Akesson, T.R., and Micevych, P.E., Estrogen-concentrating cells within cell groups of the medial preoptic area: sex differences and co-localization with galanin-immunoreactive cells, *Brain Res.*, 1992, 301–308, 1992.

93. Okamura, H., Yokosuka, M., and Hayashi, S., Induction of substance P-immunoreactivity by estrogen in neurons containing estrogen receptors in the anteroventral periventricular nucleus of female but not male rats, *J. Neuroendocrinol.*, 6(6), 609–615, 1994.

94. Breedlove, S.M., Hormonal control of the anatomical specificity of motoneuron-to-muscle innervation in rats, *Science*, 227, 1357–1359, 1985.

95. Sohal, G.S., The role of target size in neuronal survival, *J. Neurobiol.*, 23(9), 1124–1130, 1992.

96. Thorn, R.S. and Truman, J.W., Sexual differentiation in the CNS of the moth, Manduca sexta. II. Target dependence for the survival of the imaginal midline neurons, *J. Neurobiol.*, 25(9), 1054–1066, 1994.

97. Breedlove, S.M., Cellular analyses of hormone influence on motoneuronal development and function, *J. Neurobiol.*, 17, 157–176, 1986.

98. Truman, J.W., Thorn, R.S., and Robinow, S., Programmed neuronal death in insect development, *J. Neurobiol.*, 23(9), 1295–1311, 1992.

99. Nishiuzuka, M., Sumida, H., Kano, Y., and Arai, Y., Formation of neurons in the sexually dimorphic anteroventral periventricular nucleus of the preoptic area of the rat: Effects of prenatal treatment with testosterone propionate, *J. Neuroendocrinol.*, 5, 569–573, 1993.

100. Davis, E. C., Shryne, J. E., and Gorski, R. A., Structural sexual dimorphisms in the anteroventral periventricular nucleus of the rat phyothalamus are sensitive to gonadal steroids perinatally, but develop peripubertally , *Neuroendocrinology,* 63, 142-148, 1996.

101. Arai, Y. and Murakami, S., Androgen enhances neuronal degeneration in the developing preoptic area: apoptosis in the anteroventral periventricular nucleus (AVPvN -POA), *Horm. Behav.*, 28(4), 313–319, 1994.

102. Arai, Y., Sekine, Y., and Murakami, S., Estrogen and apoptosis in the developing sexually dimorphic preoptic area in female rats, *Neurosci. Res.*, 25, 403–407, 1996.

103. Simerly, R.B., Prodynorphin and proenkephalin gene expression in the anteroventral periventricular nucleus of the rat: sexual differentiation and hormonal regulation, *Mol. Cell. Neurosci.*, 2, 473–484, 1991.

104. Simerly, R.B., Swanson, L.W., Handa, R.J., and Gorski, R.A., The influence of perinatal androgen on the sexually dimorphic distribution of tyrosine hydroxylase-immunoreactive cells and fibers in the anteroventral periventricular nucleus of the rat, *Neuroendocrinology,* 40, 501–510, 1985.

105. Simerly, R.B., Hormonal control of the development and regulation of tyrosine hydroxylase expression within a sexually dimorphic population of dopaminergic cells in the hypothalamus, *Mol. Brain Res.*, 6, 297–310, 1989.

106. Corbier, P., Edwards, D.A., and Roffi, J., The neonatal testosterone surge: a comparative study, *Arch. Int. Physiol. Biochim. Biophys.*, 100, 127–131, 1992.

107. Ibanez, M.A., Zee, J., Crabtree, M., and Simerly, R.B., Developmental critical period for sexual differentiation of dopaminergic neurons in the anteroventral periventricular nucleus (AVPV), *Soc. Neurosci.*, 24, 1546, 1998.

108. Gorski, R.A. and Jacobson, C.D., Sexual differentiation of the brain, in *Clinics in Andrology,* Kogan, S.J. and Hafez, E.S.E., Eds., Martinus Nijhoff Publishers, The Hague, 1981, 109–134.

109. McCarthy, M.M., Schlenker, E.H., and Pfaff, D.W., Enduring consequences of neonatal treatment with antisense oligodeoxynucleotides to estrogen receptor messenger ribonucleic acid on sexual differentiation of rat brain, *Endocrinology,* 133, 433–439, 1993.

110. Barraclough, C.A. and Leathem, J.H., Infertility induced in mice by a single injection of testosterone propionate, *Proc. Soc. Exp. Biol. Med.*, 85, 673–674, 1954.

111. Olsen, K.L., Sex and the mutant mouse: strategies for understanding the sexual differentiation of the brain, in *The Development of Sex Differences and Similarities in Behavior*, Haug, M., Whalen , R.E., Aron, C., and Olsen, K.L., Eds., Kluwer Academic Publishers, Dordrecht, 1993, 255–278.

112. Etgen, A.M., Steroid hormone antagonists, brain receptor systems and behavior, in *Receptor Mediated Antisteroid Action*, Walter de Gruyter, Berlin, 1987.

113. Wood, R.L. and Newman, S.W., Androgen and estrogen receptors coexist within individual neurons in the brain of the Syrian hamster, *Neuroendocrinology*, 62, 487–497, 1995.

114. MacLusky, N.J., Luine, V.N., Gerlach, J.L., Fischette, C., Naftolin, F., and McEwen, B.S., The role of androgen receptors in sexual differentiation of the brain: effects of the testicular and feminization (Tfm) gene on androgen metabolism, binding, and action in the mouse, *Psychobiology*, 16, 381–397, 1988.

115. Lubahn, D.B., Moyer, J.S., Golding, T.S., Couse, J.F., Korach, K.S., and Smithies, O., Alteration of reproduction function but not prenatal sexual development after insertional disruption of the mouse estrogen receptor gene, *Proc. Natl. Aca. Sci. U.S.A.*, 90, 11162–11166, 1993.

116. Korach, K.S., Couse, J.F., Curtis, S.W., Washburn, T.F., Lindzey, J., Kimbro, K.S., Eddy, E.M., Migliaccio, S., Snedeker, S.M., Lubahn, D.B., Schomberg, D.W., and Smith, E.P., Estrogen receptor gene disruption: molecular characterization and experimental and clinical phenotypes, *Rec. Prog. Horm. Res.*, 51, 159–188, 1996.

117. Couse, J.F., Curtis, S.W., Washburn, T.F., Golding, T.S., Lubahn, D.B., Smithies, O., and Korach, K.S., Analysis of transcription and estrogen insensitivity in the female mouse after targeted disruption of the estrogen receptor gene, *Mol. Endocrinol.*, 9, 1441–1454, 1995.

118. Ogawa, S., Lubahn, D.B., Korach, K.S., and Pfaff, D.W., Behavioral effects of estrogen gene disruption in male mice, *Proc. Natl. Acad. Sci. U.S.A.*, 94, 1476–1481, 1997.

119. Rissman, E.F., Wersinger, S.R., Taylor, J.A., and Lubahn, D.B., Estrogen receptor function as revealed by knockout studies: neuroendocrine and behavioral aspects, *Horm. Behav.*, 31, 232–243, 1997.

120. Simerly, R.B., Zee, M.C., Pendleton, J.W., Lubahn, D.B., and Korach, K.S., Estrogen receptor-dependent sexual differentiation of dopaminergic neurons in the preoptic region of the mouse, *Proc. Natl. Acad. Sci. U.S.A.*, 94, 14077–14082, 1997.

121. Shughrue, P.J., Lane, M.V., and Merchenthaler, I., Comparative distribution of estrogen receptor-α and -β mRNA in the rat central nervous system, *J. Comp. Neurol.*, 388, 507–525, 1997.

122. Simerly, R.B., Carr, A.M., Zee, M.C., and Lorang, D., Ovarian steroid regulation of estrogen and progesterone receptor messenger ribonucleic acid in the anteroventral periventricular nucleus of the rat, *J. Neuroendocrinol.*, 8, 45–56, 1996.

123. Krege, J.H., Hodgin, J.B., Couse, J.F., Enmark, E., Warner, M., Mahler, J.F., Sar, M., Korach, K.S., Gustafsson , J.A., and Smithies, O., Generation and reproductive phenotypes of mice lacking estrogen receptor beta, *Proc. Natl. Acad. Sci. U.S.A.*, 95, 15677–15682, 1998.

124. Fisher, C.R., Graves, K.H., Parlow, A.F., and Simpson, E.R., Characterization of mice deficient in aromatase (ArKO) because of targeted disruption of the cyp19 gene, *Proc. Natl. Acad. Sci. U.S.A.*, 95, 6965–6970, 1998.

125. Simpson, E.R., Genetic mutations resulting in estrogen insufficiency in the male, *Mol. Cell. Endocrinol.*, 145, 55–59, 1998.

126. Gu, Q., Korach, K.S., et al., Rapid action of 17β-estradiol on Kainate-induced currents in hippocampal neurons lacking intracellular estrogen receptors, *Endocrinology*, 140, 660–666, 1999.

127. Moss, R.L., Gu, Q., et al., Estrogen: nontranscriptional signaling pathway, *Recent Prog. Horm. Res.*, 52, 33–68, 1997.

12

Sexual Differentiation of Neuronal Circuitry in the Hypothalamus

Akira Matsumoto, Yoshie Sekine, Shizuko Murakami, and
Yasumasa Arai

CONTENTS

I. Introduction

It is well established that sex steroids play a crucial role in reproductive neuroendocrine functions. Effects of sex steroid hormones on the sex steroid–sensitive neuronal structures and functions have been classically subdivided into *organizational* and *activational* ones.[1,2] Based on their findings that female guinea pigs exposed to androgen before birth do not exhibit feminine sexual behavior in the adult, Phoenix et al.[3] first proposed that androgen permanently organizes the developing neural tissues involved in sexual behavior. To exert the organizational effects, the importance of aromatization of androgen to estrogen has been pointed out.[4] Estrogen or aromatizable androgen plays a significant role in modulating neuronal development and

0-8493-1165-9/00/$0.00+$.50

neuronal circuit formation during the perinatal period.[1,2,5-10] These organizational actions of sex steroids can induce permanent sexual dimorphism in nuclear volume,[11,12] in neuronal number,[13,14] in distribution pattern of dendrites,[15-19] in neuronal membrane organization,[20] in synaptic formation,[21-26] and in neuronal connectivity[27,28] in the hypothalamus. Sexually dimorphic neuroendocrine and behavioral functions such as regulatory mechanisms of gonadotropin secretion and sexual behavior are considered to be organized by the exposure of developing brain to sex steroids during the perinatal period.[29,30] These functional sexual differences are thought to be correlated with the sexual differences in neuronal structures of the hypothalamus.

In contrast, activational effects of sex steroids on adult neuroendocrine brain are considered to be impermanent and reversible, and involving changes in neurophysiological and/or neurochemical events. Evidence suggests, however, that even impermanent effects of sex steroids can cause structural alterations in neural circuits responsible for sex steroid–sensitive neuroendocrine functions in adulthood.[1,2,31-35]

In the present report, as one step to clarify the organizational effects of sex steroids on neuronal circuit formation, the focus is mainly placed on synaptogenic action of sex steroids on developing neural substrates of the neuroendocrine hypothalamus that is thought to be involved in regulation of gonadotropin secretion and sexual behavior.[36,37]

II. Sex Differences in Nuclear Volume in the Hypothalamus: Regulation of Neuronal Number by Sex Steroids

Gorski et al.[11] found an intensely staining neuron group with a striking sex difference in the rat medial preoptic area (POA) that is called the sexually dimorphic nucleus of the POA (SDN-POA). The sex difference in nuclear volume is due to the presence of a greater number of neurons in the SDN-POA of males than in that of females.[13] Neonatal castration of males reduces the volume of the SDN-POA to a level comparable with that of females. Conversely, the nuclear volume of females is increased by exposure to androgen in the early postnatal period. The functional significance of the SDN-POA still remains unclear, but it is presumably concerned with the regulation of male sexual behavior.[38] On the other hand, a sexually dimorphic cell group which is larger and more densely cellular in females than in males is identified in the periventricular gray of the POA just caudal to the organum vasculosum of the lamina terminalis.[39] This neuron group is called the anteroventral periventricular nucleus of the POA (AVPvN-POA). This structure is thought to play a critical role in regulating the cyclic release of pituitary gonadotropins in female rats.[40] Its volume in female rats[41] and guinea pigs[42] is decreased by perinatal exposure to androgen. According to Weisz and Ward,[43] plasma androgen titers are much higher in males than in females during the perinatal period. These results suggest that the sexual dimorphism in these nuclei is not determined genetically at birth, but rather is dependent on the perinatal sex steroid environment.

Actual mechanisms by which sex steroids cause sexual differentiation of these nuclei during development have not been clarified. In the SDN-POA, the results of previous studies have not supported a role of gonadal steroids in the regulation of neurogenesis[44] or the pattern of neuronal migration from the neuroepithelium to the nucleus perinatally.[45] However, the loss of cells in the SDN-POA in females is prevented by administration of androgen perinatally.[46] On the contrary, androgen or estrogen administered perinatally enhances degeneration of AVPvN-POA neurons, inducing a significant reduction in size of the nucleus.[47] Therefore, it seems likely that the mechanism of developmental events by gonadal steroids leading to sexual dimorphism in the nuclear size of the SDN-POA and AVPvN-POA could be cell death.

There is recent evidence suggesting that gonadal steroids regulate apoptotic cell death in these sexually dimorphic neuron groups. Estrogen or androgen given perinatally has facilitatory effect on the incidence of apoptotic cell death in the AVPvN-POA[48] (Figure 12.1) but inhibitory effect in the SDN-POA.[48,49] As shown in Figure 12.2, a single injection of estradiol

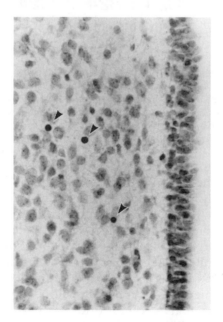

FIGURE 12.1

A photomicrograph of nuclear staining of the AVPvN-POA with the TUNEL (terminal deoxynucleotydyl transferase (TdT)-mediated dUTP-biotin nick end-labeling) method. TUNEL-positive cells (arrowheads) were recognized in the nucleus of a female pup 24 h after estradiol benzoate. ×290. (From Arai, Y., Sekine, Y., and Murakami, S., *Neurosci. Res.*, 25, 403, 1996. With permission.)

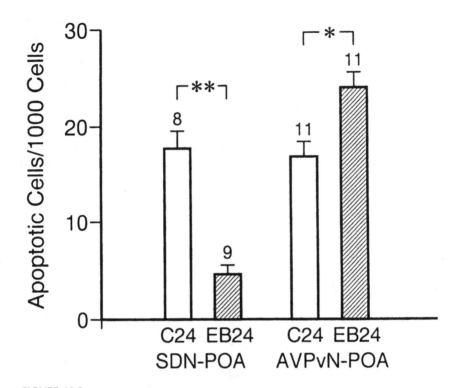

FIGURE 12.2

Effect of a single injection of 25 μg of EB on apoptotic cell death in the developing SDN-POA and AVPvN-POA of Day 5 female pups. The hatched bars represent the mean (±SEM) of TUNEL-positive cells/1000 cells in the pups sacrificed 24 h (EB24) after EB injection. The open bars represent that of control pups for EB24 (C24). The number at the top of the bars indicates the number of animals used in each group. $*p < 0.005$, $**p < 0.0002$. (From Arai, Y., Sekine, Y., and Murakami, S., *Neurosci. Res.*, 25, 403, 1996. With permission.)

benzoate (EB) given to female rats at 5 days of age (Day 5) is capable of facilitating apoptosis in the AVPvN-POA neurons within 24 h after the injection, whereas EB effectively inhibits apoptotic cell death of the SDN-POA neurons. These results suggest that the sexual dimorphism in these nuclei is not determined genetically at birth. One possible mechanism underlying the development of sexual dimorphism is regulation of apoptosis by perinatal gonadal steroids, and the regulatory mechanism seems to be different between the SDN-POA and AVPvN-POA.

According to Garnier et al.,[50] in human neuroblastoma cells SK ER3 in which the classical estrogen receptor (ERα) gene is transfected, estrogen induces an increase in mRNA expression of growth-associated nuclear protein prothymosin-α (PTMA). Conversely, mRNA expression of the Bcl-2-interacting protein Nip2 is decreased by estrogen treatment. Although biological significance of PTMA still remains unclear, it is suggested that PTMA may be correlated with proliferative activities.[51-53] On the other hand, Nip2 has been reported to show a significant homology with members of the interleukin-1β-converting enzyme (ICE)/Ced-3 protease which positively modulate apoptotic cell death.[54,55] Based on the homology between Nip2 and ICE/Ced-3 protease, it is possible to speculate that the decreased levels of Nip2 observed in SK ER3 cells at a short time after estrogen treatment might protect these cells from apoptotic cell death. With respect to Bcl-2 that prevents natural and induced neural apoptosis,[56] it is noteworthy that the number of Bcl-2-immunoreactive neurons in the hypothalamic arcuate nucleus (ARCN) in female rats is significantly increased on the day of estrus compared with proestrus and diestrus and is decreased by ovariectomy.[57] Therefore, it is plausible that estrogen may influence cell growth and survival in the above-mentioned sexually dimorphic nuclei through regulation of gene expression such as PTMA, Nip2, and/or Bcl-2 genes.

III. Chemical Nature of Sex Differences in the Hypothalamus

Immunohistochemical studies have shown that there is sexual dimorphism in the distribution pattern of neuronal cell bodies and fibers containing several types of neurotransmitters and neuromodulators in the hypothalamus. These include vasopressin in the POA,[58] cholecystokinin in the POA,[59-62] opioid peptides in the periventricular and medial POA,[62-66] galanin in the medial POA,[67-71] calcitonin gene-related peptide (CGRP) in the medial POA[72,73] and ARCN,[74] gonadotropin-releasing hormone (GnRH) in the POA,[75-77] corticotropin-releasing hormone in the AVPvN-POA and medial POA,[78] serotonin in the medial POA,[79,80] and tyrosine hydroxylase (TH) in the AVPvN-POA,[14,81] acetylcholinesterase in the medial POA-anterior hypothalamus (AH),[82] and γ-aminobutyric acid[83,84] and glutamate[83] in the medial POA and hypothalamic ventromedial nucleus (VMN). Furthermore, *in situ* hybridization histochemical studies have revealed sexual differences in mRNA expression of somatostatin in the hypothalamic periventricular nucleus,[85,86] galanin in the POA,[87] CGRP in the medial POA,[88] neuropeptide Y in the ARCN,[89] proGnRH (a precursor common to GnRH and GnRH-associated peptide) in the POA-AH,[90] neurotensin/neuromedin N,[91] proenkephalin and prodynorphin,[92] and TH in the AVPvN-POA.[93]

The distribution pattern of vasopressin-immunoreactive fibers in the POA-AH,[58] enkephalin-immunoreactive fibers in the periventricular POA,[63,64] and serotonin-immunoreactive fibers in the medial POA[79,80] is sexually dimorphic, and the distribution of these neuropeptides depends on perinatal exposure to sex steroids. Similarly, some neuronal groups of the hypothalamus have the chemical characteristics that are determined by the perinatal influence of sex steroids. The numbers of CGRP-immunoreactive[72,73] and CGRP mRNA-expressing neurons[88] in the medial POA, somatostatin mRNA-expressing neurons in the hypothalamic periventricular nucleus,[86] and proenkephalin mRNA-expressing neurons in the AVPvN-POA[92] are significantly greater in male rats than in females. Furthermore, neonatal castration of male rats reduces the numbers of these neurons to those observed in females. Conversely, neonatal exposure of female rats to androgen induces a masculine pattern in the density of these neurons. On the contrary, the numbers of TH-immunoreactive[14,81] and TH mRNA–[93] and prodynorphin mRNA–expressing neurons[92] in the AVPvN-POA is greater in female rats than in males, and the differences are dependent on the neonatal sex steroid environment. According to Merchenthaler et al.,[69] the number of GnRH-immunoreactive neurons coexpressing galanin in the medial POA/diagonal band of Broca is four to five times higher in female rats than in males. Moreover, neonatal castration of males reduces the incidence of galanin-GnRH colocalization to that found in females. Expression levels of galanin mRNA in the GnRH neurons are higher in normal females and in males castrated neonatally than in normal males and in females treated with androgen neonatally.[87] Because CGRP-, somatostatin-, galanin-, and TH-immunoreactive and their mRNA-expressing neurons have been reported to contain ERα,[86,94-96] ERα is thought to mediate the effect of estrogen on sexually dimorphic development of their expression.

GnRH neurons of the POA represent the final common pathway regulating gonadotropin secretion by the anterior pituitary. The activity of GnRH neurons is considered to be regulated by intra- and extrahypothalamic neuronal elements as well as the steroidal environment.[97] Because GnRH neurons do not accumulate estrogen,[98] it seems to be plausible that other estrogen-sensitive (ERα-containing) interneurons in the POA may integrate steroidal signals that in turn modify the activity of GnRH neurons. Immunohistochemical studies have suggested that other neurochemicals may also modulate ERα-immunoreactive neurons in the POA.[99] This is supported by immunoelectron microscopic studies indicating that neuropeptide Y-,[99] GnRH-,[100] and enkephalin-immunoreactive terminals[101] make synaptic connections with ERα-immunoreactive neurons. Langub and Watson[101] have pointed out that more enkephalin-immunoreactive terminals (synaptic and non-synaptic) contact the cell body of ERα-immunoreactive neurons in female rats than those in males. Neurons located in the POA may be influenced by both estrogen and neurotransmitters/neuromodulators via, respectively, nuclear receptors and synaptic inputs. With respect to the GnRH neuronal system, a sexual dimorphism exists in the synaptic inputs to GnRH neurons in the rat POA.[102] Thus, GnRH neurons in females have approximately twice the number of synapses as do those of males. β-Endorphin-immunoreactive terminals contribute to this dimorphism. Because a subset of β-endorphin-immunoreactive neurons has been reported to accumulate estrogen[103] and to contain immunoreactive ERα,[104] it is possible that physiological differences in the regulation of gonadotropin secretion may also be reflected in a sexually dimorphic connectivity of the GnRH system in which this neuronal subset may be involved.

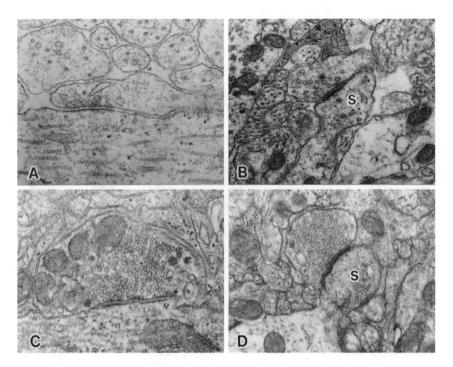

FIGURE 12.3

Axodendritic shaft and spine synapses in the ARCN. (A) Axodendritic shaft synapse in the ARCN of a 6-day-old female rat. Note the various amounts of extracellular space. Axon terminal contains only a small number of synaptic vesicles, and pre- and postsynaptic membrane specializations are poorly developed. ×32,600. (B) Axodendritic spine synapse in the ARCN of a 6-day-old female rat. S = spine. ×23,300. (C) Axodendritic shaft synapse in the ARCN of a 100-day-old female rat. ×24,800. (D) Axodendritic spine synapse in the ARCN of a 100-day-old female rat. S = spine; ×24,800. (From Matsumoto, A., and Arai, Y., *Neuroendocrinology*, 33, 166, 1981. With permission.)

IV. Synaptogenesis in the Hypothalamus and Sex Steroids

The presence of synaptic structures is one of the characteristic features of neural tissues. Synapses are considered to be the sites of functional contacts between axon terminals and other neural elements. Axon terminals containing a number of synaptic vesicles make synaptic contact with dendrites (axodendritic synapse) or somata (axosomatic synapse). Two types of axodendritic synapses, one made on the dendritic shaft (Figure 12.3A, C) and the other made on the dendritic spine (Figure 12.3B, D), are identified.

According to autoradiographic studies on the development of fetal hypothalamus, the final cell divisions of the neuroblasts which give rise to neurons in the ARCN and VMN nuclei occur by Day 17 of gestation in the rat.[105,106]

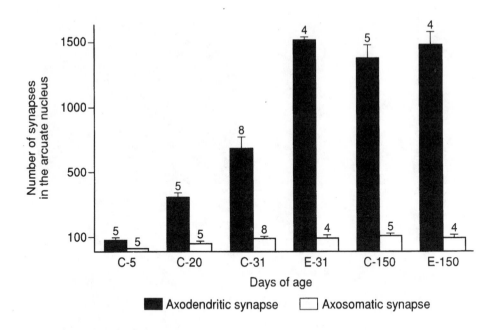

FIGURE 12.4
The number of axodendritic and axosomatic synapses in the ARCN at different ages. C-5, 20, 31 and 150 indicate normal female rats (controls) sacrificed at 5, 20, 31 and 150 days of age. E-31 and 150 indicate estrogenized rats sacrificed at 31 and 150 days of age. These rats were treated with 2, 4, and 8 µg of EB for each 10-day period from the day of birth to day 30. Vertical lines indicate SEM. Numbers on vertical lines refer to the number of rats examined. (From Matsumoto, A., *Psychoneuroendocrinology,* 16, 25, 1991. With permission.)

However, the synaptic density in the ARCN and VMN is progressively increased during the course of development. The number of axodendritic and axosomatic synapses in the ARCN is very small at the neonatal period (Figure 12.4), and reaches a plateau around the onset of puberty.[24,107] As shown in Figure 12.5, the synaptic density in the VMN (VL-VMN, the ventrolateral part of the VMN; DM-VMN, the dorsomedial part of the VMN, see below) is also small at neonatal period (Day 5) and increases remarkably within the first 20 days of age (Day 20).[26] The number of shaft and spine synapses reaches almost 80% of that at 45 days of age (Day 45), when it reaches the maximal level. More increase in the number of synapses does not occur at 100 days of age (Day 100). This tendency has also been detected in the POA.[108,109]

The neuropil matrix of the ARCN in the neonatal rat brain is still an immature state which is characterized by the presence of various degrees of extracellular space, the presence of growth cones, and the paucity of synapses.[24,107,110-112] These features have also been detected in the POA[104,105] and VMN.[26] Therefore, the neuropil matrix of these neuronal structures seems to be subjected to the organizational action of sex steroids. The formation of

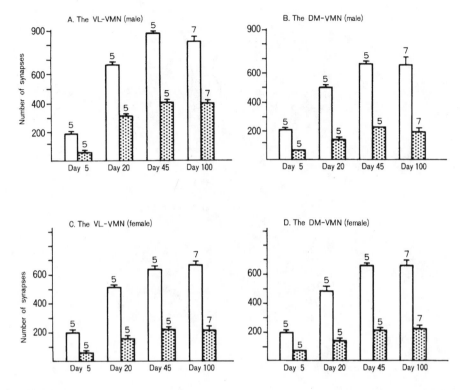

FIGURE 12.5

The number of axodendritic shaft (open bars) and spine (shaded bars) synapses in the VMN. Day 5, 5 days of age; Day 20, 20 days of age; Day 45, 45 days of age; Day 100, 100 days of age. Vertical bars indicate SEM. Numbers on vertical bars refer to the number of rats examined. (From Matsumoto, A. and Arai, Y., *Neurosci. Lett.*, 68, 165, 1986. With permission.)

major neural circuits driving postpubertal neuroendocrine and/or behavioral functions can be influenced under these circumstances.

There is evidence indicating that estrogen can act as a neurotrophic factor on neonatal brain tissue, stimulating axonal and dendritic growth and synapse formation. Estrogen markedly enhances axodendritic synapse formation in the ARCN during the neonatal period (see Figure 12.4).[110,113] In addition, estrogen exerts a stimulatory influence on the development of neuronal structures in the POA tissues transplanted into the third ventricle of adult female rats.[114] In this experiment, medial POA or parietal cortical tissues of newborn female rats were transplanted into the third ventricle of adult ovariectomized female rats. At four weeks after transplantation, most of the grafts were highly vascularized and well established in the third ventricle of the host (Figure 12.6). All of the POA and cortical grafts showed an appearance similar to normal neural tissue (Figure 12.7). The volume of the POA grafts exposed to estrogen via hosts is significantly greater than that without estrogen treatment (Figure 12.8). The number of shaft and spine

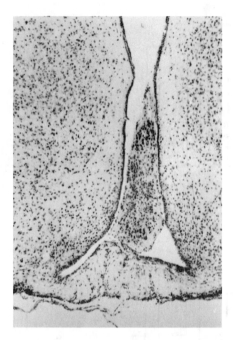

FIGURE 12.6
POA graft in the third ventricle of an estrogen-treated ovariectomized female rat. ×59. (From Matsumoto, A., Murakami, S., and Arai, Y., *Cell Tissue Res.*, 252, 33, 1988. With permission.)

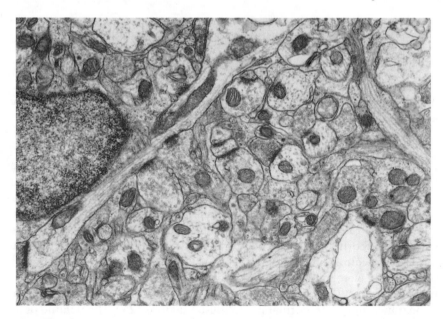

FIGURE 12.7
Neuropil in POA graft in an estrogen-treated ovariectomized female rat. ×24,900. (From Matsumoto, A., Murakami, S., and Arai, Y., *Cell Tissue Res.*, 252, 33, 1988. With permission.)

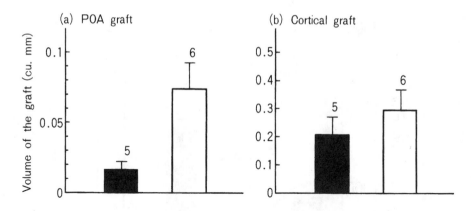

FIGURE 12.8
Volume of POA (a) and cortical grafts (b). From the transplantation, Silastic capsules containing estradiol-17β were placed subcutaneously into host animals for 4 weeks. Solid bars = Control; open bars = estrogen-treated rats. Vertical lines indicate SEM. Numbers on vertical lines refer to the number of rats examined. (From Matsumoto, A., Murakami, S., and Arai, Y., *Cell Tissue Res.*, 252, 33, 1988. With permission.)

synapses in the POA grafts is similarly influenced by estrogen (Figure 12.9). Because there is no difference in cortical graft volume between control and estrogen-treated hosts, it is reasonable to assume that stimulatory effects of estrogen are rather specific to estrogen-sensitive POA. These findings are in good agreement with the evidence that the proliferation of neuronal processes of the explants from newborn mouse POA tissues[115] and dissociated POA cells[116] is markedly stimulated by the addition of estrogen to culture medium. These brain regions at neonatal period have been reported to contain sex steroid–sensitive neurons.[117-121] It is probable, therefore, that sexually undifferentiated neuropil matrix of the neonatal brain could be subjected to organizational action of sex steroids.

Sexual dimorphism in synaptic patterns has been found in several regions. In the ARCN, the number of spine synapses is approximately twice as great in female rats as in males, whereas the number of somatic synapses in females is approximately twice that in males (Table 12.1).[23] There is no sex difference in the number of shaft synapses. This sexually dimorphic pattern of synaptic distribution is similar to that found in the dorsal part of POA.[11] In the ventrolateral part of the VMN, the number of shaft and spine synapses is significantly greater in males than in females (Figure 12.10).[25,26] In the suprachiasmatic nucleus, the incidence of spine synapses is higher in males than in females.[122,123] All of these nuclei except the suprachiasmatic nucleus contain a number of sex steroid–sensitive neurons. These findings suggest that synaptic organization may vary according to the genomic responses of the individual nuclei to organizational action of sex steroids.

There is a regional difference in distribution pattern of sex steroid–sensitive neurons in the VMN.[124-126] This may be correlated with a regional difference

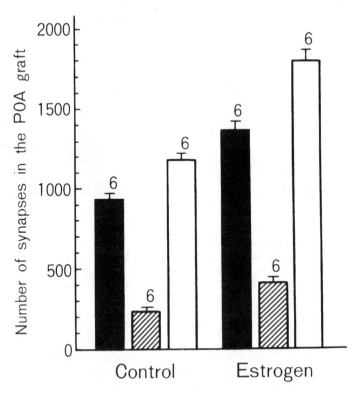

FIGURE 12.9
Number of shaft, spine and total synapses per 10,000 μm² of POA grafts in control (Control) and estrogen-treated (Estrogen) female rats. Solid bars = shaft synapses; hatched bars = spine synapses; open bars = total synapses. Vertical lines indicate SEM. Numbers on vertical lines refer to the number of rats examined. (From Matsumoto, A., Murakami, S., and Arai, Y., *Cell Tissue Res.*, 252, 33, 1988. With permission.)

in synaptic pattern in the VMN.[25,26] As shown in Figure 12.10, the number of shaft and spine synapses in the ventrolateral part of the VMN which contains abundant sex steroid–sensitive neurons is significantly greater than that in the dorsomedial part of the VMN which contains only a few ones. The number of shaft and spine synapses in the male VL-VMN is significantly greater than that in the female VL-VMN, whereas the DM-VMN shows no sex difference. The evidence indicates that the presence of sexually dimorphic synaptic organization is only restricted in the VL-VMN, which contains a number of sex steroid–sensitive neurons. Neonatal castration of males reduces the number of shaft and spine synapses in the VL-VMN to the level comparable with that of normal females. The number of shaft and spine synapses in the female VL-VMN is increased to the level comparable with that of normal males by neonatal exposure of androgen. These findings reinforce the significance of the sex steroid environment at the neonatal period for the development of sexually dimorphic synaptic organization in the neuroendocrine brain.

TABLE 12.1

Number of Axodendritic and Axosomatic Synapses in the Arcuate Nucleus (ARCN) of Normal and Neonatally Androgenized or Castrated Rats

		Axodendritic Synapses[a]		
Group	Number of Rats	Shaft Synapses	Spine Synapses	Axosomatic Synapses[b]
Normal females	8	1655 ± 94*	242 ± 30	2.11 ± 0.15 (166)[c]
Androgenized females	7	1507 ± 79	174 ± 31	3.84 ± 0.26 (130)
Normal males	7	1607 ± 138	144 ± 24	3.86 ± 0.20 (160)
Castrated males	7	1462 ± 62	257 ± 24	1.97 ± 0.25 (150)

* Mean ± SEM.
[a] Axodendritic synapses were counted per 18,000 μm² in the ARCN.
[b] Number of axosomatic synapses per cell body. For counting synapses, only cell bodies whose profiles could be seen were randomly selected.
[c] Number of neurons in parenthesis.
Source: From Matsumoto, A. and Arai, Y., *Brain Res.*, 190, 238, 1980. With permission.

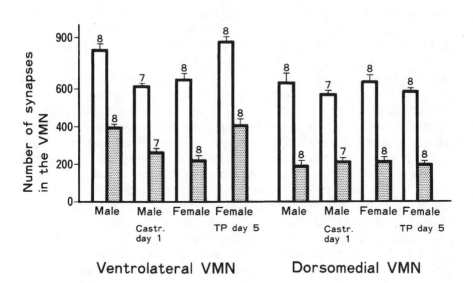

FIGURE 12.10

Number of shaft and spine synapses in the ventrolateral and dorsomedial parts per 10,000 μm² in the VMN of normal males (Male) and females (Female), males castrated on Day 1 (Male Castr. day 1) and females treated with 1.25 mg of testosterone propionate (TP) on day 5 (Female TP day 5). Open bars = shaft synapse; shaded bars = spine synapse. Vertical lines indicate SEM. Number of vertical lines refers to the number of rats examined. (From Matsumoto, A. and Arai, Y., *Neuroendocrinology*, 42, 232, 1986. With permission.)

V. Molecular Basis of Sexual Differentiation of Neuronal Circuitry

Recent studies have provided evidence of an increase in expression of specific structural proteins and their mRNAs in neuronal plasticity processes such as neuronal growth and synapse formation.[127,128] Changes in the expression of mRNAs encoding certain structural components may underlie the mechanism of plastic response of neurons. Although the molecular mechanisms underlying the sexual differentiation of neuronal circuitry in the hypothalamus are poorly understood, evidence is now accumulating that sex steroids play a significant role in regulating expression of certain structural proteins and their mRNAs in sex steroid–sensitive tissues.[129-136]

Tubulin is a main component of microtubules that are one of the major cytoskeletal proteins in neurons of the central nervous system.[137] Growth-associated protein 43 kDa (GAP-43) is concentrated in axonal growth cones and has been implicated in axonal elongation and synaptogenesis.[127] It has been reported that the amount of mRNA coding for tubulin isolated from neonatal rat POA-hypothalamus is higher in males than in females,[129,138] and that its amount in females is increased by neonatal exposure of androgen or estrogen.[129,133] Expression levels of GAP-43 mRNA[132,134,135] and synaptosomal-associated protein 25 kDa mRNA[132] in the developing POA and hypothalamus of the rat are sexually dimorphic, and are regulated by neonatal exposure of estrogen or androgen.[132,134,135] Estrogen has a facilitatory effect on expression of tau protein, one of the important microtubule-associated proteins, in dissociated cells from fetal rat hypothalamus.[131] In addition, estrogen and androgen induce neuritic growth in cultured rat pheochromocytoma PC 12 cells in which ERα and androgen receptor genes are transfected, respectively.[139,140] Estrogen can arrest cell division of human neuroblastoma cells SK ER3 in which the ERα gene is transfected.[136] Following arrest of cell division, the cells extend neuritic processes and express tau and synaptophysin. It is also reported that in the developing hypothalamus, androgen modifies the monoclonal antibody immunoreactivity in radial glia[130] that may be responsible for neuronal organization. It is plausible, therefore, that activation of genes encoding certain structural proteins induced by sex steroids is a possible mechanism by which sex steroid–dependent sexual differentiation occurs in synaptic connections in the hypothalamus.

As mentioned above, sex steroids can act as a neurotrophic factor on developing hypothalamic tissue, stimulating axonal and dendritic growth, synapse formation, and gene expression of certain structural proteins that would be involved in neuropil organization. These organizational actions of sex steroids seem to induce permanent sexual dimorphism in synaptic connections. Although it is not known how sex steroids act on developing hypothalamic tissues, resulting in sexual difference in neuronal circuitry, it is possible to

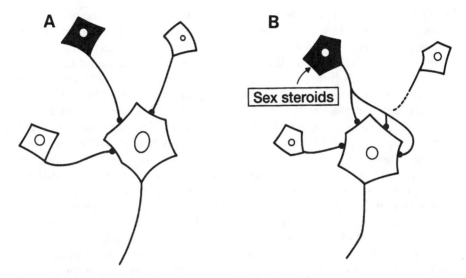

FIGURE 12.11

A speculative schema of the effect of sex steroids on the developing hypothalamic neurons. Different synaptic connections would be established between the absence (A) and presence (B) of sex steroids. Sex steroid–sensitive neuron is inked in black.

speculate that axons of sex steroid–sensitive neurons within the hypothalamus exposed to sex steroids during the developmental period would grow and branch more rapidly than axons of sex steroid–insensitive neurons (Figure 12.11). They could establish synaptic connections on their target neurons before the other sex steroid–insensitive axons could. The other ones developing more slowly would arrive at the target neurons at a time when much of the available synaptic sites are already occupied by the axon terminals which are stimulated by sex steroids. Thus, different synaptic connections seem to be established between the presence and absence of sex steroids.

VI. Concluding Remarks

This chapter provides morphological evidence suggesting that estrogen or aromatizable androgen play significant roles in modulating neural circuit formation in sex steroid–sensitive hypothalamus. Moreover, attempts have been made to clarify the molecular mechanisms underlying sexual differentiation of synaptic connections. Estrogen has been shown to regulate gene expression first by binding to ERα and then by interaction of estrogen–ERα complex with estrogen responsive elements (ERE) of the DNA that act as transcriptional enhancers.[141-143] Therefore, expression of ERα gene during development seems to be one of the important processes for sexual differentiation of the brain. The conjecture is supported by the recent evidence that administration of antisense oligonucleotide to ERα mRNA into the hypothalamus of neonatal female rats effectively blocks the masculinizing effect of testosterone on the brain morphology and behavioral functions.[144] Moreover, in ERα knockout male mice, the number of TH-immunoreactive neurons in the AVPvN-POA was three times that of wild-type ones, suggesting that disruption of the ERα gene results in feminization of the number of TH-immunoreactive neurons.[96] As mentioned above, estrogen or aromatizable androgen modifies the expression of certain structural proteins and their mRNAs in the developing hypothalamic neurons. Estrogen also modifies mRNA expression of PTMA and Nip2 in human neuroblastoma cells SK ER3,[50] and neurotrophins and their receptors in the developing forebrain.[145] To verify regulatory mechanisms of estrogen for the expression of these genes, it is important to identify EREs in each gene. There is close homology between the consensus palindromic RER sequence and the sequences in PTMA,[146] nerve growth factor, brain-derived neurotrophic factor, p75[NTR] and trkA genes.[145] Since the EREs have not yet been detected in the rat β-tubulin[147] and GAP-43 genes,[148] it is possible that expression of the genes is mediated by some gene products such as *c-fos* and *c-jun* that are under influence of estrogen–ERα complex. It is of interest to note that an increase in *c-fos* mRNA and Fos protein in human neuroblastoma cells SK ER3 is observed after short-term treatment with estrogen.[149] Thus, sex steroids may be involved in organization of neuronal substrates of the developing hypothalamus, directly or indirectly through regulation of expression of the genes involved in neural development and synapse formation.

Acknowledgments

The research was supported in part by grants from the Ministry of Education, Culture and Science of Japan.

Reference

1. Arnold, A.P. and Breedlove, S.M., Organizational and activational effects of sex steroids on brain and behavior, *Horm. Behav.*, 19, 469, 1985.
2. Matsumoto, A., Synaptogenic action of sex steroids in developing and adult neuroendocrine brain, *Psychoneuroendocrinology*, 16, 25, 1991.
3. Phoenix, C.H., Goy, R.W., Gerall, A.A., and Young, W.C., Organizing action of prenatally administered testosterone propionate on the tissues mediating mating behavior in the female guinea pig, *Endocrinology*, 65, 369, 1959.

4. Naftolin, F., Ryan, K.J., Davies, I.J., Reddy, V.V., Flores, F., Petro, Z., Kuhn, M., White, R.J., Takaoka, Y., and Wolin, L., The formation of estrogen by central neuroendocrine tissues, *Recent Prog. Horm. Res.*, 31, 295, 1975.

5. Goy, R.W. and McEwen, B.S., *Sexual Differentiation of the Brain*, MIT Press, Cambridge, MA, 1980.

6. MacLusky, N.J. and Naftolin, F., Sexual differentiation of the central nervous system, *Science*, 211, 1294, 1981.

7. Arnold, A.P. and Gorski, R.A., Gonadal steroid induction of structural sex differences in the central nervous system, *Annu. Rev. Neurosci.*, 7, 423, 1984.

8. Matsumoto, A. and Arai, Y., Morphological evidence for sexual dimorphism in wiring pattern in the neuroendocrine brain, in *Pars Distalis of the Pituitary Gland — Structure, Function and Regulation*, Yoshimura, F. and Gorbman, A., Eds., Elsevier, Amsterdam, 1986, 239.

9. Arai, Y., Matsumoto, A., and Nishizuka, N., Synaptogenesis and neuronal plasticity to gonadal steroids: implications for the development of sexual dimorphism in the neuroendocrine brain, in *Current Topics in Neuroendocrinology*, Vol. 7, *Morphology of Hypothalamus and Its Connections*, Ganten, D. and Pfaff, D., Eds., Springer-Verlag, Berlin, 1986, 291.

10. Matsumoto, A. and Arai, Y., Sexual differentiation of neuronal circuitry in the neuroendocrine hypothalamus, *Biomed. Rev.*, 7, 5, 1997.

11. Gorski, R.A., Gordon, J.H., Shryne, J.E., and Southam, A.M., Evidence for a morphological sex difference within the medial preoptic area of the rat brain, *Brain Res.*, 148, 333, 1978.

12. Matsumoto, A., and Arai, Y., Sex difference in volume of the ventromedial nucleus of the hypothalamus in the rat, *Endocrinol. Japan*, 30, 277, 1983.

13. Gorski, R.A., Harlan, R.E., Jacobson, C.D., Shryne, J.E., and Southam, A.M., Evidence for the existence of a sexually dimorphic nucleus in the preoptic area of the rat, *J. Comp. Neurol.*, 193, 529, 1980.

14. Simerly, R.B., Swanson, L.W., Handa, R.J., and Gorski, R.A., Influence of perinatal androgen on the sexually dimorphic distribution of tyrosine hydroxylase-immunoreactive cells and fibers in the anteroventral periventricular nucleus of the rat, *Neuroendocrinology*, 40, 501, 1985.

15. Greenough, W.T., Carter, C.S., Steerman, C., and DeVoogd, T.J., Sex differences in dendritic patterns in hamster preoptic area, *Brain Res.*, 126, 63, 1977.

16. Ayoub, D.M., Greenough, W.T., and Juraska, J.M., Sex differences in dendritic structure in the preoptic area of the juvenile macaque monkey brain, *Science*, 219, 197, 1983.

17. Hammer, R.P. and Jacobson, C.D., Sex difference in dendritic development of the sexually dimorphic nucleus of the preoptic area in the rat, *Int. J. Dev. Neurosci.*, 2, 77, 1984.

18. Tobet, S.A., Zahniser, D.J., and Baum, M.J., Sexual dimorphism in the pre-optic/anterior hypothalamic area of ferrets: effects of adult exposure to sex steroids, *Brain Res.*, 364, 249, 1986.

19. Cherry, J.A., Tobet, S.A., DeVoogd, T.J., and Baum, M.J., Effects of sex steroids and androgen treatment on dendritic dimensions of neurons in the sexually dimorphic preoptic/anterior hypothalamic area of male and female ferrets, *J. Comp. Neurol.*, 323, 577, 1992.

20. Garcia-Segura, L.M., Baetens, D., and Naftolin, F., Sex differences and maturational changes in arcuate nucleus neuronal plasma membrane organization, *Dev. Brain Res.*, 19, 146, 1985.

21. Raisman, G. and Field, P.M., Sexual dimorphism in the preoptic area of the rat, *Science*, 173, 731, 1971.

22. Raisman, G. and Field, P.M., Sexual dimorphism in the neuropil of the preoptic area of the rat and its dependence on neonatal androgen, *Brain Res.*, 54, 1, 1973.

23. Matsumoto, A. and Arai, Y., Sexual dimorphism in "wiring pattern" in the hypothalamic arcuate nucleus and its modification by neonatal hormone environment, *Brain Res.*, 190, 238, 1980.

24. Matsumoto, A. and Arai, Y., Effect of androgen on sexual differentiation of synaptic organization in the hypothalamic arcuate nucleus: an ontogenic study, *Neuroendocrinology*, 33, 166, 1981.

25. Matsumoto, A. and Arai, Y., Male–female difference in synaptic organization of the ventromedial nucleus of the hypothalamus in the rat, *Neuroendocrinology*, 42, 232, 1986.

26. Matsumoto, A. and Arai, Y., Development of sexual dimorphism in synaptic organization in the ventromedial nucleus of the hypothalamus in rats, *Neurosci. Lett.*, 68, 165, 1986.

27. Dyer, R.G., MacLeod, N.K., and Ellendorff, F., Electrophysiological evidence for sexual dimorphism and synaptic convergence in the preoptic and anterior hypothalamic areas of the rat, *Proc. R. Soc. Lond. B*, 193, 421, 1976.

28. Sakuma, Y. and Pfaff, D.W., Electrophysiological determination of projections from ventromedial hypothalamus to midbrain central gray: differences between female and male rats, *Brain Res.*, 225, 184, 1981.

29. Gorski, R.A., Gonadal hormones and the perinatal development of neuroendocrine functions, in *Frontiers in Neuroendocrinology*, Martini, L. and Ganong, W.F., Eds., Oxford University Press, New York, 1971, 237.

30. Arai, Y., Sexual differentiation and development of the hypothalamus and steroid-induced sterility, in *Neuroendocrine Control*, Yagi, K. and Yoshida, S., Eds., University of Tokyo Press, Tokyo, 1973, 27.

31. Matsumoto, A., Hormonally induced synaptic plasticity in the adult neuroendocrine brain, *Zool. Sci.*, 9, 679, 1992.

32. Matsumoto, A., Sex steroid induction of synaptic reorganization in adult neuroendocrine brain, *Rev. Neurosci.*, 3, 287, 1992.

33. Matsumoto, A., Hormonally induced neuronal plasticity in the adult motoneurons, *Brain Res. Bull.*, 44, 539, 1997.

34. Matsumoto, A., Arai, Y., Urano, A., and Hyodo, S., Androgen regulates gene expression of cytoskeletal proteins in adult rat motoneurons, *Horm. Behav.*, 28, 357, 1994.

35. Matsumoto, A., Arai, Y., Urano, A., and Hyodo, S., Molecular basis of neuronal plasticity to gonadal steroids, *Funct. Neurol.*, 10, 59, 1995.

36. Kordon, C., Drouva, S.V., Martinez de la Escalera, G., and Weiner, R.I., Role of classic and peptide neuromodulators in the neuroendocrine regulation of luteinizing hormone and prolactin, in *The Physiology of Reproduction*, 2nd ed., Vol. 1, Knobil, E., Neill, J.D., Greenwald, G.S., Markert, C.L., and Pfaff, D.W., Eds., Raven Press, New York, 1994, 1621.

37. Pfaff, D.W., Schwartz-Giblin, S., McCarthy, M.M., and Kow, L-M., Cellular and molecular mechanisms of female reproductive behaviors, in *The Physiology of Reproduction*, 2nd ed., Vol. 2 , Knobil, E., Neill, J.D., Greenwald, G.S., Markert, C.L., and Pfaff, D.W., Eds., Raven Press, New York, 1994, 107.

38. De Jonge, F.H., Louwerse, A.L., Ooms, M.P., Evers, P., Endert, E., and van de Poll, N.E., Lesions of the SDN-POA inhibit sexual behavior of male Wistar rats. *Brain Res. Bull.*, 23, 91, 1989.

39. Bleier, R., Byne, W., and Siggelkow, I., Cytoarchitectonic sexual dimorphisms of the medial preoptic and anterior hypothalamic areas in guinea pig, rat, hamster, and mouse, *J. Comp. Neurol.*, 212, 118, 1982.

40. Terasawa, E., Wiegand, S.J., and Bindon, W.E., A role for medial preoptic nucleus on afternoon proestrus in female rats, *Am. J. Physiol.*, 238E, 533, 1980.

41. Ito, S., Murakami, S., Yamanouchi, K., and Arai, Y., Perinatal androgen decreases the size of the sexually dimorphic medial preoptic nucleus in the rat, *Proc. Japan. Acad. Ser. B*, 62, 408, 1986.

42. Byne, W. and Bleier, R., Medial preoptic sexual dimorphisms in the guinea pig. I. An investigation of their hormonal dependence, *J. Neurosci.*, 7, 2668, 1987.

43. Weisz, J. and Ward, I.L., Plasma testosterone and progesterone titers of pregnant rats, their male and female fetuses and neonatal offspring, *Endocrinology*, 106, 306, 1980.

44. Dodson, R.E., Shryne, J.E., and Gorski, R.A., Hormonal modification of the number of total and late-arising neurons in the central part of the medial preoptic nucleus of the rat, *J. Comp. Neurol.*, 275, 623, 1988.

45. Jacobson, C.D., Davis, F.C., and Gorski, R.A., Formation of the sexually dimorphic nucleus of the preoptic area: neuronal growth, migration and changes in cell number, *Dev. Brain Res.*, 21, 7, 1985.

46. Dodson, R.E. and Gorski, R.A., Testosterone propionate administration prevents the loss of neurons within the central part of the medial preoptic nucleus, *J. Neurobiol.*, 24, 80, 1993.

47. Murakami, S. and Arai, Y., Neuronal cell death in the developing sexually dimorphic periventricular nucleus of the preoptic area in the female rat: effect of neonatal androgen treatment, *Neurosci. Lett.*, 102, 185, 1989.

48. Arai, Y., Sekine, Y., and Murakami, S., Estrogen and apoptosis in the developing sexually dimorphic preoptic area in female rats, *Neurosci. Res.*, 25, 403, 1996.

49. Davis, E.C., Popper, P., and Gorski, R.A., The role of apoptosis in sexual differentiation of the rat sexually dimorphic nucleus of the preoptic area, *Brain Res.*, 734, 10, 1996.

50. Garnier, M., Di Lorenzo, D., Albertini, A., and Maggi, A., Identification of estrogen-responsive genes in neuroblastoma SK-ER3 cells, *J. Neurosci.*, 17, 4591, 1997.

51. Dominguez, F., Magdalena, C., Cancio, E., Roson, E., Paredes, J., Loidi, L., Zalvide, J., Fraga, M., Forteza, J., Regueiro, B.J., and Puente, J.L., Tissue concentrations of prothymosin alpha: a novel proliferative index of primary breast cancer, *Eur. J. Cancer*, 29A, 893, 1993.

52. Bustelo, X.R., Otero, A., Gómez-Márquez, J., and Freire, M., Expression of the rat prothymosin α gene during T-lymphocyte proliferation and liver regeneration, *J. Biol. Chem.*, 266, 1443, 1991.

53. Sburlati, A.R., Manrow, R.E., Krug, M.S., and Berger, S.L., Prothymosin α antisense oligomers inhibit myeloma cell division, *Proc. Natl. Acad. Sci. U.S.A.*, 88, 253, 1991.

54. Thornberry, N.N., Bull, H.G., Calaycay, J.R., Chapman, K.T., Howard, A.D., Kostura, M.J., Miller, D.K., Molineaux, S.M., Weidner, J.R., and Aunins, J., A novel heterodimeric cysteine protease is required for interleukin-1 beta processing monocytes, *Nature*, 356, 768, 1992.

55. Schwartz, L.M., and Milligan, C.E., Cold thoughts of death: the role of ICE proteases in neuronal cell death, *Trends Neurosci.*, 19, 555, 1996.
56. Merry, H.H. and Korsmeyer, S.J., Bcl-2 gene family in the nervous system, *Annu. Rev. Neurosci.*, 20, 245, 1997.
57. Garcia-Segura, L.M., Cardona-Gomez, P., Naftolin, F., and Chowen, J.A., Estradiol upregulates Bcl-2 expression in adult brain neurons, *NeuroReport*, 9, 593, 1998.
58. Crenshaw, B.J., De Vries, G.J., and Yahr, P., Vasopressin innervation of sexually dimorphic structures of the gerbil forebrain under various hormonal conditions, *J. Comp. Neurol.*, 322, 589, 1992.
59. Frankfurt, M., Siegel, R.A., Sim, I., and Wuttke, W., Cholecystokinin and substance P concentrations in discrete areas of the rat brain: Sex differences, *Brain Res.*, 358, 53, 1985.
60. Larriva-Sahd, J., Gorski, R.A., and Micevych, P.E., Cholecystokinin synapses in the sexually dimorphic central part of the medial preoptic nucleus, *Exp. Neurol.*, 92, 639, 1986.
61. Micevych, P.E., Park, S.S., Akesson, T.R., and Elde, R., Distribution of cholecystokinin-immunoreactive cell bodies in the male and female rat. I. Hypothalamus, *J. Comp. Neurol.*, 255, 124, 1987.
62. Simerly, R.B., Young, B.J., Capozza, M.A., and Swanson, L.W., Estrogen differentially regulates neuropeptide gene expression in a sexually dimorphic olfactory pathway, *Proc. Natl. Acad. Sci. U.S.A.*, 86, 4766, 1989.
63. Watson, R.E., Hoffman, G.E., and Wiegand, S.J., Sexually dimorphic opioid distribution in the preoptic area: Manipulation by gonadal steroids, *Brain Res.*, 398, 157, 1986.
64. Watson, R.E., Wiegand, S.J., and Hoffman, G.E., Ontogeny of a sexually dimorphic opioid system in the preoptic area of the rat, *Dev. Brain Res.*, 44, 49, 1988.
65. Simerly, R.B., McCall, L.D., and Watson, S.J., Distribution of opioid peptides in the preoptic region: immunohistochemical evidence for a steroid-sensitive enkephalin sexual dimorphism, *J. Comp. Neurol.*, 276, 442, 1988.
66. Ge, F., Hammer, R.P., Jr., and Tobet, S.A., Ontogeny of leu-enkephalin and β-endorphin innervation of the preoptic area in male and female rats, *Dev. Brain Res.*, 73, 273, 1993.
67. Liposits, Z., Reid, J.J., Negro-Vilar, A., and Merchenthaler, I., Sexual dimorphism in copackaging of luteinizing hormone-releasing hormone and galanin into neurosecretory vesicles of hypophysiotrophic neurons: Estrogen dependency, *Endocrinology*, 136, 1987, 1995.
68. Merchenthaler, I., Lopez, F.J., Lennard, D.E., and Negro-Vilar, A., Sexual differences in the distribution of neurons coexpressing galanin and luteinizing hormone-releasing hormone in the rat brain, *Endocrinology*, 129, 1977, 1991.
69. Merchenthaler, I., Lopez, F.J., Lennard, D.E., and Negro-Vilar, A., Neonatal imprinting predetermines the sexually dimorphic estrogen-dependent expression of galanin in luteinizing hormone-releasing hormone neurons, *Proc. Natl. Acad. Sci. U.S.A.*, 90, 10479, 1993.
70. Bloch, G.J., Eckersell, C., and Millis, R., Distribution of galanin-immunoreactive cells within sexually dimorphic components of the medial preoptic area of the male and female rat, *Brain Res.*, 620, 259, 1993.

71. Park, J.-J., Baum, M.J., and Tobet, S.A., Sex difference and steroidal stimulation of galanin immunoreactivity in the ferret's dorsal preoptic area/anterior hypothalamus, *J. Comp. Neurol.*, 389, 277, 1997.

72. Herbison, A.E., Identification of a sexually dimorphic neural population immunoreactive for calcitonin gene-related peptide (CGRP) in the rat medial preoptic area, *Brain Res.*, 591, 289, 1992.

73. Herbison, A.E. and Dye, S., Perinatal and adult factors responsible for the sexual dimorphic calcitonin gene-related peptide-containing cell population in the rat preoptic area, *Neuroscience*, 54, 991, 1993.

74. Leclercq, P. and Herbison, A.E., Sexually dimorphic expression of calcitonin gene-related peptide (CGRP) immunoreactivity by rat mediobasal hypothalamic neurons, *J. Comp. Neurol.*, 367, 444, 1996.

75. King, J., Elkind, K.E., Gerall, A.A., and Millar, R.P., Investigation of the LH-RH system in the normal and neonatally steroid-treated male and female rat, in *Brain Endocrine Interaction III*, Scott, D.E., Kozlowski, G.P. and Weindl, A., Eds., Karger, Basel, 1978, 97.

76. King, J., Kugel, G., Zahniser, D., Wooledge, K., Damassa, D., and Alexsavich, G., Changes in population of LH-RH immunoreactive cell bodies following gonadectomy, *Peptides*, 8, 721, 1987.

77. Langub, M.C., Jr., Maley, B.E., and Watson, R.E., Jr., Ultrastructural evidence for luteinizing hormone-releasing hormone neural control of estrogen responsive neurons in the preoptic area, *Endocrinology*, 128, 27, 1991.

78. McDonald, A.J., Mascagni, F., and Wilson, M.A., A sexually dimorphic population of CRF neurons in the medial preoptic area, *NeuroReport*, 5, 653, 1994.

79. Simerly, R.B., Swanson, L.W., and Gorski, R.A., Demonstration of a sexual dimorphism in the distribution of serotonin-immunoreactive fibers in the medial preoptic nucleus of the rat, *J. Comp. Neurol.*, 225, 151, 1984.

80. Simerly, R.B., Swanson, L.W., and Gorski, R.A., Reversal of the sexually dimorphic distribution of serotonin-immunoreactive fibers in the medial preoptic nucleus by treatment with perinatal androgen, *Brain Res.*, 340, 91, 1985.

81. Simerly, R.B., Swanson, L.W., and Gorski, R.A., The distribution of monoaminergic cell and fibers in a periventricular preoptic nucleus involved in the control of gonadotropin release: immunohistochemical evidence for a dopaminergic sexual dimorphism, *Brain Res.*, 330, 55, 1985.

82. Commins, D. and Yahr, P., Adult testosterone levels influence the morphology of sexually dimorphic area in the mongolian gerbil brain, *J. Comp. Neurol.*, 224, 132, 1984.

83. Frankfurt, M., Fuchs, E., and Wuttke, W., Sex differences in γ-aminobutyric acid and glutamate concentrations in discrete rat brain nuclei, *Neurosci. Lett.*, 50, 245, 1984.

84. Gratten, D.R. and Selmanoff, M., Sex differences in the activity of γ-aminobutyric acidergic neurons in the rat hypothalamus, *Brain Res.*, 775, 244, 1997.

85. Chowen-Breed, J.A., Steiner, R.A., and Clifton, D.K., Sexual dimorphism and testosterone-dependent regulation of somatostatin gene expression in the periventricular nucleus of the rat brain, *Endocrinology*, 125, 357, 1989.

86. Simonia, S.X., Murray, H.E., Gillies, G.E., and Herbison, A.E., Estrogen-dependent ontogeny of sex differences in somatostatin neurons of the hypothalamic periventricular nucleus, *Endocrinology*, 139, 1420, 1998.

87. Finn, P.D., McFall, T.B., Clifton, D.K., and Steiner, R.A., Sexual differentiation of galanin gene expression in gonadotropin-releasing hormone neurons, *Endocrinology*, 137, 4767, 1996.

88. Herbison, A.E. and Spratt, D.P., Sexually dimorphic expression of calcitonin gene-related peptide (CGRP) mRNA in rat medial preoptic nucleus, *Mol. Brain Res.*, 34, 143, 1995.

89. Urban, J.H., Bauer-Dantoin, A.C., and Levine, J.E., Neuropeptide Y gene expression in the arcuate nucleus: sexual dimorphism and modulation by testosterone, *Endocrinology*, 132, 139, 1993.

90. Jakubowski, M., Blum, M., and Roberts, J.L., Postnatal development of gonadotropin-releasing hormone and cyclophilin gene expression in the female and male brain, *Endocrinology*, 128, 2702, 1991.

91. Alexander, M.J., Kiraly, Z.J., and Leeman, S.E., Sexually dimorphic distribution of neurotensin/neuromedin N mRNA in the rat preoptic area, *J. Comp. Neurol.*, 311, 84, 1991.

92. Simerly, R.B., Prodynorphin and proenkephalin gene expression in the anteroventral periventricular nucleus of the rat: sexual differentiation and hormonal regulation, *Mol. Cell. Neurosci.*, 2, 473, 1991.

93. Simerly, R.B., Hormonal control of the development and regulation of tyrosine hydroxylase expression within a sexually dimorphic population of dopaminergic cells in the hypothalamus, *Mol. Brain Res.*, 6, 297, 1989.

94. Bloch, G.J., Kurth, S.M., Akesson, T.R., and Micevych, P.E., Estrogen-concentrating cells within cell groups of the medial preoptic area: sex differences and co-localization with galanin-immunoreactive cells, *Brain Res.*, 595, 301, 1992.

95. Herbison, A.E. and Theodosis, D.T., Immunocytochemical identification of oestrogen receptors in preoptic neurons containing calcitonin gene-related peptide in the male and female rat, *Neuroendocrinology*, 56, 761, 1992.

96. Simerly, R.B., Zee, M.C., Pendleton, J.W., Lubahn, D.B., and Korach, K.S., Estrogen receptor-dependent sexual differentiation of dopaminergic neurons in the preoptic region of the mouse, *Proc. Natl. Acad. Sci. U.S.A.*, 94, 14077, 1997.

97. Fink, G., Gonadotropin secretion and its control, in *The Physiology of Reproduction*, Knobil, E., Neill, J.D., Ewing, L.L., Greenwald, G.S., Markert, C.L., and Pfaff, D.W., Eds., Raven Press, New York, 1988, 1349.

98. Shivers, B.D., Harlan, R.E., Morrell, J.I., and Pfaff, D.W., Absence of oestradiol concentration in cell nuclei of LHRH immunoreactive neurons, *Nature*, 304, 345, 1983.

99. Kallo, I., Liposits, Z.S., Flerco, B., and Coen, C.W., Immunocytochemical characterization of afferents to estrogen receptor-containing neurons in the medial preoptic area of the rat, *Neuroscience*, 50, 299, 1992.

100. Langub, M.C., Jr., Maley, B.E., and Watson, R.E., Jr., Ultrastructural evidence for luteinizing hormone-releasing hormone neural control of estrogen responsive neurons in the preoptic area, *Endocrinology*, 128, 27, 1991.

101. Langub, M.C., Jr. and Watson, R.E., Jr., Estrogen receptor neurons in the preoptic area of the rat are postsynaptic targets of a sexually dimorphic enkephalinergic fiber plexus, *Brain Res.*, 573, 61, 1992.

102. Chen, W-P., Witkin, J.W., and Silverman, A.W., Sexual dimorphism in the synaptic input to gonadotropin releasing hormone, *Endocrinology*, 126, 695, 1990.

103. Morrell, J.I., McGinty, J.F., and Pfaff, D.W., A subset of β-endorphin- or dynorphin-containing neurons in the medial basal hypothalamus accumulates estradiol, *Neuroendocrinology*, 41, 417, 1985.

104. Lehman, M.N. and Karsch, F.J., Do gonadotropin-releasing hormone-, tyrosine hydroxylase-, and β-endorphin-immunoreactive neurons contain estrogen receptors? A double-label immunocytochemical study in the Stuffolk ewe, *Endocrinology*, 133, 887, 1993.

105. Ifft, J.D., An autoradiographic study of the time of final division of neurons in rat hypothalamic nuclei, *J. Comp. Neurol.*, 144, 193, 1972.

106. Altman, J., and Bayer, S.A., Development of the diencephalon in the rat. I. Autoradiographic study of the time of origin and settling patterns of neurons of the hypothalamus, *J. Comp. Neurol.*, 182, 945, 1978.

107. Matsumoto, A. and Arai, Y., Developmental changes in synaptic formation in the hypothalamic arcuate nucleus of female rats, *Cell Tissue Res.*, 169, 143, 1976.

108. Reier, P.J., Cullen, M.J., Froelich, J.S., and Rothchild, I., The ultrastructure of the developing medial preoptic nucleus in the postnatal rat, *Brain Res.*, 122, 415, 1977.

109. Lawrence, J.M. and Raisman, G., Ontogeny of synapses in a sexually dimorphic part of the preoptic area in the rat, *Brain Res.*, 183, 466, 1980.

110. Arai, Y. and Matsumoto, A., Synapse formation of the hypothalamic arcuate nucleus during post-natal development in the female rat and its modification by neonatal estrogen treatment, *Psychoneuroendocrinology*, 3, 31, 1978.

111. Koritsanszky, S., Cyto- and synaptogenesis in the arcuate nucleus of the rat hypothalamus during fetal and early postnatal life, *Cell Tissue Res.*, 200, 135, 1979.

112. Walsh, R.J. and Brawer, J.R., Cytology of the arcuate nucleus in the newborn male and female rats, *J. Anat.*, 128, 121, 1979.

113. Matsumoto, A. and Arai, Y., Effect of estrogen on early postnatal development of synaptic formation in the hypothalamic arcuate nucleus of female rats, *Neurosci. Lett.*, 2, 79, 1976.

114. Matsumoto, A., Murakami, S., and Arai, Y., Neurotropic effects of estrogen on the neonatal preoptic area grafted into the adult rat brain, *Cell Tissue Res.*, 252, 33, 1988.

115. Toran-Allerand, C.D., Sex steroids and the development of the newborn mouse hypothalamus and preoptic area in vitro: implications for sexual differentiation, *Brain Res.*, 106, 407, 1976.

116. Uchibori, M., and Kawashima, S., Effects of sex steroids on the growth of neuronal processes in neonatal rat hypothalamus-preoptic area and cerebral cortex in primary culture, *Intern. J. Dev. Neurosci.*, 3, 169, 1985.

117. Sheridan, P.J., Sar, M., and Stumpf, W.E., Autoradiographic localization of ^{3}H-estradiol or its metabolites in the central nervous system of the developing rat, *Endocrinology*, 94, 1386, 1974.

118. Sheridan, P.J., Sar, M., and Stumpf, W.E., Autoradiographic localization of ^{3}H-testosterone or its metabolites in the neonatal rat brain, *Am. J. Anat.*, 140, 589, 1974.

119. Sibug, R.M., Stumpf, W.E., Shughrue, P.J., Hochberg, R.B., and Drews, U., Distribution of estrogen target sites in the 2-day-old mouse forebrain and pituitary gland during the "critical period" of sexual differentiation, *Dev. Brain Res.*, 61, 11, 1991.

120. DonCarlos, L.L. and Handa, R.J., Developmental profile of estrogen receptor mRNA in the preoptic area of male and female neonatal rats, *Dev. Brain Res.*, 79, 283, 1994.

121. Khnemann, S., Brown, T.J., Hochberg, R.B., and MacLusky, N.J., Sexual differentiation of estrogen receptor concentrations in the rat brain: effects of neonatal testosterone exposure, *Brain Res.*, 691, 229, 1995.

122. Güldner, F.H., Sexual dimorphism of axo-spine synapses and postsynaptic density material in the suprachiasmatic nucleus of the rat, *Neurosci. Lett.*, 28, 145, 1982.

123. Le Blond, C.B., Morris, S., Karakiulakis, G., Powell, R., and Thomas, P.J., Development of sexual dimorphism in the suprachiasmatic nucleus of the rat, *J. Endocrinol.*, 95, 137, 1982.

124. Stumpf, W.E., Estrogen-neurons and estrogen-neuron system in the periventricular brain, *Am. J. Anat.*, 129, 207, 1970.

125. Pfaff, D.W. and Keiner, M., Atlas of estradiol-concentrating cells in the central nervous system of the female rat, *J. Comp. Neurol.*, 151, 121, 1973.

126. Simerly, R.B., Chang, C., Muramatsu, M., and Swanson, L.W., Distribution of androgen and estrogen receptor mRNA-containing cells in the rat brain. An *in situ* hybridization study, *J. Comp. Neurol.*, 294, 76, 1990.

127. Cambray-Deakin, M.A., Cytoskeleton of the growing axon, in *The Neuronal Cytoskeleton*, Burgoyne, R.D., Ed., Wiley-Liss, New York, 1991, 233.

128. Baudry, M., Thompson, R.F., and Davis, J.L., *Synaptic Plasticity. Molecular, Cellular, and Functional Aspects*, The MIT Press, Cambridge, MA, 1993.

129. Stanley, H.F. and Fink, G., Synthesis of specific brain protein is influenced by testosterone at mRNA level in the neonatal rat, *Brain Res.*, 370, 223, 1986.

130. Tobet, S.A. and Fox, T.O., Sex- and hormone-dependent antigen immunoreactivity in developing rat hypothalamus, *Proc. Natl. Acad. Sci. U.S.A.*, 86, 382, 1989.

131. Ferreria, A. and Caceres, A., Estrogen-enhanced neurite growth: evidence for a selective induction of tau and stable microtubules, *J. Neurosci.*, 11, 392, 1991.

132. Lustig, R.H., Hua, P., Wilson, M.C., and Federoff, H.J., Ontogeny, sex dimorphism, and neonatal sex hormone determination of synapse-associated messenger RNAs in rat brain, *Mol. Brain Res.*, 20, 101, 1993.

133. Rogers, L.C., De Boer, I., Junier, M-P., and Ojeda, S.R., Estradiol increases neural-specific class II β-tubulin mRNA levels in the developing female hypothalamus by regulating mRNA stability, *Mol. Cell. Neurosci.*, 4, 424, 1993.

134. Shughrue, P.J. and Dorsa, D.M., Gonadal steroids modulate the growth-associated protein GAP-43 (neuromodulin) mRNA in postnatal rat brain, *Dev. Brain Res.*, 73, 123, 1993.

135. Shughrue, P.J. and Dorsa, D.M., Estrogen modulates the growth-associated protein GAP-43 (neuromodulin) mRNA in the rat preoptic area and basal hypothalamus, *Neuroendocrinology*, 57, 439, 1993.

136. Ma, Z.Q., Spreafico, E., Pollio, G., Santagati, S., Conti, E., Cattaneo, E., and Maggi, A., Activated estrogen receptor mediates growth arrest and differentiation of a neuroblastoma cell line, *Proc. Natl. Acad. Sci. U.S.A.*, 90, 3740, 1993.

137. Burgoyne, R.D., *The Neuronal Cytoskeleton*, Wiley-Liss, New York, 1991.

138. Rogers, L.C., Junier, M-P., Farmer, S.R., and Ojeda, S.R., A sex-related difference in the developmental expression of class II β-tubulin messenger RNA in rat hypothalamus, *Mol. Cell. Neurosci.*, 2, 130, 1991.

139. Lustig, R.H., Hua, P., Yu, W., Ahmad, F.J., and Bass, P.W., An in vitro model for the effects of estrogen on neurons employing estrogen receptor-transfected PC12 cells, *J. Neurosci.*, 14, 3945, 1994.

140. Lustig, R.H., Hua, P., Smith, L.S., Wang, C., and Chang, C., An in vitro model for the effects of androgen on neurons employing androgen receptor-transfected PC12 cells, *Mol. Cell. Neurosci.*, 5, 587, 1994.

141. Yamamoto, K.R., Steroid receptor regulated transcription of specific genes and gene networks, *Annu. Rev. Genet.*, 19, 209, 1985.

142. Evans, R.M., The steroid and thyroid hormone receptor superfamily, *Science*, 240, 889, 1988.

143. Truss, M., and Beato, M., Steroid hormone receptors: Interaction with deoxyribonucleic acid and transcription factors, *Endocrine Rev.*, 14, 459, 1993.

144. McCarthy, M.M., Schlenker, E.H., and Pfaff, D.W., Enduring consequences of neonatal treatment with antisense oligodeoxynucleotide to estrogen receptor messenger ribonucleic acid on sexual differentiation of rat brain, *Endocrinology*, 133, 433, 1993.

145. Toran-Allerand, C.D., The estrogen/neurotrophin connection during neural development: is co-localization of estrogen receptors with the neurotrophins and their receptors biologically relevant? *Dev. Neurosci.*, 18, 36, 1996.

146. Szabo, P., Panneerselvam, C., Clinton, M., Frangou-Lazaridis, M., Weksler, D., Whittington, E., Macera, M.J., Grzeschik, K.-H., Selvakumar, A., and Horecker, B., Prothymosin α gene in humans: organization of its promoter region and localization to chromosome 2, *Hum. Genet.*, 90, 629, 1993.

147. Bond, J.F., Robioson, G.S., and Farmer, S.R., Differential expression of two neural cell-specific, β-tubulin mRNAs during rat brain development, *Mol. Cell. Biol.*, 4, 1313, 1984.

148. Karns, L.R., Ng, S.-G., Freeman, J.A., and Fishman, M.C., Cloning of complementary DNA for GAP-43, a neuronal growth-related protein, *Science*, 236, 597, 1987.

149. Santagati, S., Ma, Z.Q., Ferrarini, C., Pollio, G., and Maggi, A., Expression of early genes in estrogen induced phenotypic conversion of neuroblastoma cells, *J. Neuroendocrinol.*, 7, 875, 1995.

13

Structural Sex Differences in the Mammalian Brain: Reconsidering the Male/Female Dichotomy

James C. Woodson and Roger A. Gorski

CONTENTS

0-8493-1165-9/00/$0.00+$.50
© 2000 by CRC Press LLC

I. Introduction

In mammals, certain brain functions such as the ability to support ovulation are irrefutably defeminized following exposure to gonadal steroids.[1-3] In this example, defeminization is demonstrated by a permanent loss of a female-specific function. However, the concept of defeminization may not apply equally well to complex behaviors that are expressed to a variable extent by both sexes. The considerable behavioral bipotential expressed by males and females of numerous species presents several conceptual and semantic challenges to theories of sexual differentiation of the brain. In this chapter, we challenge the dichotic categorization of the brain as *globally* "male" or "female." Retention of the potential for sex-atypical behavior by developmentally normal animals is best explained by the regionally and perhaps temporally independent sexual differentiation of separate brain regions. Differences in the onset, termination, and duration of critical periods may be major factors underlying the independent sexual differentiation of separate nuclei within the brain. Applying the concept of developmental independence to known structural sex differences in the human brain improves the explanatory power of current theories regarding the ontogeny of sexual orientation. We caution that conceiving of the brain as globally "male" or "female" may promote a false dichotomy in human studies, when male homosexual orientation is assumed to presuppose the global demasculinization of the brain or, conversely, when female homosexual orientation is assumed to presuppose a globally masculinized brain.

II. Organization, Activation, and Expression

Sexual ability, attraction, and expression are all influenced to different degrees by both biology and the environment. Arguably, sexual attraction to a partner of a particular sex may be the single most defining characteristic of sexual orientation. The stage of development at which sexual attraction is established, or is most greatly influenced by hormone action, is unknown. However, sex differences in brain structure and sexual ability are traditionally thought to be organized by the relatively permanent, genomic actions of gonadal steroid exposure during critical periods of development.[4] The motivational control of sex-typical behavior in rats[5-8] and men[9] is modulated in adulthood by "activational" genomic effects of gonadal hormones and by experience, both of which interact with neurotransmitter systems including dopaminergic,[10,11] noradrenergic,[12] and peptidergic systems.[13-17] The study of how rapid, nongenomic actions of steroids, occurring at cell membrane receptors,[18-20] change in response to sexual stimuli may soon provide clearer insight into the biological mechanisms that underlie the *expression* of sexual behavior.

It is important to note that hormonal influences are undeniably a biological mediator of sexuality, and inextricably link nature and nurture in the production of gender identity, sexual orientation, and behavior. Although a great deal of investigation has been conducted on the social and environmental factors influencing sexual expression in human beings, further consideration of this area is beyond the scope of this discussion. In this chapter, we will focus on the temporally and hormonally mediated processes underlying the development of structural sex differences in the brain. Obviously, appropriate moral and ethical restrictions preclude the manipulation of hormonal systems during human development. So, to understand the significance of human sex differences in the brain, it becomes essential to include comparative findings from laboratory animals in which true experimentation can take place. First, we will illustrate how during multiple critical periods, the hormonal environment alters the morphology of the reproductive system in a permanent fashion. Then, we will focus on the hormonally influenced mechanisms thought to produce structural sex differences in neural systems that underlie sexual behavior or reproductive function in laboratory animals.

A. Human Fetal Development Demonstrates Independent Processes of Masculinization and Feminization

The traditional dogma of sexual differentiation has been that testosterone (T) and its metabolites act to masculinize and defeminize both the body and the brain of what would, in their absence, default to a female morphology.[21] In most cases, the external genitalia provide a clear clue to the genetic sex of an individual, and to their basic reproductive capacity (but see Money[22]). During fetal development, the presence of gonadal factors, including T and its metabolites, triggers the morphological differentiation of shared genital precursor organs into male or female genitalia. However, describing the separate processes underlying the masculinization, feminization, demasculinization, and defeminization of the reproductive organs suggests why global categorization of the entire brain as "male" or "female" is a false truth.

Sexual differentiation of the internal genitalia begins at about 4 to 5 weeks of gestation and extends to about Week 14, regardless of the embryonic presence of an ovary. Both the mesonephric (Wolffian) ducts and paramesonephric

(Müllerian) ducts are present initially, illustrating the primordial bipotential of the reproductive system. In the presence of locally produced T, the mesonephric ducts differentiate into the epididymis, ductus deferens, seminal vesicle, and the ejaculatory duct,[21] otherwise, these ducts regress, probably via apoptosis.[23] Wolffian duct regression is clearly an excellent example of demasculinization, the complete and permanent loss of male-specific structures. In addition, the embryonic testes produce Müllerian inhibiting hormone (MIH) that actively defeminizes, causing the permanent loss of female-specific structures via regression of the paramesonephric ducts. In the absence of MIH, the differentiating reproductive system is said to become feminized as the paramesonephric ducts each become a fallopian tube, half of the uterus, and half of the deepest part of the vagina. Exactly what signals trigger the organizational feminization of these structures is not well known, but the fact that feminization occurs in the absence of a fetal ovary has led to the belief that Nature's blueprint for the developing fetus is female in the absence of perturbation by testicular hormones.

Sexual differentiation of the external genitalia occurs later, during the third month of human gestation. Following the formation of the basic excretory structures common to both sexes (the anterior urogenital groove and the posterior anorectal canal), high steroid levels in the male cause the genital tubercle to lengthen, pulling forward and fusing the urethral folds to form the penile urethra. At a slower pace, in the absence of high steroid levels in the female, the genital tubercle lengthens only slightly to form the clitoris, while the urethral folds remain to become the labia minora. Organizationally, lengthening of the genital tubercle to become the penis is a good example of masculinization, whereas the fusion of the urethral folds can be considered a loss of female potential, or organizational defeminization.

Importantly, these two processes have separate critical periods of sensitivity to organizational steroid influences. After 14 weeks, steroid exposure fails to defeminize the urethral folds, yet continues to lengthen and enlarge (masculinize) the clitoris effectively, leaving a child with female labia minora and a penislike phallus. Case studies of babies born with ambiguous genitalia and their clinical treatment are discussed at length by Money.[22] A similar independent timing of critical periods appears to exist for the brain. Rather than being entirely masculinized or not, it appears that the complexly interconnected and modular brain is capable of being functionally and structurally masculinized in one or more areas while retaining female-typical structure and function elsewhere.

B. Males Appear to Retain Brain Structures Enabling Female-Typical Behavior

Physical defeminization is clearly seen in the regression of the Müllerian system during development. Similarly, defeminization is illustrated in the brain by the dramatic and permanent loss of the ability to support ovulation because of a functionally defeminized portion of the male hypothalamic–pituitary–gonadal axis. However, with regard to brain structures mediating higher-level behaviors, it is possible that defeminization may not be reflected in a physical loss or regression of brain structure. In the rat, behavioral sex differences in the expression of female-typical sexual receptivity appear to result from the neural inhibition of lordosis circuitry that is retained in the adult male brain, rather than resulting from the developmental loss or regression of these structures.[24] Thus, in the normal male, some female-typical behaviors appear to be suppressed actively by the ontogenetic development of other, inhibitory brain structures (i.e., by masculinization). This suppression is not complete, as normal male rats can and do lordose, albeit at a much lower frequency and intensity than normal female rats. In 1924, Stone described female-typical receptive (lordotic) behavior in unprimed male rats who otherwise copulated normally when given access to estrous females.[25] In 1938, Beach[26,27] reported on a similar male, and he later concluded that the neural substrates for both behavioral potentials were retained in the normal adult of both sexes. Obviously, the argument against dichotomizing behavior as either masculine or feminine is not new.

To a limited extent, development of male-typical behavioral capacities occurs in the normal female rat as well. By the time of birth in the normal female, the brain appears to have been exposed to a limited yet influential level of steroid hormones, whether from the shared uterine environment or production by the female's own endocrine glands.[28-30] Anogenital distance, sexual behavior, and even the sexually dimorphic nucleus of the preoptic area (SDN-POA) volume appear to be influenced by the sex of a rat's gestational neighbors.[31] Thus it appears that in females of many species, partial androgenization of the brain may well underlie the well-documented occurrence of behavior traditionally categorized as male-typical, including mounting,[32] intromission patterns,[33] and ejaculation patterns,[34] often reported to occur regardless of the presence of exogenous T. While a female-typical pattern of structure and behavior results from manipulations that remove or interfere with gonadal hormones in early development, we suggest that a semantic distinction should be maintained between the terms *inherently female* and *indifferent* or *bipotential* when describing the brain at various, rather arbitrary, developmental stages.

III. Structural Sex Differences in the Brain

The existence of most sexually dimorphic brain regions in both sexes implies a retention of sex-atypical abilities, albeit at a diminished level. To provide insight into the meaning of structural sex differences in the human brain, consideration of studies of laboratory animals is essential. Physical differences between males and females in brain structure have been documented in many vertebrate species. Tables 13.1 and 13.2 list currently reported volumetric sex differences in the mammalian central nervous system. Differences in shape or asymmetry are also listed. While most of these regions are volumetrically larger in males, several exceptions exist. Notably, most differences are not absolute with one sex possessing a structure that does not exist in the other sex. Rather, a smaller nucleus usually remains in the other sex, and frequently appears to contain fewer of the neurons that comprise the larger nucleus in the first sex. Less frequently, a sex difference in volume could result from a similar number of large vs. small cells. Brain regions found to be sexually dimorphic in laboratory animals may be sexually dimorphic in humans as well, as is the case with several hypothalamic nuclei that may or may not be actually homologous with those in the rat.

A. Sex-Specific Differs From Sexually Dimorphic

The anterior hypothalamus/POA (AH/POA) continuum has been known for decades as an integrative center controlling mammalian copulatory behavior in a variety of species. Following the localization of ovulatory control centers to the hypothalamic region,[1,3] and the subsequent report of ultrastructural sex differences in the preoptic neuropil,[62] the AH/POA became one region in which to look for gross structural sex differences in the mammalian brain. In some cases, volumetric sex differences in the central nervous system were directly linked to sex-specific behaviors. The volumetric differences in the vocal control nuclei of the songbird discovered by Nottebohm and Arnold[90] provide a direct neural explanation for the sex-specific singing patterns exhibited only by males of the species. (Please see Chapter 9 by Arnold for an elaboration.) Likewise, in the rat, the spinal nucleus of the bulbocavernosus muscle (SNB) discovered by Breedlove and Arnold innervates a male-specific penile muscle.[45] (Please see Chapter 10 by Christensen, Breedlove, and Jordan for more on the SNB.) The brains of male and female rats have been found to regulate gonadotropin secretion differently, with males bringing about the release of luteinizing hormone (LH) only in a tonic fashion, whereas females can produce a cyclical release of LH, an ability caused by a positive feedback action of estrogen (E). The permanent loss of this sex-specific ability (cyclical LH release due to positive E feedback) from a single injection perinatally of T propionate or E clearly illustrates an organizational defeminization.[1,2] However, the effects of hormone manipulations perinatally are less clear when measuring complex behaviors that exhibit overlap in their normal expression by each sex.

B. The Sexually Dimorphic Nucleus of the Preoptic Area

Figure 13.1 schematically illustrates the SDN-POA of the rat and four hypothalamic nuclei in the human hypothalamus, two of which have potential homology with the rat SDN-POA (detailed later in this chapter). In the rat,

TABLE 13.1

Reported Structural Sex Differences in the Mammalian Brain

Mammal	Structure	Ref.
Volume: Male > Female		
Rat	SDN-POA	35, 36
	Accessory olfactory bulb	37
	Corpus callosum	38, 39
	Bed nucleus of the accessory olfactory tract	40
	Bed nucleus of the olfactory tract	37
	BNST	41, 42
	Medial nucleus of the amygdala	43, 42
	Medial preoptic nucleus (MPN)	44
	Spinal nucleus of the bulbocavenosus	45
	Supraoptic nucleus (dependent on body weight)	46
	Ventromedial nucleus (VMN)	47
	Vomeronasal organ	48
Volume: Female > Male		
Rat	Anterior commissure (rostral part)	49
	AVPV[a]	50, 51
	Locus coerulus	52
	Parastriatal nucleus	53
Asymmetry		
Rat	Cortex	54
Connectivity/Synaptic Morphology		
Rat	Arcuate nucleus	55
	AVPV	56
	Hippocampal dentate gyrus	57
	Hippocampal CA3 pyramidal cells	58
	Lateral septum	59
	Medial nucleus of the amygdala	60, 61
	MPN	62
	MPN/SDN-POA	44, 63
	VMN	64
	Visual cortex	65
Volume Male > Female		
Guinea Pig	MPOA	66
	BNST	66
Greater Volume Possibly Due to Larger Neurons		
Guinea Pig	Spinal motoneurons innervating bulbocavernosus/Levator ani and ishiocavernosus muscles	67

TABLE 13.1 (CONTINUED)

Reported Structural Sex Differences in the Mammalian Brain

Mammal	Structure	Ref.
Volume: Female > Male		
Pig	Vasopressin and oxytocin containing nucleus	68
	Supraoptic nucleus	68
Volume Male > Female		
Ferret	Male nucleus of the POA/AH	69
Gerbil	SDA-POA	70
Volume: Female > Male		
Gerbil	AVPV[a]	50
Connectivity		
Gerbil	Suprachiasmatic nucleus	71
Volume: Female > Male		
Hamster	AVPV[a]	50
Dendritic		
Pattern		
Hamster	POA	72
Volume: Female > Male		
Mouse	AVPV[a]	50
Volume: Male > Female		
Rhesus monkey	Dorsocentral component of the AH	73

[a] Bleier, Byne, and Siggelkow[50] referred to AVPV as MPN.

Gorski and colleagues[35] first identified the SDN-POA as being, on average, five to seven times larger in volume in normal males than in females. The SDN-POA can be sex-reversed by removal of, or exposure to, hormones around the time of birth in the male and female, respectively; see Reference 91. Several alternatives exist to explain volumetric structural dimorphisms, including differences in neuron size, density, or number. In the case of the SDN-POA, differences in neuron size or density are not responsible for the volumetrically larger nucleus in males.[36] Rather, the larger nucleus consists of more cells in the adult male than in the adult female. The increased number of cells comprising the SDN-POA in the male or androgenized female, in comparison with control females or males castrated on the day of birth (feminine males or "fales"[92]), probably does not result from an increased production of neurons during the perinatal period.[93] Rather, this dramatic structural

TABLE 13.2

Reported Structural Sex Differences in the Human Brain

	Magnitude of Difference	Ref.
Volume: Male > Female		
INAH-1		
INAH-1, "SDN-POA"	2.5 × (size) 2.2 × (# cells)	74
INAH-2	2 ×	75
INAH-3	2.8 ×	75, 76, 73
BNST-darkly staining posteromedial component	2.5 ×	77
BNST-central component	44% larger in heterosexual men	78
Cerebral volume	9%	79
Globus pallidus	Value not reported	79
Volume: Female > Male		
Anterior commissure (midsaggital area)	12%	80, 81
Massa intermedia (midsaggital area + incidence)	53.3%	81
Corpus callosum (midsaggital area)	Not listed, possibly related to handedness	82, 83
Isthmus of the corpus callosum (midsaggital area[a])	Not listed, possible related to handedness	84
Superior temporal cortex (part of Wernicke's area)	17.8%	85
Planum temporale	29.8%	85
Inferior frontal gyrus of dominant hemisphere (Broca's area)	20.4%	85
Caudate	Value not reported	79
Shape		
Splenium of corpus callosum	More bulbous in females	82, 86
Suprachiasmatic nucleus	Longer in females	87
Dimorphic with Sexual Orientation		
Anterior commissure	Homosexual men: 18% > females, homosexual men: 34% > presumed heterosexual men	88
INAH-3	Homosexual men: 2 × < presumed heterosexual men	76
Suprachiasmatic nucleus	1.7 × (size) 2.1 × (# cells); homosexual men > presumed heterosexual men	89

[a] Compared to right-handed men.

sex difference appears to result at least in part from the hormonal inhibition of apoptotic cell death.[94]

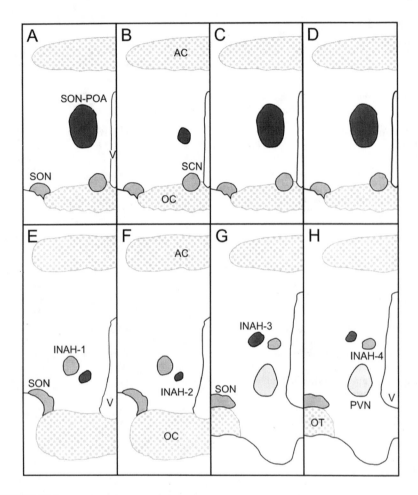

FIGURE 13.1
Highly schematic representation of the adult rat's SDN-POA in the intact male (A), intact female (B), and intact females exposed perinatally to testosterone propionate (C) or DES (D). The lower panels present highly schematic representations of the INAH 1 to 4 of the human brain as reported by Allen et al.,[75] The heavily shaded nuclei are sexually dimorphic, the intermediate shading indicates nuclei which were not found to be sexually dimorphic and the lightest nucleus (PVN) was not studied. Note that INAH-1 is the SDN-POA of Swaab and Fliers.[74] Abbreviations: AC, anterior commissure; OC, optic chiasm; OT, optic tract; PVN, paraventricular nucleus; SCN, suprachiasmatic nucleus; SON, supraoptic nucleus; V, third ventricle.

Within the SDN-POA, exposure to endogenous steroids in the male or exogenous steroids in the female, most likely aromatization-derived E, prevents the decrease in the number of neurons seen between postnatal Day 4 and 10 in the untreated female.[95] Gorski and his colleagues[94] determined that apoptosis, one form of programmed cell death, is at least partially responsible for the increased number of cells dying in females and fales. T injections strongly inhibit apoptosis in the SDN-POA when administered on postnatal Day 5 (PN5).[94] However, the neurons comprising the SDN-POA may be sensitive to the removal of T for a much longer period. There is one recent report that castration as late as 29 days after birth reduces SDN-POA volume.[96]

Importantly, exposure to steroid signals during its critical period affects neuron fate in the SDN-POA differently from another sexually dimorphic region in the brain that is likewise sensitive to hormone signaling perinatally, but unlike the SDN-POA develops its volumetric sexual dimorphism peripubertally. The anteroventral periventricular nucleus (AVPV) is sensitive to castration on PN1, but does not develop its structural sex difference until about PN30 to PN40.[97,98]

Hormone-sensitive cell death is not unique to the SDN-POA. Differences in SNB motor neuron number also arise in part because neuron loss is greater in females than in males during the early postnatal period. During a critical period similar to that of the SDN-POA, the application of androgen can inhibit motoneuron loss.[99] Aromatization of T to E by the enzyme aromatase is not required for masculinization of the SNB.[100] In contrast, the aromatization of T into E appears to account for most of the organizational effects of T on sexually dimorphic brain structures in mammals. Alpha-fetoprotein is a molecule present in fetal rat blood perinatally that binds E produced by the ovaries, placenta, or of maternal origin.[101] The role of alpha-fetoprotein as either a protective or site-specific delivery agent is still a matter of debate (see Reference 102), but most estrogenic actions in the developing brain are thought to occur through the process of aromatization, and not by direct exposure to plasma E.

C. Structural Sex Differences in the Human Hypothalamus

About 7 years after the discovery of the SDN-POA in the rat, a human hypothalamic nucleus was found to be sexually dimorphic, and was also named the SDN-POA.[74] According to studies by Swaab[103] and colleagues, the human SDN-POA is approximately twice as large in males as in females, and consists of twice as many cells. This large difference is not consistent across the life span, possibly explaining why this reported dimorphism has not been replicated by three other laboratories, including ours (see below). Swaab's studies show that the sex difference does not occur until the fourth year of life, when a decrease in cell number begins in girls. In contrast, the number of cells in the nucleus remain stable in men until about 50 years of age, when the neurons suffer a rapid reduction in number for a decade, and then level out again from about 60 years onward. In women at 50, a second phase of cell loss is incurred, stabilizes for a few years, and then accelerates to almost a 5% loss per year until the cell numbers are only 10 to 15% of the peak value by about 85 years of age.[103]

1. Possibilities Exist for Homology Between the SDN-POA of the Rat and Hypothalamic Nuclei in Human Beings

The human SDN-POA has also been referred to as the "intermediate nucleus."[104] It was renamed yet a third time by our laboratory, upon the discovery of three other, previously undescribed and adjacent cell groups in the human AH/POA, two of which were found to be sexually dimorphic (see Figure 13.1). In 1989, Allen, Hines, Shryne, and Gorski[75] suggested the term *interstitial nuclei of the anterior hypothalamus* (INAH 1-4), and found sex differences in INAH-2 and INAH-3. INAH-3 was 2.8 times larger in the male brain than in the female brain regardless of age, while INAH-2 was twice as large in the male brain, but in women appeared to be related to circulating steroid hormone levels. INAH-2 was 3.7 times larger in women age 20 to 32 than in women who were postmenopausal. In fact, three of the women of childbearing age had INAH-2s of similar volume to those of the men in the study.[75] INAH-1 (the same nucleus Swaab and Fliers named the SDN-POA[74]), INAH-4, and the supraoptic nucleus were not found to be sexually dimorphic. However, these authors did find, like Swaab's group, that INAH-1 decreased significantly with age. Differences between the studies have been discussed at length elsewhere, and may relate to technical differences or perhaps more likely to the age range of subjects studied.[75,105-107] Confirmation of the sexual dimorphism of INAH-3 has been obtained by two independent laboratories without any intervening failures.[73,76] Both of these studies found no sex differences in INAH-1, INAH-2, or INAH-4.

Although questions remain the sexual dimorphism of INAH-1, neurochemical evidence supports a possible homology with the SDN-POA of the rat based on the expression of mRNA for the GABA synthesizing enzyme, glutamic acid decarboxylase (GAD). In both the human and the rat, staining for GAD_{65} and GAD_{67} mRNA is heavy enough to distinguish the SDN-POA from the rest of the preoptic area.[108] Production of GAD suggests that neurons in the SDN-POA are involved the GABAergic inhibition of regions to which they project, which may provide a clue to function. Unfortunately, the exact projections are not well known. No similar GAD expression exists in INAH-3 (Moore, personal communication).

Therefore, although the parallels in GAD expression appear to support homology with INAH-1, when using morphological criteria it currently appears that INAH-3 is also a reasonable candidate for homology with the rat SDN-POA. While this judgment is largely based on location[73] and the greater element of certainty surrounding the sexual dimorphism of INAH-3, further evidence must be obtained by demonstrations of similarities in ontogeny, connectivity, and perhaps function.

2. "Male" Brains, "Female" Brains, and "Gay" Brains?

Excitement about the INAHs in the press and the lay public exploded in 1991 following the *Science* article by LeVay.[76] Included in LeVay's study was a group of homosexual men, all of whom had died from AIDS-related complications.

The INAHs of the homosexual men were compared to the INAHs of presumed heterosexual men and women. LeVay found that the mean volume of INAH-3 was twice as large in heterosexual men than in women or homosexual men, whose mean volumes did not differ significantly from each other.[76] The mean differences notwithstanding, it should be emphasized that the second *largest* INAH-3 volume in the study belonged to a homosexual male who died from AIDS. This fact helps to contest the argument that LeVay's findings were biased by AIDS-related factors, and draws attention to the great deal of variance in INAH volume observed when grouping subjects by sex or by apparent sexual orientation. Although LeVay was reasonably prudent in his own interpretation of his findings, INAH-3 was subsequently interpreted by many as a biological substrate contributing to sexual orientation in humans.

In 1990, Swaab and Hofman[89] provided evidence against the assumption that homosexual men must possess globally feminized or demasculinized brains. They reported that the mean volume of the suprachiasmatic nucleus (SCN) in homosexual men was 1.7 times larger and contained 2.1 times as many cells as that of a heterosexual reference group of both sexes. No differences in either SDN-POA (INAH-1) volume or cell number were found between heterosexual and homosexual males. These data imply a regional specificity in the processes that led to the enlarged SCN in homosexual men, and do not support a role for the SDN-POA of the human in homosexual orientation.[89]

Obviously, the human subjects in these studies were not experimentally manipulated. The nuclear volumes of the SCN and INAH-3 were only correlated with homosexuality, disallowing any causal link between the observed differences in the nuclei and sexual orientation. Resolution limitations currently preclude the use of functional brain imaging to examine the INAHs of living humans. Thus, it remains possible that activities engaged in by the subjects in these studies altered the sizes of their brain nuclei. There has been a recent report from animal studies implicating a potentially deleterious effect of high sexual activity on the size of the SNB.[109] Unfortunately, in all studies to date, inadequate information is available regarding human subjects' specific sexual behaviors to enable discrimination between the effects of high and low levels of sexual activity or other factors (e.g., stress) on the human nervous system.

D. Hypothalamic Nuclei and Aggressive Behavior

There may be other behavioral correlates of sex that could have a biological substrate in hypothalamic nuclei, for example, male-typical aggression. The SDN-POA is disproportionately large in a Hawaiian strain of Long Evans rats bred to be socially aggressive, and lesioning their SDN-POAs results in a reduction of their aggressive behavior (Gorski and Hori, unpublished observations). Like the SDN-POA to which it connects, the bed nucleus of the stria terminalis (BNST) is sexually dimorphic,[66,71] and is involved in both sexual,[110,111] and aggressive behaviors.[112-114] Allen and Gorski[77] discovered that the volume of the "darkly staining posteromedial" component of the BNST in the human brain was 2.5 times greater in males than in females.[77] Information regarding the existence of sexually aggressive or sexually receptive tendencies in these subjects is unavailable. It is well known, however, that sexual and aggressive behaviors in numerous species including humans are modulated by serotonin (5-hydroxytryptophan; 5-HT), and the medial preoptic nucleus (MPN) may be one locus of the influence of 5-HT on aggressive behaviors.[115] Each subdivision (lateral, l; medial, m; and central, c) of the sexually dimorphic MPN contains a unique pattern of 5-HT immunoreactive fibers, and a male-typical pattern of 5-HT staining can be obtained in female rats by exposing them to T during the perinatal critical period.[116] In each sex, the MPN is surrounded by a low to medium density of 5-HT staining, whereas the MPNl contains a dense concentration of 5-HT immunoreactive fibers.[44] The MPNm, surrounding the MPNc, contains a medium density. In contrast, the MPNc, which corresponds to a *portion* of the SDN-POA, is virtually devoid of 5-HT-positive fibers. Thus, a large proportion of SDN-POA cell bodies appear to be virtually devoid of the serotonergic influence found in the other subdivisions of the MPN. Because both the MPNm and the MPNc are larger in the male, a larger region of very low 5-HT-stained fiber density is found in the male. Therefore, it is possible that the larger region of reduced 5-HT input to the male MPN may be related to the higher levels of aggression exhibited by males. As mentioned earlier, the SDN-POA has been found to be approximately 1.9 times larger in an aggressive strain of Long Evans rats that were compared with a normal Long Evans strain from our laboratory (Gorski and Hori, unpublished observations). To date, no studies have been published linking any INAH volumes to violent or sexually aggressive behavior in humans.

E. The SDN-POA and Sexual Behavior

The SDN-POA is a locus of hormone/peptide interactions influencing sexual behavior. Approximately 13% of the E-concentrating cells in the rat SDN-POA are galanin positive, suggesting that an interaction between galanin and gonadal steroids may be another mechanism by which cells in the SDN-POA regulate reproductive function.[15] Unilateral microinjection of galanin into the region facilitates copulatory behavior in male rats in a dose-dependent manner, increasing the percentage of males that display sexual behaviors and decreasing mount and intromission latencies.[14] While the medial POA (MPOA) has previously been categorized as primarily being a copulatory center,[5,6,117,118] the SDN-POA may be involved to a variable extent in both consummatory and motivational aspects of sexuality. The transplantation of MPOA brain tissue including the SDN-POA from neonatal males into developing female rats results in an increase in both male-typical and female-typical sexual behaviors.[119] Decreases in mount and intromission latencies following MPOA tissue transplantation or galanin microinjection might be interpreted as a heightening of sexual arousal, thereby implicating the SDN-POA in sexual motivation. Ejaculation is a rewarding aspect of copulation that feeds back positively on motivated behaviors,[120] and unpublished electrical brain stimulation experiments originating in our laboratory suggest that neurons in the SDN-POA are also involved in ejaculatory behavior (Gorski and Hori, unpublished observations). Unilateral electrical stimulation of the anterior pole of the SDN-POA caused one male rat to ejaculate eight times in a single, 20-min test. Three other subjects with electrodes located in the SDN-POA also exhibited higher ejaculation frequencies and lower postejaculatory intervals than rats with electrodes placed outside the region.

IV. Possible Masculinizing and Demasculinizing Influences During Development

While the developmental influences resulting in sex differences in the brains of human beings remain difficult to investigate experimentally, laboratory studies of animals again provide cautionary insight about clinical treatment of pregnant women. Perinatal administration of androgens or estrogens, including diethylstylbestrol (DES), increase SDN-POA volume in the rat. Although DES is no longer used to treat pregnant women, antiestrogens and aromatase inhibitors are currently being used to control certain types of cancer. Caution with these substances is warranted because the SDN-POA of the rat can also be developmentally demasculinized by perinatal exposure to aromatase inhibitors[121] or exposure postnatally to the antiestrogen Tamoxifen.[122] Another concern regarding prenatal care involves the use of alcohol or drugs during pregnancy. Although an effect on the SDN-POA has not been studied, barbiturate treatment inhibits functional masculinization by exogenous androgen in female rats.[123] Administration of alcohol to pregnant rats has been reported to reduce SDN-POA volume significantly.[124] Chronic exposure perinatally to cocaine from the fifteenth day of gestation (G15) to PN10 has also been found to reduce SDN-POA volume,[125] probably by inhibiting both T and E incorporation into the hypothalamus.[126] Administration prenatally of cocaine from G15 to G20 inhibits hypothalamic T and E incorporation by approximately 50%. Although most dietary factors have not been directly addressed with regard to SDN-POA development, subcutaneous injection of high doses (4 mg/g) of monosodium l-glutamate to rat pups on days PN1 and PN3 reduces SDN-POA volume at 6 months of age.[127] The mechanism behind this reduction is not understood, but excitotoxicity may be a possibility. The calcium-binding proteins calbindin and calretinin appear to be expressed at higher levels around the day of birth in male rats compared with female littermates at several hypothalamic sites including portions of the SDN-POA.[128] It is tempting to speculate that the sexually dimorphic control of cytosolic calcium levels in this fashion might protect against an excitotox-icity-based loss of neurons during the critical period, and thereby contribute to differences in SDN-POA volume.

A. Prenatal Stress Affects the Brain

Several studies of prenatal stress-induced demasculinization in rodents have shown a link between stress, SDN-POA volume, and heterotypical sexual behavior. Prenatal stress inhibits aromatase activity,[129] and alters plasma T in males by causing a premature shift in the prenatal T surge to day G17.[130,131] Using a similar stress paradigm in 1985, Anderson et al.[132] were able to dem-onstrate that repeated restraint, light and heat stress during the third trimes-ter of gestation in Sprague-Dawley rats produces a significant decrease (51.2%) in SDN-POA cross-sectional area measured at 20 days of age. The same treatment of littermates caused 46.4% decrease in SDN-POA area mea-sured at 60 days. A subsequent study using the same stressor regime exam-ined SDN-POA volume rather than cross-sectional area, and reported similar reductions in nuclear volume.[133] The effect was independently replicated by Ward's laboratory.[134] However, the deleterious effect of stress on nuclear vol-ume is not global, nor does it affect all SDNs that are larger in the male. Sex-ually dimorphic areas of the medial amygdala are resistant to the same stressors.[135] The mechanism underlying this resistance is unknown.

B. Human Studies of Prenatal Stress

Based in part on the probably erroneous assumption that male homosexuality presupposes global brain demasculinization, prenatal stress has been weakly implicated as a "cause" of male homosexuality in human offspring by retro-spective studies involving their mothers.[136-138] However, a different study by Bailey et al.[139] found no relationship between self-reports of stress during preg-nancy and homosexual orientation in offspring, but did find a significant rela-tionship between self-reports of stressful pregnancy and femininity scores of offspring, regardless of their sexual orientation. These results favor the argu-ment that independent masculinization of separate brain regions occurs in the human as in the rat, by clearly demonstrating a functional independence between effeminate behavior and male homosexual orientation.

V. Conceptual Challenges

"False truths" often develop for good reason. When communicating the meaning of empirical findings to students and the public, it is often necessary to group the results of extensive research under broad headings that have intuitive appeal. However, in the simplification process, conceptual validity may be unintentionally compromised. Such is the case with the dichotic conceptualization of the brain as globally "male" or "female." Rather, it appears that cells utilizing different criteria for initiating apoptosis may underlie the independent differentiation of sexually dimorphic brain regions. Because apoptosis is a protein transcription–dependent event, the discovery of at least partial apoptotic determination of nuclear volume of the SDN-POA places renewed emphasis on genetically mediated mechanisms of sexual differentiation. Formerly characterized as the proximate mechanism underlying the development of sexually dimorphic brain structures, gonadal hormones now appear to be a primary signaling factor, perhaps among several, upon which a decision to undergo apoptosis or not is made by the cell. The temporal window during which the apoptotic decision is made appears to be amenable to scientific investigation. The apoptotic "decision window" is a likely mechanism delimiting the critical or sensitive period of a given brain region. Its parameters can probably be determined by evaluating the effectiveness of removing or supplying critical signaling factors to the region at different stages of development. For example, whether or not neural growth factors or other neuropeptides are signaling factors involved in the apoptotic organization of SDN-POA volume is currently under investigation in this laboratory.

A. How Can the Same Hormonal Signal Have Opposite Effects on Apoptosis?

In contrast to the demonstration by Davis and colleagues[94] that estrogen regulates the neuronal number by inhibiting apoptosis in the developing SDN-POA, Arai et al.[140] found that estrogen *facilitates* apoptotic cell death in the developing AVPV, a sexually dimorphic nucleus larger in females than in males. Thus, it appears that the organizational actions of steroid hormones on the brain are paradoxical, in one region forming larger nuclei in males (e.g., SDN-POA) and in another nearby region causing a nucleus to become smaller (e.g., AVPV). This is very strong evidence for the independent differentiation of separate brain regions, even though both changes are in a male-typical direction. However, an important question is raised by the phenomenon. How can exposure to the same steroid inhibit cell death in one region and promote or permit it in another? A conceptual mechanism by which the same steroid exposure can have opposite effects in different brain regions is elaborated below.

B. Hormonally Signaled "Logical Decision Rules" May Mediate Apoptosis in Sexually Dimorphic Brain Regions

Even though many established sex differences have not yet been shown to be mediated by apoptosis, the idea that genetically programmed *logical decision rules* might govern apoptosis provides a potential answer to this long-standing puzzle of structural sexual differentiation. It is tempting to speculate that genetically programmed logical decision rules could function as cellular "if–then statements" sensitive to hormonal or neurotrophic signals during critical periods, thus allowing the same signal to have opposite effects in separate brain regions. For example, it appears that during the critical period of the SDN-POA, the neurons comprising the nucleus respond to the presence of aromatization-derived estrogen according to a logical rule. *If* the required signal is present, *then* inhibit apoptosis and survive; *if* not, *then* initiate a protein synthesis-dependent program of cell death. The mechanism underlying estrogenic inhibition of this developmentally common cellular process is unknown, and it remains possible that steroids interfere directly with the cell death process. However, the fact that the presence of the same steroids facilitates apoptosis in adjacent hypothalamic regions (e.g., the AVPV) makes direct steroidal interference seem less likely. In the AVPV, then, it would seem likely that the apoptotic decision rules are different, or rather, inverted. The converse rule might then be stated: *If* the required signal is absent, *then* inhibit apoptosis and survive, if present, then initiate cell death. Figure 13.2 illustrates the concept of logical decision rules governing cell fate.

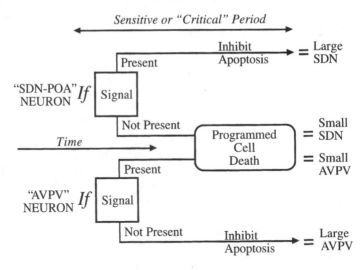

FIGURE 13.2
Theoretical "logical decision rules" potentially mediating apoptosis, which results in the sexually dimorphic development of brain regions in male and female mammals. With regard to the SDN-POA and AVPV of the hypothalamus, the signal most well understood is E, probably derived from T by aromatization.

V. Conclusions

1. Brain sex is regional and specific, not global and general.

We contend here that it is essential to conceptualize the brain as a heterogeneous component of the reproductive system, rather than as a homogeneous "male" or "female" structure. The "sex of the brain" in any given individual is likely to be a mixture of independently differentiated structures underlying the independent development of male-typical and female-typical traits along separate dimensions. It appears that brain regions displaying structural sex differences may independently be subject to multiple organizational periods of steroid influence, rather than a single critical period during which the brain is differentiated in a global fashion.

2. Characterizing the undifferentiated brain as "inherently female" may be inaccurate.

Because most sexually dimorphic parameters of the brain develop in a female-typical fashion in the absence of the influence of gonadal steroids, the developing brain has been characterized as being inherently feminine. However, the categorical emphasis placed on the sexual state of the brain at particular developmental stages is rather arbitrary and varies from study to study. Rather, it may prove useful to consider the brain as inherently bipotential with many separate stages of sexual development. Additionally, the normal female brain does not develop in the absence of gonadal steroids. Rather, it is exposed to them *in utero*, for example, by sharing the gestational environment with male siblings. Even in adulthood, the female brain appears to retain much of its inherent bipotential. A similar retention of bisexual capacity exists in the developmentally normal male rat, which is capable of exhibiting lordosis. Expression of receptive sexual behavior in the rat is decreased by circulating T, and the removal of T may reduce the activity of inhibitory brain centers preventing female-typical behavior. We suggest that, at least in rats, the "defeminization" of sexual receptivity may be promoted by the hormonally influenced development of inhibitory structures (i.e., masculinization) rather than by the loss of brain functions related to lordotic ability or "defeminization" per se.

References

1. Barraclough, C.A. and Gorski, R.A., Evidence that the hypothalamus is responsible for androgen-induced sterility in the female rat, *Endocrinology*, 68, 68, 1961.
2. Gorski, R.A., Modification of ovulatory mechanisms by postnatal administration of estrogen to the rat, *Am. J. Physiol.*, 205, 842, 1963.
3. Gorski, R.A., Localization and sexual differentiation of the nervous structures which regulate ovulation, *J. Reprod. Fert.* (Suppl.), 1, 67, 1966.
4. Phoenix, C.H., Goy, R.W., Gerall, A.A., and Young, W.C., Organizing action of prenatally administered testosterone proprionate on the tissues mediating mating behavior in the guinea pig, *Endocrinology*, 65, 369, 1959.

5. Everitt, B.J. and Stacey, P., Studies of instrumental behavior with sexual reinforcement in male rats (*Rattus norvegicus*): II. Effects of preoptic area lesions, castration, and testosterone, *J. Comp. Psychol.*, 101, 407, 1987.

6. Everitt, B.J., Sexual motivation: a neural and behavioural analysis of the mechanisms underlying appetitive and copulatory responses of male rats, *Neurosci. Biobehav. Rev.*, 14, 217, 1990.

7. McGinnis, M.Y., Mirth, M.C., Zebrowski, A.F., and Dreifuss, R.M., Critical exposure time for androgen activation of male sexual behavior in rats, *Physiol. Behav.*, 46, 159, 1989.

8. McGinnis, M.Y. and Dreifuss, R.M., Evidence for a role of testosterone-androgen receptor interactions in mediating masculine sexual behavior in male rats, *Endocrinology*, 124, 618, 1989.

9. Davidson, J.M., Camargo, C.A., and Smith, E.R., Effects of androgen on sexual behavior in hypogonadal men, *J. Clin. Endocrinol. Metab.*, 48, 955, 1979.

10. Clark, J.T., Effects of a novel dopamine-receptor agonist RDS-127 (2-*N,N-di-n*-propylamino-4,7-dimethoxyindane), on hormone levels and sexual behavior in the male rat, *Physiol. Behav.*, 29, 1, 1982.

11. Clark, J.T., Stefanick, M.L., Smith, E.R., and Davidson, J.M., Further studies on alterations in male rat copulatory behavior induced by the dopamine-receptor agonist RDS-127, *Pharmacol., Biochem. Behav.*, 19, 781, 1983.

12. Clark, J.T., Smith, E.R., and Davidson, J.M., Enhancement of sexual motivation in male rats by yohimbine, *Science*, 225, 847, 1984.

13. Angulo, J.A. and McEwen, B.S., Molecular aspects of neuropeptide regulation and function in the corpus striatum and nucleus accumbens, *Brain Res. Rev.*, 19, 1, 1994.

14. Bloch, G.J., Butler, P.C., Kohlert, J.G., and Bloch, D.A. Microinjection of galanin into the medial preoptic nucleus facilitates copulatory behavior in the male rat, *Physiol. Behav.*, 54, 615, 1993.

15. Bloch, G.J., Eckersell, C., and Mills, R., Distribution of galanin-immunoreactive cells within sexually dimorphic components of the medial preoptic area of the male and female rat, *Brain Res.*, 620, 259, 1993.

16. Clark, J.T., Benextramine, a putative neuropeptide Y receptor antagonist, attenuates the termination of receptivity, *Physiol. Behav.*, 52, 965, 1992.

17. Micevych, P.E. and Block, G.J., Estrogen regulation of a reproductively relevant cholecystokinin circuit in the hypothalamus and limbic system of the rat, in *The Neuropeptide Cholecystokinin (CCK)*, Hughes, J., Dockray, G., and Woodruff, G., Eds., Halsted Press, New York, 1989, 68.

18. Nemere, I. and Farach-Carson, M.C. Membrane receptors for steroid hormones: a case for specific cell surface binding sites for vitamin D metabolites and estrogens, *Biochem. Biophys. Res. Commun.*, 248, 443, 1998.

19. Dufy, B., Vincent, J.D., Fleury, H., Du Pasquier, P., Gourdji, D., and Tixier-Vidal, A., Membrane effects of thyrotropin-releasing hormone and estrogen shown by intracellular recording from pituitary cells, *Science*, 204, 509, 1979.

20. Towle, A.C. and Sze, P.Y., Steroid binding to synaptic plasma membrane: differential binding of glucocorticoids and gonadal steroids, *J. Steroid Biochem.*, 18, 135, 1983.

21. Jost, A., Hormonal factors in the sex differentiation of the mammalian fetus, *Philos. Trans.*, 259, 119, 1970.

22. Money, J., Sex errors of the body and related syndromes in *A Guide to Counseling Children, Adolescents, and Their Families*, 2nd ed., Paul H. Brookes, Baltimore, MD, 1994, 132.

23. Lee, D.M., Osathanondh, R., and Yeh, J., Localization of Bcl-2 in the human fetal Müllerian tract, *Fertil. Steril.*, 70, 135, 1998.

24. Yamanouchi, K. and Arai, Y., Presence of a neural mechanism for the expression of female sexual behaviors in the male rat brain, *Neuroendocrinology*, 40, 393, 1985.

25. Stone, C.P., A note on "feminine" behavior in adult male rats, *Am. J. Physiol.*, 68, 39, 1924.

26. Beach, F.A., Sex reversals in the mating pattern of the rat, *J. Genet. Psychol.*, 1938

27. Beach, F.A., Female mating behavior shown by male rats after administration of testosterone proprionate, *Endocrinology*, 29, 409, 1941.

28. Clemens, L., Gladue, B.A., and Coniglio, L.P., Prenatal endogenous androgenic influences on masculine sexual behavior and genital morphology in male and female rats, *Horm. Behav.*, 10, 40, 1978.

29. Gladue, B.A. and Clemens, L.G., Development of feminine sexual behavior in the rat: androgenic and temporal influences, *Physiol. Behav.*, 29, 263, 1982.

30. Meisel, R.L. and Ward, I.L., Fetal female rats are masculinized by littermates located caudally in the uterus, *Science*, 213, 239, 1981.

31. Faber, K.A. and Hughes, C.L., Anogenital distance at birth as a predictor of volume of the sexually dimorphic nucleus of the preoptic area of the hypothalamus and pituitary responsiveness in castrated adult rats, *Biol. Reprod.*, 46, 101, 1992.

32. Whalen, R.F. and Edwards, D.A., Hormonal determinants of the development of masculine and feminine behavior in male and female rats, *Anat. Rec.*, 157, 173, 1967.

33. Sachs, B.D., Pollak, E.I., Krieger, M.S., and Barfield, R.J., Sexual behavior: Normal male patterning in androgenized female rats, *Science*, 181, 770, 1973.

34. Emery, D.E. and Sachs, B.D., Ejaculatory pattern in female rats without androgen treatment, *Science*, 190, 484, 1975.

35. Gorski, R.A., Gordon, J.H., Shryne, J.E., and Southam A.M., Evidence for a morphological sex difference within the medial preoptic area of the rat brain, *Brain Res.*, 148, 333, 1978.

36. Gorski, R.A., Harlan, R.E., Jacobson, C.D., Shryne, J.E., and Southam, A.M., Evidence for the existence of a sexually dimorphic nucleus in the preoptic area of the rat, *J. Comp. Neurol.*, 193, 529, 1980.

37. Guillamón, A. and Segovia, S., Sex differences in the vomeronasal system, *Brain Res. Bull.*, 44, 377, 1997.

38. Berrebi, A.S., Fitch, R.H., Ralphe, D.L., Denenberg, J.O., Friedrich, V.L., Jr., and Denenberg, V.H., Corpus callosum: region-specific effects of sex, early experience and age, *Brain Res.*, 438, 216, 1988.

39. Ojima, K., Abiru, H., and Fukui, Y., Effects of postnatal exposure to cocaine on the development of the rat corpus callosum, *Reprod. Toxicol.*, 10, 221, 1996.

40. Collado, P., Valencia, A., Del Abril, A., Rodríguez-Zafra, M., Pérez-Laso, C., Segovia, S., Guillamón, A., Effects of estradiol on the development of sexual dimorphism in the bed nucleus of the accessory olfactory tract in the rat, *Brain Res. Dev. Brain Res.*, 75, 285, 1993.

41. del Abril, A., Segovia, S., and Guillamón, A., The bed nucleus of the stria terminalis in the rat: regional sex differences controlled by gonadal steroids early after birth, *Brain Res.*, 429, 295, 1987.

42. Hines, M., Allen, L.S., and Gorski, R.A., Sex differences in subregions of the medial nucleus of the amygdala and the bed nucleus of the stria terminalis of the rat, *Brain Res.*, 579, 321, 1992.

43. Mizukami, S., Nishizuka, M., and Arai, Y., Sexual difference in nuclear volume and its ontogeny in the rat amygdala, *Exp. Neurol.*, 79, 569, 1983.

44. Simerly, R.B., Swanson, L.W., and Gorski, R.A., Demonstration of a sexual dimorphism in the distribution of serotonin-immunoreactive fibers in the medial preoptic nucleus of the rat, *J. Comp. Neurol.*, 225, 151, 1984.

45. Breedlove, S.M. and Arnold, A.P., Hormone accumulation in a sexually dimorphic motor nucleus of the rat spinal cord, *Science*, 210, 564, 1980.

46. Madeira, M.D., Sousa, N., Cadete-Leite, A., Lieberman, A.R., and Paula-Barbosa, M.M., The supraoptic nucleus of the adult rat hypothalamus displays marked sexual dimorphism which is dependent on body weight, *Neuroscience*, 52, 497, 1993.

47. Matsumoto, A. and Arai, Y., Sex difference in volume of the ventromedial nucleus of the hypothalamus in the rat, *Endocrinol. Jpn.*, 30, 277, 1983.

48. Segovia, S. and Guillamón, A., Effects of sex steroids on the development of the vomeronasal organ in the rat, *Brain Res.*, 281, 209, 1982.

49. Jones, H.E., Ruscio, M.A., Keyser, L.A., Gonzalez, C., Billack, B., Rowe, R., Hancock, C., Lambert, K.G., and Kinsley, C.H., Prenatal stress alters the size of the rostral anterior commissure in rats, *Brain Res. Bull.*, 42, 341, 1997.

50. Bleier, R., Byne, W., and Siggelkow, I., Cytoarchitectonic sexual dimorphisms of the medial preoptic and anterior hypothalamic areas in guinea pig, rat, hamster, and mouse, *J. Comp. Neurol.*, 212, 118, 1982.

51. Sumida, H., Nishizuka, M., Kano, Y., and Arai, Y., Sex differences in the anteroventral periventricular nucleus of the preoptic area and in the related effects of androgen in prenatal rats, *Neurosci. Lett.*, 151, 41, 1993.

52. Guillamón, A., de Blas, M.R., and Segovia, S., Effects of sex steroids on the development of the locus coeruleus in the rat, *Brain Res.*, 468, 306, 1988.

53. del Abril, A., Segovia, S., and Guillamón, A., Sexual dimorphism in the parastrial nucleus of the rat preoptic area, *Brain Res. Dev. Brain Res.*, 52, 11, 1990.

54. Diamond, M.C., Hormonal effects on the development or cerebral lateralization, *Psychoneuroendocrinology*, 16, 121, 1991.

55. Matsumoto, A. and Arai, Y., Sexual dimorphism in "wiring pattern" in the hypothalamic arcuate nucleus and its modification by neonatal hormonal environment, *Brain Res.*, 190, 238, 1980.

56. Simerly, R.B., Swanson, L.W., Handa, R.J., and Gorski, R.A., Influence of perinatal androgen on the sexually dimorphic distribution of tyrosine hydroxylase-immunoreactive cells and fibers in the anteroventral periventricular nucleus of the rat, *Neuroendocrinology*, 40, 501, 1985.

57. Parducz, A. and Garcia-Segura, L.M., Sexual differences in the synaptic connectivity in the rat dentate gyrus, *Neurosci. Lett.*, 161, 53, 1993.

58. Gould, E., Westlind-Danielsson, A., Frankfurt, M., and McEwen, B.S., Sex differences and thyroid hormone sensitivity of hippocampal pyramidal cells, *J. Neurosci.* 10, 996, 1990.

59. DeVries, G.J., Sex differences in neurotransmitter systems, *J. Neuroendocrinol.*, 2, 1, 1990.

60. Nishizuka, M. and Arai Y., Sexual dimorphism in synaptic organization in the amygdala and its dependence on neonatal hormone environment, *Brain Res.*, 212, 31, 1981.

61. Nishizuka, M. and Arai, Y., Male-female differences in the intra-amygdaloid input to the medial amygdala, *Exp. Brain Res.*, 52, 328, 1983.

62. Raisman, G. and Field, P.M., Sexual dimorphism in the neuropil of the preoptic area of the rat and its dependence on neonatal androgen, *Brain Res.*, 54, 1, 1973.

63. Larriva-Sahd, J., Ultrastructural evidence of a sexual dimorphism in the neuropil of the medial preoptic nucleus of the rat: a quantitative study, *Neuroendocrinology*, 54, 416, 1991.

64. Matsumoto, A. and Arai, Y., Development of sexual dimorphism in synaptic organization in the ventromedial nucleus of the hypothalamus in rats, *Neurosci. Lett.*, 68, 165, 1986.

65. Muñoz-Cueto, J. A., Garcia-Segura, L.M., and Ruiz-Marcos, A., Regional sex differences in spine density along the apical shaft of visual cortex pyramids during postnatal development, *Brain Res.*, 540, 41, 1991.

66. Hines, M., Davis, F.C., Coquelin, A., Goy, R.W., and Gorski, R.A., Sexually dimorphic regions in the medial preoptic area and the bed nucleus of the stria terminalis of the guinea pig brain: a description and an investigation of their relationship to gonadal steroids in adulthood, *J. Neurosci.*, 5, 40, 1985.

67. Freeman, L.M. and Breedlove, S.M., Motoneurons innervating guinea pig perineal muscles are sexually dimorphic in size but not number, *Brain Res.*, 690, 1, 1995.

68. van Eerdenburg, F.J. and Swaab, D.F., Postnatal development and sexual differentiation of pig hypothalamic nuclei, *Psychoneuroendocrinology*, 19, 471, 1994.

69. Tobet, S.A., Zahniser, D.J., and Baum, M.J., Sexual dimorphism in the preoptic/anterior hypothalamic area of ferrets: effects of adult exposure to sex steroids, *Brain Res.*, 364, 249, 1986.

70. Commins, D. and Yahr, P., Lesions of the sexually dimorphic area disrupt mating and marking in male gerbils, *Brain Res. Bull.*, 13, 185, 1984.

71. Crenshaw, B.J., De Vries, G.J., and Yahr, P., Vasopressin innervation of sexually dimorphic structures of the gerbil forebrain under various hormonal conditions, *J. Comp. Neurol.*, 322, 589, 1992.

72. Greenough, W.T., Carter, C.S., Steerman, C., and DeVoogd, T.J., Sex differences in dendritic patterns in hamster preoptic area, *Brain Res.*, 126, 63, 1977.

73. Byne, W., The medial preoptic and anterior hypothalamic of the Rhesus monkey: a comparison with the human and evidence for sexual dimorphism, *Brain Res.*, 793, 346, 1998.

74. Swaab, D.F. and Fliers, E., A sexually dimorphic nucleus in the human brain, *Science*, 228, 1112, 1985.

75. Allen, L.S., Hines, M., Shryne, J.E. and Gorski, R.A., Two sexually dimorphic cell groups in the human brain, *J. Neurosci.*, 9, 497, 1989.

76. LeVay, S., A difference in hypothalamic structure between heterosexual and homosexual men, *Science*, 253, 1034, 1991.

77. Allen, L.S. and Gorski, R.A., Sex difference in the bed nucleus of the stria terminalis of the human brain, *J. Comp. Neurol.*, 302, 697, 1990.

78. Zhou, J.N., Hofman, M.A., Gooren, L.J., and Swaab, D.F., A sex difference in the human brain and its relation to transsexuality, *Nature*, 378, 68, 1995.

79. Giedd, J.N., Castellanos, F.X., Rajapakse, J.C., Vaituzis, A.C., and Rapoport, J.L., Sexual dimorphism of the developing human brain, *Prog. Neuro-psychopharmacol. Biol. Psychiatr.*, 21, 1185, 1997.

80. Allen, L.S. and Gorski, R.A., Sexual dimorphism of the human anterior commissure, *Anat. Rec.*, 214, 3, 1986.

81. Allen, L.S. and Gorski, R.A., Sexual dimorphism of the anterior commissure and massa intermedia of the human brain, *J. Comp. Neurol.*, 312, 97, 1991.

82. DeLacoste-Utamsing, C. and Holloway, R.L., Sexual dimorphism in the human corpus callosum, *Science*, 216, 1431, 1982.

83. Steinmetz, H., Staiger, J.F., Schlaug, G., Huang, Y., Corpus callosum and brain volume in women and men, *NeuroReport*, 6, 1002, 1995.

84. Witelson, S.F., Neural sexual mosaicism: Sexual differentiation of the human temporo-parietal region for functional asymmetry. Special Issue: Neuroendocrine effects on brain development and cognition, *Psychoneuroendocrinology*, 16, 131, 1991.

85. Harasty, J., Double, K.L., Halliday, G.M., Kril, J.J., and McRitchie, D.A., Language-associated cortical regions are proportionally larger in the female brain, *Arch. Neurol.*, 54, 171, 1997.

86. Allen, L.S., Richey, M.F., Chai, Y.M., and Gorski, R.A., Sex differences in the corpus callosum of the living human being, *J. Neurosci.*, 11, 933, 1991.

87. Swaab, D.F., Fliers, E., and Partiman, T.S., The suprachiasmatic nucleus of the human brain in relation to sex, age and senile dementia, *Brain Res.*, 342, 37, 1985.

88. Allen, L.S. and Gorski, R.A., Sexual orientation and the size of the anterior commissure in the human brain, *Proc. Nat. Acad. Sci. U.S.A.*, 89, 7199, 1992.

89. Swaab, D.F. and Hofman, M.A., An enlarged suprachiasmatic nucleus in homosexual men, *Brain Res.*, 537, 141, 1990.

90. Nottebohm, F. and Arnold, A.P., Sexual dimorphism in vocal control areas of the songbird brain, *Science*, 194, 211, 1976.

91. Gorski, R.A., Gonadal hormones and the organization of brain structure and function, in *The Lifespan Development of Individuals: Behavioral, Neurobiological, and Psychosocial Perspectives: A Synthesis*, E. David Magnusson et al., Ed., Cambridge University Press, New York, 1997, 315.

92. Gorski, R.A., Localization of the neural control of luteinization in the feminine male rat (FALE), *Anat. Rec.*, 157, 63, 1967.

93. Dodson, R.E., Shryne, J.E., and Gorski, R.A., Hormonal modification of the number of total and late-arising neurons in the central part of the medial preoptic nucleus of the rat, *J. Comp. Neurol.*, 275, 623, 1988.

94. Davis, E.C., Popper, P., and Gorski, R.A., The role of apoptosis in sexual differentiation of the rat sexually dimorphic nucleus of the preoptic area, *Brain Res.*, 734, 10, 1996.

95. Dodson, R.E. and Gorski, R.A., Testosterone propionate administration prevents the loss of neurons within the central part of the medial preoptic nucleus, *J. Neurobiol.*, 24, 80, 1993.

96. Davis, E.C., Shryne, J.E., and Gorski, R.A., A revised critical period for the sexual differentiation of the sexually dimorphic nucleus of the preoptic area in the rat, *Neuroendocrinology*, 62, 579, 1995.

97. Arai, Y., Nishizuka, M., Murakami, S., Miyakawa, M., Machida, M., Takeuchi, H., and Sumida, H., Morphological correlates of neuronal plasticity to gonadal steroids: Sexual differentiation of the preoptic area, in *The Development of Sex Differences and Similarities in Behavior*, Huag, M., Ed., Kluwer Academic, Dordrecht, 1993, 311.

98. Davis, E.C., Shryne, J.E., and Gorski, R.A., Structural sexual dimorphisms in the anteroventral periventricular nucleus of the rat hypothalamus are sensitive to gonadal steroids perinatally, but develop peripubertally, *Neuroendocrinology*, 63, 142, 1996.

99. Nordeen, E.J., Noordeen, K.W., Sengelaub, D.R., and Arnold, A.P., Androgens prevent normally occurring cell death in a sexually dimorphic spinal nucleus, *Science*, 229, 671, 1985.

100. Breedlove, S.M., Jacobson, C.D., Gorski, R.A., and Arnold, A.P., Masculinization of the female rat spinal cord following a single neonatal injection of testosterone propionate but not estradiol benzoate, *Brain Res.*, 237, 173, 1982.

101. Toran-Allerand, C.D., On the genesis of sexual differentiation of the general nervous system: morphogenetic consequences of steroidal exposure and possible role of alpha-fetoprotein, *Prog. Brain Res.*, 61, 63, 1984.

102. Gorski, R.A., The 13th J.A.F. Stevenson memorial lecture. Sexual differentiation of the brain: possible mechanisms and implications, *Can. J. Physiol. Pharmacol.*, 63, 577, 1985.

103. Swaab, D.F., Aging of the human hypothalamus, *Horm. Res.*, 43, 8, 1995.

104. Braak, H. and Braak, E., The hypothalamus of the human adult: chiasmatic region, *Anat. Embryol.*, 175, 315, 1987.

105. Swaab, D. F., Zhou, J., Fodor, M., and Hofman, M.A., Sexual differentiation of the human hypothalamus: differences according to sex, sexual orientation, and transsexuality, in *Sexual Orientation: Toward Biological Understanding*, E. Lee Ellis and E. Linda Ebertz, Eds., Praeger Greenwood, Westport, CT, 1997, 129.

106. Swaab, D.F. and Hofman, M.A., Sexual differentiation of the human hypothalamus in relation to gender and sexual orientation, *Trends Neurosc.*, 18, 264, 1995.

107. Swaab, D.F., Gooren, L.J.G., and Hofman, M.A., The human hypothalamus in relation to gender and sexual orientation, *Prog. Brain Res.*, 93, 205, 1992.

108. Gao, B. and Moore, R.Y., The sexually dimorphic nucleus of the hypothalamus contains GABA neurons in rat and man, *Brain Res.*, 742, 163, 1996.

109. Breedlove, S.M., Sex on the brain, *Nature*, 389, 801, 1997.

110. Yahr, P., Finn, P.D., Hoffman, N.W., and Sayag, N., Sexually dimorphic cell groups in the medial preoptic area that are essential for male sex behavior and the neural pathways needed for their effects, *Psychoneuroendocrinology*, 19, 463, 1994.

111. Emery, D.E. and Sachs, B.D., Copulatory behavior in male rats with lesions in the bed nucleus of the stria terminalis, *Physiol. Behav.*, 17, 803, 1976.

112. Adamec, R.E., The role of the temporal lobe in feline aggression and defense. Special Issue: Ethoexperimental psychology of defense: behavioral and biological processes, *Psychol. Rec.*, 41, 233, 1991.

113. Albert, D.J., Petrovic, D.M., Walsh, M.L., and Jonik, R.H., Medial accumbens lesions attenuate testosterone-dependent aggression in male rats, *Physiol. Behav.*, 46, 625, 1989.

114. Shaikh, M.B., Brutus, M., Siegel, H.E., and Siegel, A., Regulation of feline aggression by the bed nucleus of stria terminalis, *Brain Res. Bull.*, 16, 179, 1986.
115. Cologer-Clifford, A., Simon, N.G., Lu, S.F., and Smoluk, S.A., Serotonin agonist-induced decreases in intermale aggression are dependent on brain region and receptor subtype, *Pharmacol. Biochem. Behav.*, 58, 425, 1997.
116. Simerly, R.B., Swanson, L.W., and Gorski, R.A., Reversal of the sexually dimorphic distribution of serotonin-immunoreactive fibers in the medial preoptic nucleus by treatment with perinatal androgen, *Brain Res.*, 340, 91, 1985.
117. Brackett, N.L. and Edwards, D.A., Medial preoptic connections with the midbrain tegmentum are essential for male sexual behavior, *Physiol. Behav.*, 79, 1984
118. Liu, Y.C., Salamone, J.D., and Sachs, B.D., Lesions in medial preoptic area and bed nucleus of stria terminalis: differential effects on copulatory behavior and noncontact erection in male rats, *J. Neurosci.*, 17, 5245, 1997.
119. Arendash, G. and Gorski, R.A., Brain tissue transplants and reproductive function: Implications for the sexual differentiation of the brain, in *Neural Transplants*, J.R. Sladek and D.M. Gash, Eds., Plenum Press, New York, 1984, 223.
120. Oldenburger, W.P., Everitt, B.J., and de Jonge, F.H., Conditioned place preference induced by sexual interaction in female rats, *Horm. Behav.*, 26, 214, 1992.
121. Ohe, E., [Effects of aromatase inhibitor on sexual differentiation of SDN-POA in rats]. Nippon Sanka Fujinka Gakkai Zasshi, *Acta Obstet. Gynaecol. Japan*, 46, 227, 1994.
122. Döhler, K.D., Srivastava, S.S., Shryne, J.E., Jarzab, B., Sipos, A., and Gorski, R.A., Differentiation of the sexually dimorphic nucleus in the preoptic area of the rat brain is inhibited by postnatal treatment with an estrogen antagonist, *Neuroendocrinology*, 38, 297, 1984.
123. Arai, Y. and Gorski, R.A., Protection against the neural organizing effect of exogenous androgen in the neonatal female rat, *Endocrinology*, 82, 1005, 1968.
124. Ahmed, I.I., Shryne, J.E., Gorski, R.A., Branch, B.J., and Taylor, A.N., Prenatal ethanol and the prepubertal sexually dimorphic nucleus of the preoptic area, *Physiol. Behav.*, 49, 427, 1991.
125. Maecker, H.L., Perinatal cocaine exposure inhibits the development of the male SDN, *Brain Res. Dev. Brain Res.*, 76, 288, 1993.
126. Raum, W.J., McGivern, R.F., Peterson, M.A., Shryne, J.E., and Gorski, R.A., Prenatal inhibition of hypothalamic sex steroid uptake by cocaine: effects on neurobehavioral sexual differentiation in male rats, *Brain Res. Dev. Brain Res.*, 53, 230, 1990.
127. Hsieh, Y.L., Hsu, C., Lue, S.I., Hsu, H.K., and Peng, M.T., The neonatal neurotoxicity of monosodium L-glutamate on the sexually dimorphic nucleus of the preoptic area in rats, *Dev. Neurosci.*, 19, 342, 1997.
128. Brager, D.H., Sickel, M.J., and McCarthy, M.M., Sex differences in calcium binding proteins in the neonatal rat hypothalamus, *Soc. Neurosci. Abstr.*, 24, 1549, 1998.
129. Weisz, J., Brown, B.L., and Ward, I.L., Maternal stress decreases steroid aromatase activity in brains of male and female rat fetuses, *Neuroendocrinology*, 35, 374, 1982.
130. Ward, I.L. and Weisz, J., Maternal stress alters plasma testosterone in fetal males, *Science*, 207, 328, 1980.
131. Ward, I.L. and Weisz, J., Differential effects of maternal stress on circulating levels of corticosterone, progesterone, and testosterone in male and female rat fetuses and their mothers, *Endocrinology*, 114, 1635, 1984.

132. Anderson, D.K., Rhees, R.W. and Fleming, D.E., Effects of prenatal stress on differentiation of the sexually dimorphic nucleus of the preoptic area (SDN-POA) of the rat brain, *Brain Res.*, 332: 113, 1985.

133. Anderson, R.H., Fleming, D.E., Rhees, R.W., and Kinghorn, E., Relationships between sexual activity, plasma testosterone, and the volume of the sexually dimorphic nucleus of the preoptic area in prenatally stressed and non-stressed rats, *Brain Res.*, 370: 1, 1986.

134. Kerchner, M. and Ward, I.L., SDN-MPOA volume in male rats is decreased by prenatal stress, but is not related to ejaculatory behavior, *Brain Res.* 581, 244, 1992.

135. Kerchner, M., Malsbury, C.W., Ward, O.B., and Ward, I.L., Sexually dimorphic areas in the rat medial amygdala: resistance to the demasculinizing effect of prenatal stress, *Brain Res.*, 672, 251, 1995.

136. Dörner, G., Geier, T., Ahrens, L., Krell, L., Münx, G., Sieler, H., Kittner, E., and Müller, H., Prenatal stress as possible aetiogenetic factor of homosexuality in human males, *Endokrinologie*, 75, 365, 1980.

137. Dörner, G., Götz, F., and Döcke, W.D., Prevention of demasculinization and feminization of the brain in prenatally stressed male rats by perinatal androgen treatment, *Exp. Clin. Endocrinol.*, 81, 88, 1983.

138. Ellis, L., Ames, M.A., Peckham, W., and Burke, D., Sexual orientation of human offspring may be altered by severe maternal stress during pregnancy, *J. Sex Res.*, 25, 152, 1988.

139. Bailey, J.M., Willerman, L., and Parks, C., A test of the maternal stress theory of human male homosexuality, *Arch. Sexual Behav.*, 20, 277, 1991.

140. Arai, Y., Sekine, Y., and Murakami, S., Estrogen and apoptosis in the developing sexually dimorphic preoptic area in female rats, *Neurosci. Res.*, 25, 403, 1996.

14

Gonadal Hormones and Sexual Differentiation of Human Behavior: Effects on Psychosexual and Cognitive Development

Melissa Hines

CONTENTS

I. Introduction and Overview

Other chapters in this volume have discussed the powerful influences of gonadal steroids on sexual differentiation of brain and behavior in a wide range of mammals. This chapter will focus on the question of whether gonadal hormones exert similar influences on human neurobehavioral development. In other mammals, true experiments are possible, where animals are

0-8493-1165-9/00/$0.00+$.50
© 2000 by CRC Press LLC

randomly assigned to have hormones manipulated or not and where animals who do not experience a hormone manipulation are treated with a placebo or sham procedure. These types of experiments are generally not possible in human beings, because of ethical considerations. Therefore, information relevant to hormone influences on human development has come from studies of naturally occurring situations where hormonal aberrations occur. These include genetic disorders as well as situations where women have been prescribed hormones during pregnancy, usually for medical reasons. Of course, data obtained from these naturally occurring situations must be viewed more cautiously than data from true experiments, and the nonhormonal consequences of the genetic disorder or other situation resulting in hormonal abnormality must be considered in interpreting data. For this reason, as each situation involving hormonal abnormality is introduced, relevant details will be provided. Because of space constraints, these details will not be exhaustive and the reader interested in more extensive information should consult a basic text in pediatric endocrinology.

This chapter will not review in detail the extensive evidence that the gonadal hormones, androgen and estrogen, direct certain aspects of brain development during early critical periods, as these are described in other chapters. However, a few particular general points about these influences are worth reviewing before beginning the discussion of human studies. First, in general, hormones have two major types of influences on brain and behavior, activational and organizational, and these influences vary in their permanence and in the developmental time period during which they occur. Activational influences are temporary, waxing and waning as hormone levels rise and fall, and generally occur in sexually mature animals. An example would be behavioral changes that occur across the estrous cycle in rodents. In contrast, organizational influences of hormones typically occur during critical periods of early (prenatal or neonatal) development and are permanent. Although the hormone is present only briefly, its effect persists across the animal's life span. This is thought to occur because hormones direct basic processes of neural development, affecting, for instance, whether nerve cells live or die, which other cells they connect with anatomically, and which neurotransmitters they use.[1] This chapter will evaluate the possibility of similar organizational influences on human behavior. A second important point is that gonadal hormones influence behaviors that show sex differences. Therefore, in evaluating the hypothesis that gonadal hormones influence human neurobehavioral development, the focus will be on behaviors that show sex differences. In human beings, such behaviors fall into three general categories: (1) core gender identity, or the sense of self as male or female; (2) sexual orientation, or the direction of one's erotic interests, in persons of the same sex, the other sex, or both and (3) what are sometimes called gender role behaviors, or behaviors that have no obvious relation to sexuality or reproduction, but which differ on average in males and females. This chapter will discuss research in each of these three categories in turn. Third, in other species, particularly rodents, administration of estrogen to females during early development has many of the same neurobehavioral influences as administration of androgen. This is because the androgen, testosterone, is normally converted to the estrogen, estradiol, within certain regions of the brain before it acts to masculinize and defeminize the normal developing male animals. For this reason exposure of girls and women to high levels of estrogen during development will be discussed with the expectation that results will resemble those seen following exposure to androgen. Finally, as other chapters have outlined, the early hormone environment influences the development of brain structures that show sex differences as well as behavioral outcomes. In addition, there are sex differences in the human brain. However, to date, there are no studies of these sexually differentiated brain structures in people who developed in atypical hormone environments. For this reason, this chapter will focus on behavioral outcomes following development in sex-atypical hormone environments.

II. Core Gender Identity

There is growing evidence that the early hormone environment contributes to the development of core gender identity and its disorders. Several studies have focused on genetic females with congenital adrenal hyperplasia (CAH), a genetic disorder that results in overproduction of adrenal androgens, beginning prenatally. The diagnosis of a physical intersex condition, including CAH, precludes the diagnosis of gender identity disorder.[2] Nevertheless, women with CAH appear to be at increased risk for the symptoms of gender identity disorder, which include a strong and persistent cross-gender identification, the desire to be, or insistence that one is, of the other sex, and a persistent discomfort with one's assigned sex and its associated gender role. For instance, a Canadian study[3] found that one of a sample of 30 genetically female (XX) individuals with CAH who had been raised as females was now living as a male. The possibility that this occurred by chance was calculated to be one in 608. A second study[4] reported on four XX CAH patients in the New York area who had been raised as females but who now chose to live as males. The possibility that this was a chance occurrence was calculated to be 1 in 420 million. Similarly, a study conducted in the Netherlands[5] found that 2 of 18 girls with CAH also met the criteria for a diagnosis of gender identity disorder. So did 5 of 29 children with other diagnoses who had been exposed to high levels of androgen prenatally and raised as females. Reasons for the androgen exposure in these 29 children included partial androgen insensitivity syndrome (PAIS), cloacal extrophy, a transverse penis, gonadal dysgenesis, and true hermaphroditism. Others also have reported gender identity disorder in a true hermaphrodite raised as a female[6] and in a person with PAIS raised as a female.[7]

So, hormones can influence core gender identity. But, can they determine it? Probably not. Most people exposed to high levels of androgen prenatally because of endocrine disorders and raised as females do not have gender identity disorder. This clearly is true of females with CAH, as evidenced by the two studies described above where 29 of 30 CAH women in one study and 16 of 18 CAH girls in a second were content in the female gender. However, the elevated androgen levels experienced by genetic females with CAH are likely to be lower and to differ in timing from those of normal males. For this reason, other disorders, such as those experienced by XY individuals with cloacal extrophy or transverse penises, may be more informative. Although information is limited to relatively few cases, it appears that for these patients also, despite an increased risk of gender identity disorder, female gender assignment and rearing usually lead to contentment with the female role.[5] This despite the XY chromosome complement and presumed exposure to male-typical levels of testicular hormones prenatally and even during the early postnatal period until the testes are removed.

Perhaps the most stringent test of possible hormonal determination of gender identity is provided by situations where no underlying disorder is present, but other circumstances lead to gender reassignment. How likely is gender identity disorder in these cases? One well-publicized case involved a pair of identical twins, one of whom had his penis damaged during a phimosis repair at the age of 8 months. This twin was reassigned as female at the age of 17 months and surgically corrected to have female-appearing genitalia. Initial reports from early childhood suggested good adjustment to the female role,[8] an outcome that was widely cited as support for socialization as the primary determinant of gender identity. As an adult, however, this twin lives as a man and was reportedly unhappy for many years with the female sex of assignment.[9] This outcome could be seen as providing evidence that early exposure to testicular hormones determines male gender identity, despite female rearing. However, such a strong conclusion is unwarranted, partly because for the first 17 months of life, this child was socialized as a boy. Also, as in most intersex cases, no actual data on the rearing environment are available. Thus, we have no way of knowing how successful the parents were at changing the socialization of a child who had been their son for almost a year and a half to socialization that would be typical for a daughter. In addition, another case of a similar nature[10] produced a somewhat different outcome. In this case, the penis of a male infant was damaged during electrocautery circumcision at the age of 2 months. By 7 months of age, when the child was hospitalized for reconstructive surgery of the genitalia, the child had been reassigned to the female sex. At the age of 16 and again at the age of 26 this woman was interviewed and found to have a female core gender identification with no sign of gender dysphoria at present or in the past. Differences between this individual and the prior case may relate to the age at sex reassignment (prior to 7 months vs. 17 months or later), or to other factors.

One question that has not been explored systematically is why some children with endocrine disorders are content with their sex of assignment and others are not. No single syndrome is particularly susceptible to problems. Factors suggested to be important include slowness to assign gender after birth, a change from the original gender assignment, ambiguity in the sex of rearing, failure to correct the genitalia surgically, poor postnatal hormonal control, problems in the parent-child relationship, and gender-atypical psychological and body image.[4,5,10] However, no studies have directly addressed these possibilities. Resolution of this issue is important not only to increase our understanding of the basic science of gender development but also to assist in clinical decision making and patient management.

III. Sexual Orientation

The two cases of ablatio penis discussed above differed not only in outcomes for core sexual identity but also in outcomes for sexual orientation. The child reassigned as female after the age of 17 months had the sexual orientation of a heterosexual male as an adult,[9] whereas the child reassigned as female by the age of 7 months was bisexual. During the period of assessment at age 26 she had sought additional vaginal surgery to facilitate sexual intercourse with a male partner. However, several months later she had switched from living with this man to living with a new partner, this time a woman.[10] Both of these cases suggest that masculine-typical levels of testicular hormones during the prenatal and neonatal period predisposes away from a sexual orientation exclusively toward men, although the contribution of early rearing as a male, or of genitalia that require surgical correction for comfortable intercourse, cannot be ruled out.

What of other "experiments of nature"? These also suggest a hormonal contribution to sexual orientation. Once again, however, hormones appear to be a contributory rather than determining factor. Data have come from studies of genetic females exposed to high levels of androgen prenatally because of CAH, genetic females and males exposed to high levels of estrogen prenatally because their mothers were prescribed the synthetic hormone, diethylstilbestrol (DES) during pregnancy, and genetic males exposed to lower than normal levels of androgen because of defective receptors or deficiencies in enzymes needed to produce androgens.

A. Genetic Females

1. CAH

This chapter will focus on XX patients treated for CAH beginning early in life. The small number of studies examining sexual orientation in CAH patients who were not treated until relatively late in life will be excluded, to avoid the problems of interpretation associated with postnatal exposure to high levels of androgen and attendant continued physical virilization.

Several studies have reported that women with CAH have decreased heterosexual interest, increased homosexual interest, or decreased sexual interest in general. An initial study from the United States compared 30 women with CAH to 27 control women, 15 of whom had complete androgen insensitivity syndrome (CAIS) and 12 of whom had Rokitansky syndrome. Women with these two syndromes were selected as controls because they were being followed at the same endocrine clinic as the CAH women and because, like the CAH women, they had an endocrine disorder requiring genital correction. Evaluations of sexual orientation, in imagery as well as behavior, were made by the women themselves and verified by investigator ratings based on clinic notes and an interview. Seven of the 30 CAH women were noncommittal about their sexual orientation. Of the remaining 23, 12 rated themselves as heterosexual. This was significantly different from controls, all 27 of whom were willing to indicate their sexual orientation and 25 of whom rated themselves as heterosexual. The remaining two women indicated that they were bisexual. A second study, from Germany,[11] compared 34 women with CAH to 14 of their unaffected sisters. Among those old enough to have had sexual experiences, 44% of the CAH group desired or had experienced homosexual relations, compared with 0% of controls. In addition, on inventories of sexual interests, the CAH group scored higher on a homosexual interest scale and lower on a heterosexual interest scale compared with the sister controls. The CAH group also reported decreased general sexual interest. A third study, from Canada,[3] compared 30 women with CAH to unaffected sisters ($n = 12$), half-sisters ($n = 1$) and female cousins ($n = 2$). All women but three were interviewed face to face. Two were interviewed by telephone and one completed questionnaires. The interviews included an assessment of sexual orientation in fantasy and experience. Participants also completed a questionnaire assessing erotic response and orientation. The CAH group showed decreased general sexual activity as well as decreased heterosexual activity. A fourth study, from Germany,[12] assessed 45 CAH women (37 with classic CAH, the disorder studied in other reports, and 8 with late-onset CAH, meaning that the disorder was not apparent at birth or in infancy, but appeared later in life) and 46 controls matched for sex, age, education, and professional background. Both groups of women completed a questionnaire and the CAH women were also interviewed. The women with CAH were less sexually active, and reported fewer relationships than controls. However, self-report of homosexuality appeared similar. The authors stated: "Two patients and one control individual stated they were lesbians and lived with a female partner. One of the women stated in the questionnaires that she was a lesbian but denied it in the personal interview. Whether this was a sign of shyness or instability in her decision remains unclear."

This raises one difficulty in studying sexual orientation. Because it is such a personal issue, research participants may be reluctant to divulge detailed or accurate information. A second complication relates to the need for surgical correction of the genitalia and differences in the quality of the surgical outcome. Mulaikal,[13] for instance, studied 80 women with CAH and found that heterosexual behavior was more frequent and lack of sexual experience less frequent in CAH patients with an adequate vagina than in those without an adequate vagina and that an adequate vagina was more likely in the simple-virilizing variant of CAH than in the salt-losing variant. He found 33 of 40 simple-virilizing patients compared to 19 of 40 salt-losing patients to have an adequate vagina. Typically, effects on gendered behavior, including sexual orientation, are more pronounced in salt-losing patients than in simple virilizers.[3,11] Is this because they have had less successful surgical repair of the vagina? Alternatively, is the salt-losing form of the disorder more severe and thus more likely to lead both to alterations in sexuality and difficulties in vaginal repair? In their 1992 study, Dittman et al.[11] asked about insecurities based on genital problems and found no relationship to sexuality. Also, to the extent that measures of sexual orientation are based on fantasy as well as actual behavior, genital adequacy should be less important. However, the possibility that genital problems contribute to alterations in sexuality in CAH cannot be ruled out.

Thus, several types of studies would improve our understanding of sexual outcomes in CAH. First, it appears to be important to assess sexuality through in-depth interviews by trained sexual interviewers who can help participants comfortably divulge personal information about their sexuality. In addition, these interviews should cover information on fantasy as well as actual behavior. It also seems important to assess at least three dimensions of sexuality (heterosexual interest, homosexual interest, and sexual interest in general) since there is some evidence of alteration in each of these areas. Finally, the relationship between medical factors, such as salt-losing vs. simple-virilizing CAH (or even late-onset CAH) and the adequacy of the surgical repair of the genitalia, merit more-detailed examination to determine if alterations in particular dimensions of sexuality are more pronounced in some subgroups than in others or relate to physical problems associated with sexual function.

2. DES

On a more theoretical level, study of other hormone-exposed groups without physical problems related to surgical repair of the genitalia could shed light on the hypothesis that hormones during early development shape brain regions involved in determining sexual orientation. One group that could be particularly informative in this regard is women exposed to the synthetic estrogen, DES. In other species, early exposure to DES has masculinizing and defeminizing influences on certain aspects of brain and behavioral development, but not on development of the genitalia.[14-17] Female offspring of these pregnancies could reveal influences of hormones on sexual orientation in the absence of genital virilization and repair.

One group of researchers has studied sexual orientation in three samples of DES-exposed women. The first sample included 30 DES-exposed women, aged 17 to 30, compared with 30 unexposed women recruited from the same gynecological clinic.[18] The controls resembled the DES-exposed women in age and in having abnormal PAP smear findings. (Although prenatal treatment with DES rarely causes genital virilization, it usually alters some aspects of genital development. A small proportion of DES-exposed women develop vaginal or cervical adenocarcinoma, and a large proportion develop vaginal adenosis. In most cases they have abnormal PAP smears.[19]) Twelve unexposed sisters of the DES-exposed women formed a second comparison group. Sexual orientation in fantasy and behavior was assessed by interview using seven-point rating scales, ranging from exclusively heterosexual (0), through bisexual to exclusively homosexual (6).[20,21] Results suggested DES increased bisexuality or homosexuality. Approximately 24% of the DES-exposed women (vs. 0% of the control group) had a lifelong bisexual or homosexual orientation. Among the 12 sister pairs, 42% of the DES-exposed women (vs. 8% of their unexposed sisters) had a life-long bisexual or homosexual orientation.

This initial study was later reported along with two further studies.[22] In one, a second sample of 30 DES-exposed women was compared with 30 demographically matched controls who did not have a history of DES exposure. Eight unexposed sisters also participated. Results resembled those of the first study. About 35% of the DES-exposed women (vs. about 13% of the matched controls) showed bisexual or homosexual responsiveness since puberty. For sister pairs, the percentages were 36 and 0%. In the third study, 37 women exposed to DES were identified from obstetrical files. Daughters of women treated with at least 1000 mg of DES were compared with age-matched daughters of untreated women identified from the files of the same obstetrical practice. Results again suggested an association between DES and homosexual or bisexual orientation, although the difference between hormone-exposed and unexposed women (16 vs. 5%) appeared less dramatic than in the first two samples.

B. Genetic Males

1. Androgen Insensitivity Syndrome (AIS)

The consequences for genetic males of developing in a hormone environment that could promote a female-typical sexual orientation are not well understood. This may be because such hormone environments are rare. AIS, a disorder in which the testes produce normal levels of androgens, but cells cannot respond to it because of receptor defects, occurs in only about 1 in 60,000 births, and very few AIS individuals have been studied. One study compared sexual orientation in CAH women to 15 AIS patients who had been raised female. The AIS group had a more female-typical sexual orientation (i.e., toward males) than the CAH group,[23] but the AIS group was not compared with female controls with no endocrine disorder.

2. Enzymatic Deficiencies

Other syndromes that provide an opportunity to study the impact of reduced levels of androgen on gender development in genetic males include deficiencies in enzymes that are required for androgen synthesis, such as 5-alpha-reductase deficiency and 17-beta-hydroxysteroid dehydrogenase deficiency. In both syndromes, the physical appearance at birth is more female than male, and affected individuals are often raised as females. However, if the individual's testes are not removed, the dramatic elevation in androgen at puberty produces physical virilization. It has been reported that these individuals then adopt a male identity and social role and the sexual orientation of a heterosexual man (i.e., toward women).[24-26] One interpretation of this outcome is that androgen exposure had programmed the brain in the male direction, despite rearing as a female.[24] Alternatively, it has been suggested that not enough data are available on rearing to know if it was unambiguously female,[27,28] or that the change to a masculine physical appearance or the social advantages of being male (or disadvantages of being a sterile female) in the societies where these syndromes have been studied account for the change.[27,29]

3. Exposure to DES or Progestins

The impact on sexual orientation of exposure of genetic males to estrogen, progestins, or a combination or estrogen and progestin has also been investigated. Neither prenatal estrogen or progesterone exposure appears to influence sexual orientation in men. One study compared two groups of men exposed prenatally to the synthetic estrogen, DES, with controls matched for sex, age, and maternal age at birth.[30] One group had been exposed to DES alone, and the other to DES plus natural progesterone. There were no differences between either hormone-exposed group and their respective controls. Of 16 DES-exposed men, 15 indicated their behavior was exclusively heterosexual, as did 16 of 16 control men. Similarly, 13 of 17 DES-exposed men and 13 of 16 controls indicated their fantasies were exclusively heterosexual. Of 21 men exposed to DES plus progesterone, 20, compared with 16 of 20 control men, indicated exclusively heterosexual behavior, while 15 of 21 in the hormone group and 14 of 21 in the control group indicated exclusively heterosexual fantasies. The study also included men exposed to natural progesterone or synthetic progestins without DES. They also showed no differences in either sexual experience or fantasy from matched controls. In the natural progesterone group, 8 of 10 were exclusively heterosexual in behavior (vs. 9 of 10 controls) and 6 of 10 were exclusively heterosexual in fantasy (vs. 6 of 10 controls). In the synthetic progestin group, 12 of 13 were exclusively heterosexual in behavior (vs. 12 of 13 controls) and 10 of 13 were exclusively heterosexual in fantasy (vs. 11 of 13 controls).

A second set of studies of DES-exposed men also found no influence on sexual orientation. One study compared 31 hormone-exposed men with 29 unexposed controls, all recruited from one obstetrical practice that had pre-scribed DES. The second included 34 DES-exposed men and 15 controls, all recruited from one urological practice. Sexual orientation in behavior and fantasy was based on interviews and measured on a 7-point heterosexual to homosexual continuum. No consistent differences were seen between the DES-exposed men and controls.[31]

The absence of an influence of estrogen and progestin on sexual orientation in men is not surprising. Animal models of hormonal influences suggest that neither estrogen nor progesterone have consistent demasculinizing or femi-ninizing influences on development in genetic males when administered early in life (see, e.g., Reference 14, but also Reference 32).

IV. Gender Role Behaviors

Gender role behaviors are sometimes defined as behaviors that are culturally fixed or assigned to one sex or the other (e.g., Reference 33). However, this definition assumes enough knowledge of the causes of behavioral sex differ-ences to determine which are culturally fixed or assigned. Perhaps a safer definition would be those behaviors, other than core sexual identity or sexual orientation, that differ on average for males and females. Even this definition, although safer, is not without problems. As Maccoby and Jacklin[34] pointed out as long ago as 1974, there is great debate and widespread misconception about psychological sex differences. In addition, they described problems associated with identifying sex differences, such as the increased probability of publishing when sex differences are significant and not when they are not, producing an overreporting of spurious effects. A second problem is a ten-dency for researchers, like others, to see behavior through the prism of their own preconceptions, a tendency that can result in observing sex differences if you expect them, but not if you do not. Although these problems cannot be avoided completely, there is general agreement that there are several sex dif-ferences in what are commonly viewed as gender role behaviors. These include juvenile play behavior (e.g., toy choices, sex of preferred play part-ners, and rough and tumble play), specific cognitive abilities (e.g., mental rotations ability, spatial perception ability, mathematical problem-solving ability, verbal fluency, and perceptual speed and accuracy), personality char-acteristics (e.g., aggression and nurturance), and manifestations of neural asymmetry (e.g., hand preferences and language lateralization). In some cases, sufficient data are available to estimate the size of these behavioral sex differences. To put them in a familiar perspective, they are typically less than half the size of the sex difference in height, which is two standard deviation units. Nevertheless, by behavioral standards, where group differences of 0.8 standard deviation units are regarded as large, they are potentially important. The next section of this chapter will discuss data regarding hormonal influences on gender role behavior. Because of space limitations, the focus will be on two areas: childhood play behavior and cognitive functioning.

A. Childhood Play Behavior

The play behavior of girls exposed to high levels of androgen prenatally, because of CAH, has been assessed in several studies and found to be more masculine or less feminine than that of controls. These results have been reported based on interviews with the girls and their mothers, on questionnaires and even in direct observations of the toy choices of girls with CAH compared with unaffected sisters and female first cousins in the same age range.[8,35-39] Questionnaire and interview data suggest that the influenced behaviors include increased rough-and-tumble play and interest in male playmates, as well as increased interest in male-typical toys, such as cars and trucks, and decreased interest in female-typical toys, such as dolls. However, the one study that involved direct observation of rough-and-tumble play[40] found no difference between 20 girls with CAH compared with 12 unaffected female relatives, despite seeing sex differences in the 12 unaffected girls compared with 15 unaffected male relative controls. Further research is needed to determine if this reflects a lack of a hormone effect on rough-and-tumble play. Alternatively, the testing situation may have been inadequate to detect an effect. Rough-and-tumble play requires a partner. Because most girls do not like this kind of play and boys prefer to play in this way with other boys, rough-and-tumble play in CAH girls could have been inhibited by the lack of a willing partner (see Reference 40 for additional discussion of this and other possibilities).

The play behavior of children whose mothers were prescribed hormones during pregnancy has also been studied. One investigation found that ten girls exposed to androgenic progestins prenatally showed increased preferences for male playmates, masculine-typical toys, and vigorous play.[8] Thus, this source of exposure to androgenic hormones appears to have similar effects on girls to those seen after prenatal androgen exposure caused by CAH. Similarly, exposure of genetic females to the synthetic progestin, medroxyprogesterone acetate (MPA), which has some anti-androgenic action, has been reported to decrease some male-typical play behaviors and increase some female-typical ones. These effects appear to be more limited, and less consistent, than those of androgen exposure,[41-43] perhaps not surprisingly since there is less room for a meaningful decrease in feminine-typical behaviors in genetic females than for a meaningful increase in masculine-typical ones.

In contrast to the effects of androgen, prenatal exposure of genetic females to the synthetic estrogen DES does not appear to influence sexual differentiation of juvenile play. Three reports on women exposed prenatally to DES used interviews and questionnaires to assess childhood play retrospectively.[44-46] The three reports were from a single research group and involved a total of 60 DES-exposed women compared with various control groups. Taken together with the results for girls exposed to androgens prenatally, these data suggest that masculine-typical play behaviors differentiate under the influence of androgen acting through androgen receptors, rather than following conversion to estrogen. This is consistent with data from experimental studies of rats, where sex differences in rough-and-tumble play, unlike sex differences in most other behaviors, have been suggested to result from direct action of androgen.[47] However, the only published study of the influences of DES on rough-and-tumble play in nonhuman primates suggests that prenatal exposure of genetic females to DES has some masculinizing effects.[17] Female rhesus monkeys exposed to a long duration of DES treatment prenatally, but not those exposed to a short duration of treatment, initiated more rough play and initiated play with male partners more frequently than untreated females. In these respects their play behavior resembled that of male monkeys. These results suggest that certain aspects of play behavior might be sensitive to prenatal estrogen exposure, or that longer duration of exposure might be influential.

It is not clear if exposure to estrogen or progestin prenatally alters the development of play behavior in males. One study suggested that exposure of boys to the antiandrogenic progestin MPA decreased some, but not all, aspects of male typical play.[43,48] However, a study of boys exposed to a different antiandrogenic progestin, 17-alpha-hydroxyprogesterone caproate (17 aHC), found no evidence of alterations in sex-typical play behavior.[49] A third study, of boys exposed to estrogen and progestin, found some evidence of decreased athleticism in a group of 6-year-olds, but no evidence of changes in other aspects of sex-typical play at this age and no evidence of changes in athleticism in a similar group of hormone-exposed boys at age 16.[50]

B. Cognitive Abilities

Early reports on cognitive function in hormone-exposed patients suggested that prenatal exposure to androgenic hormones enhanced general intelligence. Patients exposed to androgenic progestins and patients with CAH had intelligence quotient (IQ) scores that were significantly higher than the population norm.[51,52] Subsequently, individuals exposed to natural progesterone were reported to be rated by teachers as smarter and to have received more scholastic honors and progressed farther in school compared with matched controls.[53,54] Because natural progesterone should have antiandrogenic activity, these results would appear to be in conflict.

Subsequent research revealed no IQ differences between CAH patients and relative controls, and no difference from predictions based on parental IQ.[55-57] Intellectual attainment in CAH patients has also been reported to resemble that of carefully matched controls.[58] Similarly, studies of children exposed prenatally to estrogen and progestins (androgenic or antiandrogenic action unspecified)[59] and of women exposed to the synthetic estrogen DES[60,61] have found no differences in general intelligence from unexposed relatives. Also, reevaluation of the data suggesting that progesterone enhanced intellectual attainments found little support for the original conclusions,[62] suggesting instead that they resulted from questionable sampling and statistical analyses. Similarly, attempts at replication by reevaluation of some of the original participants, as well as new ones, found no evidence of an association between prenatal progesterone and academic achievement.[63] Thus, it now appears that exposure to high levels of sex hormones does not influence general intelligence. This is consistent with the absence of sex differences in general intelligence.

Although general intelligence appears to be similar for males and females, there are some sex differences in specific aspects of cognitive function. These include male advantages on tasks requiring mental rotation of two- or three-dimensional objects, spatial perception tasks, and mathematical problem solving and female advantages on verbal fluency and perceptual speed and accuracy. Although these differences have sometimes been conceptualized as a male advantage on spatial and mathematical tasks and a female advantage on verbal tasks, this is an overgeneralization. In all three areas there are some tasks that do not show sex differences. (See References 64 through 69 for meta-analyses in these areas.)

Thus the sex differences are specific to subtypes of ability. In addition, they vary in magnitude. The largest is that in three-dimensional mental rotations for which the difference between men and women is 0.92 standard deviation units (or "d" units). Effect sizes for group differences can be classified as large ($d = 0.8$ or greater), moderate ($d =$ about 0.5), or small ($d =$ about 0.2).[70] By using this approach, sex differences in mathematical problem solving, verbal fluency, and perceptual speed are moderate, those in spatial perception are small to moderate, and those in two-dimensional mental rotation are small (see, Reference 71) for further discussion and additional references regarding effect sizes).

The evidence that prenatal hormone levels influence sex differences in cognitive function is equivocal. One study reported enhanced performance on a three-dimensional mental rotations task in 17 CAH girls compared with 13 unaffected sisters and female first cousins.[72] Similar differences between CAH and control girls were also seen on a two-dimensional mental rotations task and on a spatial visualization task which would not normally show a sex difference. There were no differences between female patients and controls on tests of perceptual speed and accuracy or on measures of verbal ability. There also were no significant differences between 8 boys with CAH and 14 unaffected male relatives on any of the cognitive measures. This is the only published study of CAH patients to date that has used the three-dimensional mental rotations task that shows a large sex difference. One other study of 7 girls and 5 boys with CAH compared to 6 unaffected sisters and 4 unaffected brothers reported enhanced performance in CAH girls and impaired performance in CAH boys on a two-dimensional mental rotations task,[73] but a separate study using the same task in a sample of 17 girls and 10 boys with CAH and 11 unaffected sisters and 16 unaffected brothers found no such differences.[55] Studies have also found no differences between CAH patients and controls on the block design subtest of the Wechsler scales, (Wechsler Adult Intelligence Scale, WAIS, Wechsler Intelligence Scale for Children, WISC, and the revised versions of these, WAIS-R and WISC-R) or on the embedded figures test (EFT),[55,74] although one found CAH girls to perform worse than controls on the block design subtest,[58] a result opposite prediction based on the idea that visuospatial abilities generally show a male advantage. However, the block design subtest of the WAIS-R shows only a small sex difference (d = 0.26) and the same subtest of the WISC-R shows a negligible sex difference (d = 0.15). (Some studies may have used unrevised versions of these tests, but sex differences on subtests appear similar in the revised and unrevised versions[75,76]). Similarly, the EFT shows variable sex differences (d = 0.18 overall and for the group EFT, but effect size may be as small as 0.01 for the children's EFT and as large as 0.42 for the individual EFT[64,66]). As a consequence these tests are not ideal markers of hormonal influences.

Most studies of CAH patients have also found no differences on tasks measuring verbal abilities or perceptual speed and accuracy. As noted above, the Resnick et al.[72] study finding enhancement in CAH females on several visuospatial measures found no differences between CAH girls or boys in perceptual speed or accuracy or on verbal measures. Baker and Ehrhardt[55] also reported no differences for CAH males or females on verbal measures. Similarly, two additional studies found no evidence of differences between CAH patients and controls on measures of verbal abilities or perceptual speed and accuracy. One included 15 female and 16 male CAH patients and matched controls.[74] The other included 7 female and 12 male patients compared with matched controls, but in this one study the results were not broken down by sex.[77]

The most consistent finding regarding CAH and cognition has involved impaired computational ability. This has been reported in three studies.[55,58,77] In one study the effect was seen for girls but not boys,[58] in one it was seen in both girls and boys separately[55] and in the third it was seen in the combined group of boys and girls.[77] This effect is puzzling. Although there is a sex difference favoring females in computational ability, it is small and apparent only in young children.[68]

If androgens play a role in the development of sex differences in human cognition, they are unlikely to do so after conversion to estrogen. Two studies of DES-exposed women have found them to be highly similar to their unaffected sisters on verbal and visuospatial tasks that show sex differences as well as on those that do not. The first study compared 25 women who had been exposed for at least 20 weeks prenatally to DES with 25 of their unexposed sisters. There were no differences on a two-dimensional mental rotations task and no differences on a verbal fluency task.[60] The second study included 42 women exposed prenatally to DES and 26 unexposed sisters.[78] The groups did not differ in performance on any of several measures that show sex differences, including a three-dimensional mental rotations task, measures of spatial perception, verbal fluency, and perceptual speed and accuracy. They also did not differ on verbal or spatial tasks that do not show sex differences. A third study of cognition following prenatal exposure to DES availed itself of a sample from a true experiment where women had been administered either DES or placebo during pregnancy to evaluate its efficacy for preventing miscarriage. American College Testing (ACT) scores were obtained for 325 female offspring, 175 exposed to DES and 150 exposed to placebo prenatally. There were no differences between the groups on any of four ACT subtests that show sex differences.[79] The absence of an effect of early estrogen exposure on cognitive development, at least in the area of spatial ability, appears to contrast with the situation in rodents where levels of estrogen during early development influence sex differences in spatial ability during later life.[80]

What of cognitive abilities in other hormone-exposed groups? A syndrome that has been studied extensively is Turner syndrome (TS). This syndrome occurs when the second member of the 23 pairs of chromosomes (the sex chromosomes) is absent or imperfect. Consequences of TS include universal short stature and other more variable outcomes including gonadal failure in the great majority of cases.[81,82] For patients with the XO karyotype this gonadal failure occurs prenatally.[83] It is possible, therefore, that TS patients experience lower than normal levels of ovarian hormones during critical developmental periods. Vocabulary scores are normal in TS females, but a variety of other cognitive abilities, including some that are typically performed better by males, some that are typically performed better by females, and some that are sex neutral, are impaired (see Reference 71). Published data are insufficient to determine if impairment is greater on tasks that show sex differences compared with tasks that do not.

Men with idiopathic hypogonadotropic hypogonadism (IHH) experience gonadal failure because of a deficiency in gonadotropins or hypothalamic releasing factors.[84] They are typically born with normal-appearing male genitalia, assumedly because maternal gonadotropins stimulate androgen production prenatally.[85] However, it is not known if prenatal androgen production is equivalent to that of normal males. In addition, men with IHH would not experience the neonatal elevation in androgen that occurs in normal males. IHH men have been reported to show visuospatial deficits on several tasks including a two-dimensional mental rotations task (the Space Relations subtest of the Differential Aptitude Test) and a measure of spatial perception (the Rod and Frame Test) as well as on tests that are less sensitive to gender, such as the EFT and Block Design, although this last result has not been found consistently.[85-87] One study also found that the severity of the disorder correlated with the degree of visuospatial impairment, that treatment with testosterone in adulthood did not improve performance, and that men who became hypogonadal after puberty did not show impaired visuospatial ability, all of which points to the importance of an androgen deficit during early development.[85] In regard to verbal abilities, IHH men do not differ from controls on tasks that do not show a sex difference, including Wechsler subtests and a vocabulary test.[85,87] However, in one study they showed impairment on the Controlled Associations Test, a measure of verbal fluency.[87]

A group of ten patients with CAIS have been reported to show deficiency on a number of spatial tasks, including Block Design and other performance subtests from the Wechsler scales compared with 26 female and 9 male relatives.[88] However, most of the tests on which impairment is seen show negligible or no sex differences. Similarly, one study of ten boys exposed to the synthetic estrogen DES prenatally found them to have reduced scores on a combination of Wechsler performance subtests compared with their unexposed brothers,[89] tests which again show small to negligible sex differences. The placebo controlled study of DES-exposed offspring found one difference between 172 DES-exposed males and 175 placebo-treated controls. The DES-exposed men scored higher on the Social Science subtest of the ACT. Since males typically score slightly higher than females on this subtest, the effect is in the direction of more masculine-typical performance. However, the effect was not predicted and, since it was the single significant finding from a number of statistical comparisons, the authors attributed it to chance.[79]

V. Summary and Conclusions

Data on cognitive function in patients exposed to atypical hormone environments prenatally is inconclusive. Only the studies of DES-exposed offspring have used sample sizes large enough to be confident of detecting effects if they exist, and these studies are notable in providing no evidence of a hormonal influence. Studies of patient groups have relied on smaller samples. For some syndromes, findings are contradictory. In addition, for these syndromes as well as those producing more consistent findings, it is not clear if cognitive impairments are specific to abilities that show sex differences (or even more pronounced for these abilities compared with those that do not show sex differences). Inconsistencies could result from small sample size and reduced power to detect effects consistently. Alternatively, as noted in the discussion of research on gender differences, there is a tendency to publish results for small samples when they are significant but not when they are not, and this tendency could be operating in studies of hormonal influences as well. Conclusive information about whether or not hormones influence sexual differentiation of human cognitive functions will require larger samples and attention to the magnitude of changes in abilities that show sex differences compared with those that do not.

Data suggesting hormonal contributions to core gender identity, sexual orientation and childhood play behavior, particularly toy choices, suggest that the early hormone environment has consistent influences in these areas. Girls exposed to high levels of androgens prenatally, either because of CAH or because their mothers were prescribed androgenic progestins during pregnancy, show increases in masculine-typical play behavior. Other research suggests that sex differences in play behavior, including toy choices, are also learned, through reinforcement and modeling (see Reference 90 for reviews). Thus, the early hormone environment appears to be one of several types of influences shaping sex differences in these childhood behaviors.

Both XX and XY individuals reared as females but exposed to levels of androgenic hormones that are higher than those experienced by normal females during prenatal and neonatal development show an increased likelihood of symptoms of gender identity disorder. Similarly, women exposed to high levels of either androgens or estrogens during development show increased homosexual or bisexual interest. However, for both sexual orientation and core gender identity, the effect of hormone exposure is not universal. Bisexuality and homosexuality are increased following prenatal exposure, as is the occurrence of symptoms of gender identity disorder. However, the majority of women exposed to high levels of hormones are heterosexual and content in their assigned gender. Thus, as for childhood play behavior, hormones appear to be one of a number of factors shaping these aspects of sexual identity.

References

1. Arnold, A.P. and Gorski, R.A., Gonadal steroid induction of structural sex differences in the central nervous system, *Annu. Rev. Neurosci.*, 7, 413–442, 1984.
2. APA, *Diagnostic and Statistical Manual of Mental Disorders*, 4th ed., American Psychiatric Association, Washington, D.C., 1994.
3. Zucker, K.J., et al., Psychosexual development of women with congenital adrenal hyperplasia, *Horm. Behav.*, 30, 1996.
4. Meyer-Bahlburg, H.F.L. et al., Gender change from female to male in classical congenital adrenal hyperplasia, *Horm. Behav.*, 30, 319–332, 1996.
5. Slijper, F.M.E. et al., Long-term psychological evaluation of intersex children, *Arch. Sexual Behav.*, 27, 125–144, 1998.
6. Zucker, K.J., S.J. Bradley, and Hughes, H.E., Gender dysphoria in a child with true hermaphroditism, *Can. J. Psychiatr.*, 32, 602–609, 1987.
7. Gooren, L. and Cohen-Kettenis, P.T., Development of male gender identity/role and a sexual orientation towards women in a 46, XY subject with an incomplete form of androgen insensitivity syndrome, *Arch. Sexual Behav.*, 20, 459–470, 1991.

8. Money, J. and Ehrhardt, A., *Man and Woman: Boy and Girl*, Johns Hopkins University Press, Baltimore, 1972.

9. Diamond, M. and Sigmundson, H.K., Sex reassignment at birth: long-term review and clinical implications, *Arch. Pediatric Adolescent Med.*, 151, 298–304, 1997.

10. Bradley, S.J., et al., Experiment of nurture: ablatio penis at 2 months, sex reassignment at 7 months and a psychosexual follow-up in young adulthood, *Pediatrics*, 102, 1998, www.pediatrics.org/cgi/content/full/102/1/eg;

11. Dittman, R.W., Kappes, M.E., and Kappes, M.H., Sexual behavior in adolescent and adult females with congenital adrenal hyperplasia, *Psychoneuroendocrinology*, 17, 153–170, 1992.

12. Kuhnle, U. and Bullinger, M., Outcome of congenital adrenal hyperplasia, *Pediatr. Surg. Int.*, 12, 511–515, 1997.

13. Mulaikal, R.M., Migeon, C.J., and Rock, J.A., Fertility rates in female patients with congenital adrenal hyperplasia due to 21-hydroxylase deficiency, *N. Eng. J. Med.*, 316, 178–182, 1987.

14. Döhler, K.-D. et al., Pre- and postnatal influence of testosterone propionate and diethylstilbestrol on differentiation of the sexually dimorphic nucleus of the preoptic area in male and female rats, *Brain Res.*, 302, 291–295, 1984.

15. Hines, M. and Goy, R.W., Estrogens before birth and development of sex-related reproductive traits in the female guinea pig, *Horm. Behav.*, 19, 331–347, 1985.

16. Hines, M. et al., Estrogenic contributions to sexual differentiation in the female guinea pig: influences of diethylstilbestrol and tamoxifen on neural, behavioral and ovarian development, *Horm. Behav.*, 21, 402–417, 1987.

17. Goy, R.W. and Deputte, B.L., The effects of diethylstilbestrol (DES) before birth on the development of masculine behavior in juvenile female rhesus monkeys, *Horm. Behav.*, 30, 379–386, 1996.

18. Ehrhardt, A.A. et al., Sexual orientation after prenatal exposure to exogenous estrogen, *Arch. Sexual Behav.*, 14, 57–77, 1985.

19. Herbst, A.L. and Bern, H.A., Eds., *Developmental Effects of Diethylstilbestrol (DES) in Pregnancy*, Thieme-Stratton, New York, 1981.

20. Kinsey, A., Pomeroy, W., and Martin, C., *Sexual Behavior in the Human Male*, Philadelphia, Saunders, 1948.

21. Kinsey, A., Pomeroy, W., and Martin, C., *Sexual Behavior in the Human Female*, Philadelphia, Saunders, 1953.

22. Meyer-Bahlburg, H.F.L. et al., Prenatal estrogens and the development of homosexual orientation, *Dev. Psychol.*, 31, 12–21, 1995.

23. Masica, D.N., Money, J., and Ehrhardt, A.A., Fetal feminization and female gender identity in the testicular feminizing syndrome of androgen insensitivity, *Arch. Sexual Behav.*, 1, 131–142, 1971.

24. Imperato-McGinley, J. et al., Androgens and the evolution of male-gender identity among male pseudohermaphrodites with 5 alpha reductase deficiency, *N. Eng. J. Med.*, 300, 1233–1237, 1979.

25. Imperato-McGinley, J. et al., Male pseudohermaphroditism secondary to 17-beta-dehydroxysteroid dehydrogenase deficiency: gender role change with puberty, *J. Clin. Endocrinol. Metab.*, 49, 391–395, 1979.

26. Rosler, A. and Kohn, G., Male pseudohermaphroditism due to 17 beta-hydroxysteroid dehydrogenase deficiency: studies on the natural history of the defect and effect of androgens on gender role, *J. Steroid Biochem.*, 19, 663–674, 1983.

27. Herdt, G.H. and Davidson, J., The Sambia "Turnim-Man": sociocultural and clinical aspects of gender formation in male pseudohermaphrodites with 5-alpha-reductase deficiency in Papua New Guinea, *Arch. Sexual Behav.*, 17, 33–56, 1988.

28. Money, J., Gender identity and hermaphroditism, *Science*, 191, 872, 1976.

29. Wilson, J.D., Sex hormones and sexual behavior, *N. Eng. J. Med.*, 300, 1269–1270, 1979.

30. Kester, P. et al., Prenatal "female hormone" administration and psychosexual development in human males, *Psychoneuroendocrinology*, 5, 269–285, 1980.

31. Meyer-Bahlburg, H.F.L. et al., Sexuality in males with a history of prenatal exposure to diethylstilbestrol (DES), in *Psychosexual and Reproductive Issues Affecting Patients with Cancer*, American Cancer Society, New York, 1987.

32. Diamond, M., Llacuna, A., and Wong, C.L., Sex behavior after neonatal progesterone, testosterone, estrogen, or antiandrogens, *Horm. Behav.*, 4, 73–88, 1973.

33. Green, R., *The "Sissy Boy Syndrome" and the Development of Homosexuality*, Yale University Press, New Haven, CT, 1987.

34. Maccoby, E.E. and Jacklin, C.N., *The Psychology of Sex Differences*, Stanford University Press, Stamford, CA, 1974.

35. Ehrhardt, A.A., Epstein, R., and Money, J., Fetal androgens and female gender identity in the early-treated adrenogenital syndrome, *Johns Hopkins Med. J.*, 122, 165–167, 1968.

36. Ehrhardt, A.A. and Baker, S.W., Fetal androgens, human central nervous system differentiation, and behavior sex differences, in *Sex Differences in Behavior*, R.C. Friedman, R.M. Richart, and R.L. van de Wiele, Eds., Wiley, New York, 33–52, 1974.

37. Dittman, R.W. et al., Congenital adrenal hyperplasia II: gender-related behavior and attitudes in female salt-wasting and simple virilizing patients, *Psychoneuroendocrinology*, 15, 421–434, 1990.

38. Dittman, R.W. et al., Congenital Adrenal Hyperplasia I: Gender-related behavior and attitudes in female patients and sisters, *Psychoneuroendocrinology*, 15, 401–420, 1990.

39. Berenbaum, S.A. and Hines, M., Early androgens are related to childhood sex-typed toy preferences, *Psychol. Sci.*, 3, 203–206, 1992.

40. Hines, M. and Kaufman, F.R., Androgen and the development of human sex-typical behavior: rough-and-tumble play and sex of preferred playmates in children with congenital adrenal hyperplasia (CAH), *Child Dev.*, 65, 1042–1053, 1994.

41. Ehrhardt, A.A., Grisanti, G.C., and Meyer-Bahlburg, H.F.L., Prenatal exposure to medroxyprogesterone acetate (MPA) in girls, *Psychoneuroendocrinology*, 2, 391–398, 1977.

42. Ehrhardt, A.A. et al., Sex-dimorphic behavior in childhood subsequent to prenatal exposure to exogenous progestogens and estrogens, *Arch. Sexual Behav.*, 13, 457–477, 1984.

43. Meyer-Bahlburg, H.F.L. et al., Perinatal factors in the development of gender-related play behavior: Sex hormones versus pregnancy complications, *Psychiatry*, 51, 260–271, 1988.

44. Ehrhardt, A.A. et al., The development of gender-related behavior in females following prenatal exposure to diethylstilbestrol (DES), *Horm. Behav.*, 23, 526–541, 1989.

45. Lish, J.D. et al., Gender-related behavior development in females exposed to diethylstilbestrol (DES) in utero: an attempted replication, *J. Am. Acad. Child Adolescent Psychiatr.*, 30, 29–37, 1991.

46. Lish, J.D. et al., Prenatal exposure to diethylstilbestrol (DES): childhood play behavior and adult gender-role behavior in women, *Arch. Sexual Behav.*, 21, 423–441, 1992.

47. Meaney, M.J. and Stewart, J., Neonatal androgens influence the social play of prepubescent rats, Horm. Behav., 15, 197–213, 1981.

48. Meyer-Bahlburg, H.F.L., Grisanti, G.C., and Ehrhardt, A.A., Prenatal effects of sex hormones on human male behavior: medroxyprogesterone acetate (MPA), *Psychoneuroendocrinology,* 2, 383–390, 1977.

49. Kester, P.A., Effects of prenatally administered 17 alpha hydroxyprogesterone caproate on adolescent males, *Arch. Sexual Behav.*, 13, 441–455, 1984.

50. Yalom, I.D., Green, R., and Fisk, N., Prenatal exposure to female hormones: effect on psychosexual development in boys, *Arch. Gen. Psychiatr.*, 28, 554–561, 1973.

51. Ehrhardt, A.A. and Money, J., Progestin-induced hermaphroditism: IQ and psychosexual identity in a study of ten girls, *J. Sex Res.*, 3, 83–100, 1967.

52. Money, J. and Lewis, V., IQ, genetics and accelerated growth: adrenogenital syndrome, *Johns Hopkins Hosp. Bull.*, 118, 365–373, 1966.

53. Dalton, K., Ante-natal progesterone and intelligence, *Br. J. Psychiatr.*, 114, 1377–1382, 1968.

54. Dalton, K., Prenatal progesterone and educational attainments, *Br. J. Psychiatr.*, 129, 438–442, 1976.

55. Baker, S.W. and Ehrhardt, A.A., Prenatal androgen, intelligence and cognitive sex differences, in *Sex Differences in Behavior,* R.C. Friedman, R.N. Richart, and R.L. Vande Wiele, Eds., Wiley, New York, 53–76, 1974.

56. McGuire, L.S. and Omenn, G.S., Congenital adrenal hyperplasia: I. Family studies of IQ, *Behav. Genet.*, 5, 165–173, 1975.

57. Wenzel, U. et al., Intelligence of patients with congenital adrenal hyperplasia due to 21-hydroxylase deficiency, their parents and unaffected siblings, *Helv. Paediatr. Acta*, 33, 11–16, 1978.

58. Perlman, S.M., Cognitive abilities of children with hormone abnormalities: Screening by psychoeducational tests, *J. Learning Disabil.*, 6, 21–29, 1973.

59. Reinisch, J.M. and Karow, W.G., Prenatal exposure to synthetic progestins and estrogens: Effects on human development, *Arch. Sexual Behav.*, 6, 257–288, 1977.

60. Hines, M. and Shipley, C., Prenatal exposure to diethylstilbestrol (DES) and the development of sexually dimorphic cognitive abilities and cerebral lateralization, *Dev. Psychol.*, 20, 81–94, 1984.

61. Hines, M. and Sandberg, E., Cognitive performance in DES-exposed women, unpublished data, 1993.

62. Lynch, A. and Mychalkiw, W., Prenatal progesterone II. Its role in the treatment of pre-eclamptic toxaemia and its effect on the offspring's intelligence: a reappraisal, *Early Hum. Dev.*, 2, 323–339, 1978.

63. Lynch, A., Mychalkiw, W., and Hutt, S.J., Prenatal progesterone I. Its effect on development and on intellectual and academic achievement, *Early Hum. Dev.*, 2, 305–322, 1978.

64. Linn, M.C. and Petersen, A.C., Emergence and characterization of sex differences in spatial ability: a meta-analysis, *Child Dev.*, 56, 1479–1498, 1985.

65. Linn, M.C. and Petersen, A.C., A meta-analysis of gender differences in spatial ability: Implications for mathematics and science achievement, in *The Psychology of Gender: Advances through Meta-analysis*, J.S. Hyde and M.C. Linn, Eds., Johns Hopkins University, Baltimore, MD, 67–101, 1986.

66. Voyer, D., Voyer, S., and Bryden, M.P., Magnitude of sex differences in spatial abilities: a meta-analysis and consideration of critical variables, *Psychol. Bull.*, 117, 250–270, 1995.

67. Halpern, D., *Sex Differences in Cognitive Ability*, Laurence Erlbaum, Hillsdale, NJ, 1992.

68. Hyde, J.S., Fennema, E., and Lamon, S.J., Gender differences in mathematics performance: a meta-analysis, *Psychol. Bull.*, 107, 139–155, 1990.

69. Hyde, J.S. and Linn, M.C., Gender differences in verbal ability: a meta-analysis, *Psychol. Bull.*, 104, 53–69, 1988.

70. Cohen, J., *Statistical Power Analysis for the Behavioral Sciences*, 2nd ed., Lawrence Erlbaum Associates, Hillsdale, NJ, 567, 1988.

71. Collaer, M.L. and Hines, M., Human behavioral sex differences: a role for gonadal hormones during early development? *Psychol. Bull.*, 118, 55–107, 1995.

72. Resnick, S.M. et al., Early hormonal influences on cognitive functioning in congenital adrenal hyperplasia, *Dev. Psychol.*, 22, 191–198, 1986.

73. Hampson, E., Rovet, J.F., and Altmann, D., Spatial reasoning in children with congenital adrenal hyperplasia due to 21-hydroxylase deficiency, *Dev. Neuropsychol.*, 14, 299–320, 1998.

74. McGuire, L.S., Ryan, K.O., and Omenn, G.S., Congenital adrenal hyperplasia II: cognitive and behavioral studies, *Behav. Genet.*, 5, 175–188, 1975.

75. Seashore, H., Wesman, A., and Doppelt, J., The standardization of the Wechsler Intelligence Scale for Children, *J. Consul. Psychol.*, 14, 99–110, 1950.

76. Snow, W.G. and Weinstock, J., Sex differences among non-brain-damaged adults on the Wechsler Adult Intelligence Scales: a review of the literature, *J. Clin. Exp. Neuropsychol.*, 12, 873–886, 1990.

77. Sinforiani, E. et al., Cognitive and neuroradiological findings in congenital adrenal hyperplasia, *Psychoneuroendocrinology*, 19, 55–64, 1994.

78. Hines, M. and Sandberg, E.C., Sexual differentiation of cognitive abilities in women exposed to diethylstilbestrol (DES) prenatally, *Horm. Behav.*, 30, 354–363, 1996.

79. Wilcox, A.J., Maxey, J., and Herbst, A.L., Prenatal hormone exposure and performance on college entrance examinations, *Horm. Behav.*, 24, 433–439, 1992.

80. Williams, C.L. and Meck, W.H., The organizational effects of gonadal steroids on sexually dimorphic spatial ability, *Psychoneuroendocrinology*, 16, 155–176, 1991.

81. Wilson, J.D. and Foster, D.W., *William's Textbook of Endocrinology*, 7th Ed., Saunders, Philadelphia, 1985.

82. Lippe, B., Turner syndrome, *Endocrinol. Metab. Clin. North Am.*, 20, 121–152, 1991.

83. Singh, R.P. and Carr, D.H., The anatomy and histology of XO human embryos and fetuses, *Anat. Rec.*, 155, 369–384, 1996.

84. Grumbach, M.M. and Styne, D.M., Puberty: ontogeny, neuroendocrinology, physiology and disorders, in *Williams Textbook of Endocrinology*, J.D. Wilson and D.W. Foster, Eds., Saunders, Philadelphia, 1992, 1139-1221.

85. Hier, D.B. and Crowley, W.F., Spatial ability in androgen-deficient men, *N. Engl. J. Med.*, 306, 1202–1205, 1982.

86. Buchsbaum, M.S. and Henkin, R.I., Perceptual abnormalities in patients with chromatin negative gonadal dysgenesis and hypogonadotropic hypogonadism, *Int. J. Neurosci.*, 11, 201–209, 1980.

87. Cappa, S.F. et al., Patterns of lateralization and performance levels for verbal and spatial tasks in congenital androgen deficiency, *Behav. Brain Res.*, 31, 177–183, 1988.

88. Imperato-McGinley, J. et al., Cognitive abilities in androgen-insensitive subjects: comparison with control males and females from the same kindred, *Clin. Endocrinol.*, 34, 341–347, 1991.

89. Reinisch, J.M. and Sanders, S.A., Effects of prenatal exposure to diethylstilbestrol (DES) on hemispheric laterality and spatial ability in human males, *Horm. Behav.*, 26, 62–75, 1992.

90. Golombok, S. and Fivush, R., *Gender Development*, Cambridge University Press, Cambridge, U.K.; 1994.

15

Sexual Differentiation of Spatial Functions in Humans

Elizabeth Hampson

CONTENTS

I. Introduction

Although the two sexes do not differ in intelligence, as measured by standard IQ tests, a number of specialized cognitive functions are sexually differentiated in humans. On average, women outperform men on tests of perceptual speed and accuracy, verbal fluency, and certain memory functions. Men outperform women in many spatial functions that require the formation of accurate mental representations of the positions or movements of objects in space. For example, men tend to excel on tasks which involve route-learning, or in which visual objects or parts of objects are mentally transformed in shape or

0-8493-1165-9/00/$0.00+$.50
© 2000 by CRC Press LLC

position, manipulated, rotated, visualized in motion or from an alternate orientation in space. In everyday life, men claim to rely more than women on dynamic mental representations to guide behavior, while women claim to engage in more static mental imagery than men.[1]

Spatial abilities evolved to enable our hominid ancestors to solve spatial problems in the natural environment. Today, they are typically assessed in the laboratory setting using a variety of psychometric tests or synthetic problems. An example item from a test of spatial ability is shown in Figure 15.1. The type of function assessed in this case is called "mental rotation" or "spatial orientation." Factor analytic studies of mental test batteries have long identified spatial orientation as a separable form of spatial ability. In everyday life, mental rotation is required in many mechanical or building-related activities and is involved in recognizing one's surroundings from different vantage points. Mental rotation tests are widely used to assess spatial ability in human research and reliably elicit a male advantage. The size of the sex difference varies, but on mental rotation tests with a high degree of difficulty, average scores for men and women differ by as much as one full standard deviation.

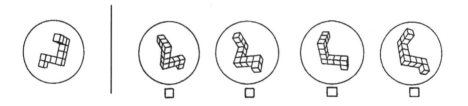

FIGURE 15.1
Example item from a test of mental rotation. The objective is to determine which two of four alternatives represent the same figure as the target figure shown at the left. (From Vandenberg, S.G., *Mental Rotations Test*, 1971. With permission.)

The proximate causes of sex differences in cognition are debated. In the early 1980s, I embarked on a series of studies to investigate whether gonadal steroids are involved in the sexual differentiation of cognitive function. At the time, there was emerging evidence in other species that sex steroids are the basis of a remarkable range of behavioral sex differences, but this was a radical or even heretical idea in humans. The popular wisdom of the day suggested that social and experiential factors, not biology, were the sole causes of cognitive sex differences. A close examination of the literature, however, provided clues that hormones might be involved, at least in spatial functions. For one thing, sex differences in a variety of spatial learning tasks had been demonstrated in nonhuman species, suggesting sexual differentiation in spatial functions is not a uniquely human phenomenon. The male advantage on spatial tasks was observed across a diverse range of human cultures with differing gender expectations and socialization practices. The observation that sex differences in spatial function are not expressed before puberty, or, if present, are weaker than the sex differences seen in mature adults, raised the possibility of an activational component. Finally, spatial proficiency had been noted in earlier research to covary predictably with individual difference variables or biological markers suggestive of a hormonal influence.[2,3] Of course, spatial functions in other species are now known to exhibit hormone sensitivity, but most of this work did not emerge until later (but see Reference 4).

In this chapter, we will review evidence from our own laboratory and elsewhere which supports the neuroendocrine hypothesis. We will also try to place hormonal modulation of spatial abilities within an evolutionary context. In general, evidence increasingly suggests that sex differences in spatial functions have a substantial basis in neuroendocrine events. Effects seen for spatial ability are important beyond these exact functions because they help build support for the more general view that the human central nervous system (CNS) is sexually differentiated and that sex steroids are important in the establishment and expression of sex-dependent neural and behavioral specializations.

II. Effects of Early Life Hormones on Spatial Functions

A difficulty facing human researchers is the inability, except under rare circumstances, to manipulate hormones experimentally in order to observe the effects on some dependent variable of interest. Researchers must rely on clinical conditions in which early life hormones differ from the norm, either being present in excess or in insufficient amounts or in which tissue responsiveness to specific steroids is reduced through a genetic error. Two of the most important sources of evidence for testing the role of early life hormones in sexual differentiation of cognitive function are people with congenital adrenal hyperplasia (CAH) and people born of pregnancies in which the mothers ingested the synthetic estrogen diethylstilbestrol (DES) during gestation.

A. Evidence from Congenital Adrenal Hyperplasia

The classical form of CAH due to 21-hydroxylase deficiency is a rare disorder of adrenal steroid biosynthesis that affects approximately 1 in 15,000 live births. In this condition, the 21-hydroxylase enzyme is deficient in the adrenal cortex as a result of a gene mutation on the short arm of chromosome 6. As a result, males or females with CAH are exposed during gestation to unusually high levels of androgens, beginning in the third month of fetal life. As soon as diagnosis is made, which usually occurs in the immediate newborn period, at least in females, replacement therapy with glucocorticoids and, if necessary, mineralocorticoids is begun. With treatment, steroid concentrations can be normalized and further virilization prevented. In early diagnosed cases who receive effective treatment, the hormonal abnormalities are confined to the prenatal and early neonatal period. An interesting question, therefore, is whether females with CAH show evidence of increased spatial abilities, compatible with their male-like gestational environment. Such an observation would suggest that early androgens are important in the organization of spatial abilities in humans, because postnatal upbringing in girls with CAH is female. Thus, environmental factors do not likely account for any observed differences between girls with CAH and unaffected girls. Partial masculinization of other behavioral traits, including sexual orientation[5] has been reported in females with CAH.

In our own research,[6] we were fortunate to have the opportunity to assess a group of young children with the classical form of CAH (N = 12) and a control group of unaffected siblings (N = 10). We used a standard paper-and-pencil spatial test, Spatial Relations, plus a nonspatial test, Perceptual Speed, taken from the same set of aptitude tests (the *Primary Mental Abilities* battery). The two tests were closely matched on mode of responding and other extraneous features. The results were striking (Figure 15.2). On Spatial Relations, a test of spatial visualization that involves mentally fitting together sets of cutout shapes, girls with CAH scored a full standard deviation above the mean for control girls. In contrast, control girls achieved the higher score on Perceptual Speed. Thus, a double dissociation was found. Perceptual speed is a skill that shows a female superiority in adult samples, so defeminization of perceptual speed in girls with CAH is not implausible.

Confirmation of our findings comes from an earlier study by Resnick et al.[7] who also found better spatial abilities in females with CAH. In the Resnick study, superior spatial scores in the CAH group relative to female controls were found on three different tests of spatial ability including two tests of mental rotation. This suggests the spatial improvement is likely to generalize to other types of spatial measures. Importantly, in both Resnick's study and our own, the effect was selective. No enhancement in the CAH group was found for other types of cognitive functions that were assessed, nor did they differ from controls in general intelligence. In Resnick's study, the effects were seen in adolescents and young adults, indicating the effects are likely to persist at older ages, although, obviously, this less clearly implicates an organizational mechanism than our own work, which was done in prepubertal children. Taken together, the two studies provide strong evidence in favor of an organizational effect of early androgens on spatial functions that ordinarily exhibit a male advantage.

Recent work by Grimshaw et al.[8] in ordinary children who do not have CAH provides convergent evidence for a relationship between spatial ability and the androgen environment *in utero*. Testosterone concentrations at 14 to 20 weeks of gestational age were measured by radioimmunoassay in specimens of amniotic fluid and correlated with the later performance of the offspring of those pregnancies on a mental rotation task at age 7. In girls,

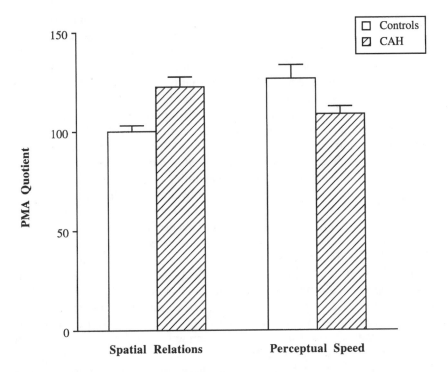

FIGURE 15.2
Girls with CAH outperformed same-sex sibling controls on a test of spatial visualization. The difference was reversed on a test of perceptual speed and accuracy.

Grimshaw et al. found a significant positive correlation ($r = 0.67$) between testosterone levels in second-trimester amniotic fluid and facility in the mental rotation task. In boys, if anything, the opposite pattern was seen ($r = -0.62$). This is consistent with our own CAH findings,[6] in that in our study boys with CAH were discovered to have *lower* spatial visualization scores than control boys. It must be stressed that the Grimshaw data do not necessarily identify Weeks 14 to 20 as the critical period when sexual differentiation of spatial functions occurs. The second trimester is often considered the time when sexual differentiation of the human brain is most likely to occur, and the Grimshaw data are certainly consistent with that prospect. But if a fetus with higher testosterone at Weeks 14 to 20 is also characterized by relatively higher levels of androgens at other points in gestation, the same pattern of correlations with spatial scores could easily emerge at other time points as well. At present, we cannot exclude this possibility.

B. Evidence from Women Exposed to Diethylstilbestrol

An important corollary source of information about the sexual differentiation of spatial functions comes from studies of people exposed to the synthetic estrogen DES during gestation. DES is a nonsteroidal estrogen that was widely prescribed for the prevention of miscarriage from the 1940s to the early 1970s. Spatial abilities along with other sexually differentiated cognitive functions have been investigated in women exposed to DES. As in the CAH work, a provisional hypothesis was that women exposed to DES might develop superior spatial abilities, relative to unexposed controls. The basis for expecting better spatial abilities in women exposed to DES is observations in laboratory rodents that exposure to DES during early development leads to masculinization and defeminization of several behavioral characteristics. Support for the masculinizing potential of DES in humans comes from work showing an increased incidence of bisexuality in women exposed to DES during gestation.[9] It is important to note, however, that DES exerts its activities by binding to estrogen receptors and is capable of exerting masculinizing effects in the CNS only in those neural regions that normally undergo sexual differentiation via conversion of testicular androgens to estradiol.

In terms of spatial abilities, Hines and Shipley[10] failed to find a difference between DES-exposed women and their unexposed sisters on a test requiring mental rotation of simple two-dimensional shapes. In a recent study, Hines and Sandberg[11] used a more extensive set of cognitive measures, including six spatial tests assessing a wide variety of different spatial functions, to investigate cognitive performance in women with a history of DES exposure. No evidence of superior spatial performance was found. DES-exposed women did not differ from their sisters who were not DES-exposed either in spatial ability or any other cognitive function. Scores were closely equivalent in the two groups of women. Both the Hines studies were well designed and well executed, so the failure to find significant differences is not likely due to methodological error. Hines concluded that prenatal DES exposure has little or no effect on women's cognitive development.

C. Conclusion

The hypothesis that there is an organizational effect of early androgens on spatial abilities in humans is supported by evidence from people with CAH and by the recent work by Grimshaw incorporating a direct assay of testosterone in amniotic fluid.[6-8] Females exposed to higher levels of androgens showed better spatial processing than females exposed to lower levels of androgens. So far, the evidence from DES studies suggests that females exposed to DES do not exhibit an enhancement on traditional measures of spatial ability, relative to unexposed female controls. Far from being disappointing in their seeming lack of support for the organizational hypothesis, the DES results are quite important theoretically. The most straightforward interpretation is that spatial functions do not masculinize via the aromatization route. However, Hines and Sandberg[11] noted that women whose DES exposure ended later in gestation scored higher on a spatial composite measure than those whose exposure ended earlier in gestation. This was true even though no group difference between the DES women and controls was found. Thus, an alternative possibility that cannot be completely ruled out at present is that sexual differentiation of brain areas mediating spatial processing does occur under the influence of estrogens, but occurs as a very late gestational event, possibly even extending into the early infant period (see References 12 and 13, for discussion of the possibility of a postnatal critical period). In that case, DES exposure might typically occur too early in gestation to have discernible effects on spatial abilities. Whether the DES data are telling us something about the sensitive period for differentiation of spatial functions or about the molecular endocrine mechanisms that subserve masculinization has yet to be resolved.

III. Reversible Effects of Sex Steroids in Adults

In the past few years, evidence has begun to accumulate suggesting that adult sex steroids might also affect the expression of spatial abilities. These effects are of considerable theoretical interest because they imply that ovarian hormones and, potentially, testicular androgens as well can act as regulators of neural function in brain regions outside the hypothalamic–pituitary area not classically thought to be steroid sensitive in adults.

A. Effects of Estrogens on Spatial Functions

Some of the earliest evidence for this position came from studies of the menstrual cycle, conducted by ourselves and others, in the mid- to late-1980s. Our work was explicitly designed to test the possibility that discernible changes in sexually differentiated cognitive functions might accompany changes in the concentrations of ovarian steroids. In an initial study, we examined a group of healthy young women at two different phases of the menstrual cycle on a battery of motor tests plus the Rod-and-Frame test. The Rod-and-Frame evaluates perceptual accuracy in aligning a rod to the true upright when it is presented against a visual background that is tilted and therefore spatially confusing. The Rod-and-Frame is considered a reliable and valid measure of a type of spatial function involving spatial perception. Males are typically slightly more accurate than females on the Rod-and-Frame, with an average sex difference of approximately 2 to 3° of error per trial. We discovered that healthy women were significantly more accurate on the Rod-and-Frame task during menses, which is characterized by low concentrations of estradiol and progesterone, than during the midluteal peak in estrogen and progesterone secretion.[14] The same women showed a relative facilitation during the luteal testing on several of the motor tasks involving fine coordination of the fingers and hands. Thus, a dissociation in the two categories of tasks was demonstrated. We followed up this work with further studies using a more extensive set of test measures to sample a wider variety of sexually differentiated cognitive functions.

Our first follow-up study assessed healthy women at the midluteal and menstrual phases using a repeated measures design.[15] Order of testing was carefully counterbalanced and, where possible, alternate but equivalent versions of the tests were given on the two occasions. Besides the original tests of motor function, we included multiple measures of several sexually differentiated cognitive abilities — these included functions that show sex differences in favor of females (e.g., verbal fluency, perceptual speed, articulatory speed and accuracy) and functions that show sex differences in favor of males (spatial abilities). The Rod-and-Frame test was supplemented with a conventional paper-and-pencil measure of spatial visualization, Space Relations, which is part of a standard aptitude test battery used in vocational placement and counseling. A test of figural disembedding, the Hidden Figures test, was included because there were hints from other work[16] that this type of spatial function, in which a simple figure must be discriminated when hidden within a more complex visual pattern, may be affected by the menstrual cycle. Thus, our set of spatial measures was diverse, deliberately being chosen to sample more than one type of spatial ability. The results of this study provided further evidence of menstrual cycle variability in spatial function. On initial exposure to the tests, women at the menstrual phase obtained higher scores on the set of spatial measures than women at the midluteal phase. Differences in accuracy between the two phases were small but consistent. Again, motor abilities including verbal articulation were if anything facilitated at higher estrogen and progesterone levels, relative to menses. The study provided modest support for the hypothesis that ovarian hormones can affect spatial functioning, but also had limitations. Notably, because estradiol and progesterone varied in parallel at the two phases of the menstrual cycle we chose to investigate, it was not possible to determine which of the two hormones was most closely associated with the cognitive and motor effects.

To remedy this, a third study was carried out. This time we assessed a new group of women twice: immediately before ovulation, when estradiol concentrations are greatly elevated, and at menses, when estradiol is low, in counterbalanced fashion.[17] The same test battery was used as before. We discovered that women's spatial scores were diminished during the preovulatory estradiol peak, relative to their achievements on the same tests during menses. Phase of cycle was confirmed by serum radioimmunoassays (RIAs). Because progesterone is still low prior to ovulation, this finding suggested that high levels of estradiol alone are sufficient to induce the effect. Importantly, in none of our three menstrual cycle studies were the cognitive effects attributable to concurrent variations in mood state. A commonly used mood inventory sensitive to alterations in mood in both ordinary individuals and psychiatric populations that was given as part of our test battery enabled us to rule out this possibility. Furthermore, among the three spatial measures, scores on two of the three showed significant albeit modest correlations with serum estradiol as quantified by the RIAs (Hidden Figures, Space Relations). The correlation for the Space Relations test was especially interesting because it was differentiated from the other spatial measures by having the form of an inverted U-shaped function.[17] Thus, a distinct relationship to serum estradiol was suggested, both by the effects of phase of cycle and by the patterns of correlations obtained.

An effect of the menstrual cycle on spatial functions has since been replicated by many laboratories (e.g., References 18 through 21), but not all (e.g., Reference 22). For example, using the Vandenberg Mental Rotations test, Silverman and Phillips[21] found strikingly consistent effects of menstrual cycle phase in studies assessing women at the menstrual and nonmenstrual phases of the cycle (Figure 15.3). It may be significant that the Vandenberg test elicits some of the largest and most reliable sex differences in the spatial abilities literature. However, not all studies have found menstrual cycle effects even when the Vandenberg Mental Rotations was used (e.g., Reference 23). In part,

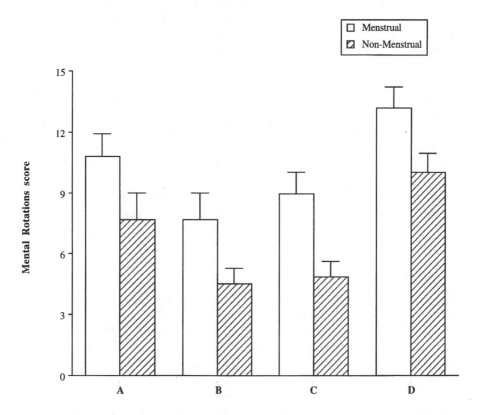

FIGURE 15.3

Several different studies by Silverman and colleagues using Vandenberg's Mental Rotations test have confirmed our earlier finding of better spatial ability in women at menses relative to high-estrogen phases of the menstrual cycle. Means shown in A, B, and C are from Studies 2, 3, and 4 in Silverman and Phillips.[21] D is from Phillips and Silverman.[19]

this reflects the relatively small size of the menstrual cycle fluctuations and their consequent vulnerability to methodological differences. Because women are notoriously unreliable in their verbal reports of their menstrual cycle length and last dates of onset, the validity of group testing is questionable. Nor can women's estimates be taken at face value without objective verification of phase of cycle, either through radioimmunoassays or indirect means (e.g., basal body temperature). Other factors might also be instrumental in determining whether menstrual cycle effects are seen but have not been systematically explored. As one example, in our own studies we typically excluded volunteers who were less than 21 years of age because the incidence of anovulatory cycles is high until women are in their early 20s[24] and because there is evidence that ovarian steroid production does not reach full adult levels until around the same age.[25] Researchers who rely on first year college students for their data may be less likely to detect significant menstrual cycle effects on cognition.

Recent support for an effect of estradiol on the expression of spatial abilities comes from studies using other methodologies. Evidence for an inhibitory influence of very high levels of estrogen or improvements in spatial abilities under reduced estrogen has come from studies of women using oral contraceptives,[21] (but see Reference 26), women who are pregnant,[27] studies of women athletes with amenorrhea,[28] and studies of male-to-female transsexuals in whom exogenous estrogen treatment resulted in diminished scores on a rotated figure test and an increase in verbal fluency.[29] The cognitive effects of intermediate estrogen levels are less clear. Our own menstrual cycle work focused on phases of the cycle where estradiol is maximized or minimized so we have little data to speak to this issue. The fact that we found an inverted U-shaped function relating serum estradiol to women's scores on the Space Relations test suggests that, on at least some types of spatial tests, a drop in spatial scores may not occur until relatively high levels of estradiol are reached.

B. Evidence From Studies of Men

The expression of spatial abilities in adult men might also be regulated by concentrations of gonadal steroids.

As far back as the 1970s, Petersen[3] and others reported that males with more masculinized somatotypes had relatively weaker spatial ability in relation to their verbal scores than males with less masculinized somatotypes. The studies were criticized on the grounds that somatotype is not reliably related to individual differences in androgen concentrations. Later studies employing direct radioimmunoassay measures of testosterone (T) or other androgens reported significant differences in spatial abilities between men with higher and lower circulating T concentrations. Young men with the highest androgens performed more poorly than young men with lower androgens on tests of spatial visualization,[30,31] and on composite scores that averaged across a mixed set of spatial[31] or spatial and mathematical tests.[30] In recent work of our own, using a saliva measure of free T, we found significant negative correlations in young male university students between free T and accuracy on the Vandenberg Mental Rotation test.[32] The fact that effects in all these studies were seen only on spatial tests and not on verbal fluency tests or other cognitive measures suggests that the effects are selective and not due to some generalized deleterious effect of very high androgens.

Negative correlations between T and spatial abilities in men have not always been found. Some studies have found positive correlations. For example, in a recent study of the !Kung San of Namibia, Christiansen[33] found positive correlations between levels of circulating androgens and scores on two spatial tests chosen to be cross-culturally valid. A salient feature of the Namibian sample was their relatively low concentrations of serum and salivary T. This may be relevant to the findings. In another group of low-T men, namely, aging men who had senescent decline in T levels, Janowsky et al.[34] found that double-blind placebo-controlled treatment with T (via the testosterone patch) produced an increase in performance after 12 weeks on a test of spatial function but not on other cognitive measures. In general, studies finding a positive relationship between T and spatial performance have involved populations in which T was relatively low (e.g., females, low T males due to age, illness, or constitutional factors), whereas studies finding a negative relationship involved young adult men in the prime of their lives. One possible way to integrate the seemingly divergent findings is to suggest that there might be a hypothetical optimum level of T for spatial functioning. In general, it might well be the case that increased T promotes better spatial ability, but that if T rises above the theoretical optimum, as it might in a proportion of young adult men, poorer rather than better spatial ability may be the result.

Because of the research designs that were used, most of the studies done so far are not able to differentiate between effects of T on spatial abilities that are due to lifelong exposure to higher vs. lower levels of T and reversible effects due to the immediate hormone environment (i.e., "activational" effects). That is, T has typically been measured at only one time point, providing a snapshot view of the relationship between T and spatial abilities across a set of individuals. Several studies now suggest that not all of the observed correlation is attributable to organizational factors. The recent studies by Janowsky et al.[34] and Van Goozen et al.[29,35] are a case in point. Van Goozen studied female-to-male transsexuals undergoing treatment with testosterone esters preparatory to surgery for sex reassignment. Relative to a control group, treated females showed a greater increase from baseline in scores on a simple mental rotation test and a corresponding decline in verbal fluency, after 12 weeks of hormone treatment. If circulating androgens do have a reversible influence on the expression of spatial abilities, we might also expect to see changes in men's spatial scores with biorhythm-based changes in levels of T secretion. In support of this possibility, we found preliminary evidence that variations in spatial function might accompany diurnal and circannual variations in T. Men tested in the spring scored almost half a standard deviation higher on tests of spatial ability than men tested in the autumn, when circulating T levels were higher.[36] This effect was seen only on spatial tests and not other cognitive measures. In another study, we found that men tested in late morning achieved higher mental rotation scores than men tested in early morning, consistent with the diurnal change in T (Figure 15.4).[32] In both studies, lower T concentrations were associated with better spatial performance, as one might expect in healthy young adult men.

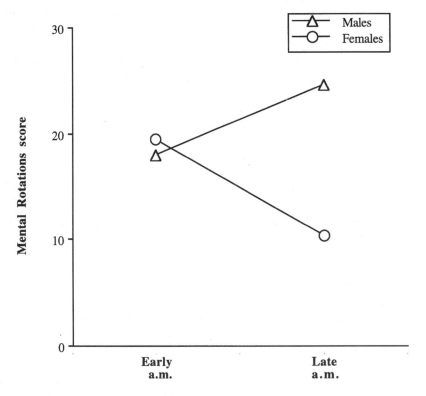

FIGURE 15.4
Diurnal changes in spatial performance in males and females. Males scored higher on the Mental Rotations test in late morning, while females scored lower. In both sexes, the pattern was consistent with the expected diurnal changes in circulating T concentrations. A salivary measure of free T confirmed that T was about 20% lower in late morning in males.

To summarize, there is preliminary evidence from a number of sources that spatial ability may vary with T levels in adult males. Although the observations must be regarded cautiously until further data come to light, so far the data suggest that increases in T when T is fairly low are associated with improvements in spatial performance, whereas increases in T when T is high may serve to diminish spatial performance. The mechanisms underlying these effects are not yet understood. One possibility is that T exerts its effects through conversion to estradiol rather than acting on androgen receptors per se at the neural level. An appealing feature of this possibility is that it would allow us to integrate the findings in males and females to form a unified theory regarding the adult effects of sex steroids on spatial abilities.

IV. Convergent Evidence from Animal Studies

To shed further light on the neuroendocrine interactions that may underlie the effects of sex steroids on spatial functions, we must turn to work in other species.

A. Effects of Perinatal Hormone Manipulations

Experimental studies in the rat have shown that the male advantage normally observed on conventional spatial learning tasks like the Morris water maze (WM) or radial-arm maze (RAM) is powerfully dependent on sex steroids. Early studies involving neonatal hormone manipulations[4,37,38] indicated a possible role for sex hormones in the organization of spatial learning. But these studies were not followed up in detail until nearly 1990. In an elegant set of experiments, Williams et al.[39] showed that exposing female rats to estradiol benzoate (EB) in the first few days of life led to male-typical levels of maze acquisition in adulthood and changes in the relative reliance on landmark vs. geometric cues for navigation in the RAM. Females exposed to neonatal estrogen showed improved performance relative to control females and increased reliance on geometry cues, a change toward the typical male pattern of performance. In later work using a different type of maze, Roof[40] found a dose-dependent improvement in both the RAM and WM in female rats treated neonatally with T propionate (TP), and also found evidence of a reverse effect in males, consistent with the optimal level hypothesis and with our own findings for spatial ability in boys with CAH. The hippocampal formation is often considered to be one important substrate for this form of spatial learning in the rat. It was therefore of interest that Roof and Havens[41] observed morphological changes in the granule cell layer (GCL) of the dentate gyrus in response to neonatal TP. Females treated with TP showed a wider GCL than control females, as is more typical of males. A larger GCL predicted better maze acquisition in females. The Roof and Williams studies implicate the first few neonatal days as a sensitive period for masculinization of spatial function in the rat and suggest that this masculinization occurs via the aromatization route.

Studies of early hormone effects have focused almost exclusively on the neonatal period. Recent evidence, however, raises the possibility of a *prenatal* contribution to sexual dimorphism in spatial function as well. Using a variety of hormonal manipulations from embryonic Day 16 to birth, Isgor and Sengelaub[42] found that TP and dihydrotestosterone propionate (DHTP) had masculinizing effects on WM acquisition in female rats. Flutamide treatment followed by gonadectomy on postnatal Day 1 demasculinized WM acquisition in males. Although females did show some evidence of masculinization by prenatal treatment with EB, EB was not as effective as androgens in altering adult levels of WM performance. As in Roof's work, some effects of prenatal hormones on hippocampal morphology were seen, but in this case changes were localized to pyramidal cells of the CA1 and CA3 subfields of the hippocampus.

B. Effects of Adult Hormone Manipulations on Spatial Learning

Animal researchers have more recently begun to focus on the question of whether adult steroids have discernible effects on spatial learning, and the extent to which these effects model the findings of human studies. Most of the evidence suggests that ovarian hormones do have an influence on spatial learning in females, although a few studies failed to find significant effects (e.g., References 43 and 44). Evidence of T effects in males is also beginning to emerge, including some work supporting a deleterious effect of high levels of T (e.g., Reference 45). Most studies of adult hormones were stimulated either by human work showing changes in spatial proficiency related to the menstrual cycle or by reports from B. McEwen's laboratory of a remarkable degree of plasticity of hippocampal synapses in response to variations in ovarian steroids in the adult rat. Woolley and McEwen[46] discovered that the density of synapses in CA1 declined by as much as 30% over the 24-h period from proestrus to estrus in the female rat, and that the changes are estrogen dependent. This stimulated a burst of studies on hippocampally dependent spatial learning, on the premise that such remarkable plasticity must be functionally significant. Many researchers assumed that impairment on hippocampal learning and memory tasks would be observed in female rats at estrus as a result of decreased synaptic connectivity, but most studies have not found this effect. In fact, the inference itself is questionable. It assumes (1) that greater numbers of hippocampal synapses will necessarily correspond to behavioral improvements and (2) that changes in other neural regions subserving spatial learning either do not occur or are insignificant in driving the behavioral response. Stewart and Kolb[47] found that estrogen deprivation in female rats increased the dendritic arbor of pyramidal neurons in parietal cortex and produced modest increases in apical dendritic spine density in that region.

So far, the bulk of the empirical evidence favors a negative influence of high levels of estradiol on spatial learning, compatible with the human findings. Several studies of naturally cycling female rats have found better spatial learning at lower levels of circulating estradiol.[48,50] In some studies, sex differences in WM acquisition were found only when males were compared with females who were in a high-estrogen state (e.g., proestrus).[48,51] In meadow voles, Galea, Kavaliers, Ossenkopp, and Hampson[51] found better maze acquisition in nonbreeding females than in females housed in breeding pairs. Plasma estradiol levels were positively correlated with latencies to find the hidden platform (i.e., with poorer performance). The vole data are of special interest because voles are induced ovulators. Increased plasma estrogen levels and a prolonged state of behavioral estrus with no detectable cycling is induced in females by pairing them with males. This helps defuse one criticism that has been raised in studies in the rat, where rapid changes in hormones across the estrous cycle make it difficult to make inferences about hormone–behavior relationships. In deer mice, Galea et al.[52] found that male WM acquisition was superior to females during the breeding season only, with female deer mice showing significant decreases in WM performance in the high-estrogen breeding season relative to the low-estrogen nonbreeding season. A few studies using hormone replacement methodologies in ovariectomized rats have reported that EB enhances performance in the RAM (e.g., Reference 53). But Korol et al.[54] found that estradiol treatment impaired acquisition in the WM. A further complication is that there might be a sex difference in the stress response to the WM which may interact in female rats with stage of the estrous cycle.[55] The role glucocorticoid responses might play in spatial learning has not been adequately explored.

One of the unresolved questions about of the effects estrogen on spatial learning and memory is whether the effects are mediated by the classical estradiol receptor, ERα, or some alternative genomic or nongenomic mechanism, perhaps involving the newly identified receptor ERβ. A recent study by Fugger et al.[56] strongly implicates the ERα system. Using a WM measure of spatial learning, Fugger et al.[56] found that acute treatment with EB impaired performance of wild-type female mice in the WM but not transgenic knockout females lacking functional copies of the ERα gene. This result suggests that impaired performance under estradiol on the WM task requires ligand-dependent ERα activation.

To summarize, clear modulatory effects of estradiol levels on a least one form of spatial behavior have been demonstrated in laboratory rodents. On the whole, the evidence favors a negative effect of high levels of estradiol, at least for WM performance, but a few studies, mostly using the RAM, have found a positive influence of estradiol on female maze proficiency. Other studies failed to find any changes at all related to the estrous cycle (e.g., Reference 43). At this preliminary stage of the research, it is not possible to tell whether the discrepancies across studies are due to the use of physiological vs. pharmacological doses of estradiol, chronicity vs. acuteness of the estradiol exposure, misidentification in some studies of estrous phase, the time lag after treatment at which the behavioral effects of EB exposure are measured, the type of maze-learning task that is employed, presence or absence of non-spatial pretraining, strain of rats or species of rodent, or other as yet unknown factors.

An important issue is the validity of comparing human and nonhuman data on spatial function. While the nonhuman data increasingly take shape, and appear to support the inferences being made from human studies, there is still an inferential problem in that the type of spatial ability being assessed in laboratory rodents almost invariably involves spatial navigational learning (e.g., the Morris WM or RAM), whereas spatial abilities in humans are typically assessed using mental rotation or other non-navigational tasks. These tasks are worlds apart. Although different spatial functions in humans likely have somewhat different neurological substrates, reliance on the hippocampal formation has not been demonstrated for tasks such as mental rotation, which, if anything, appear to rely on parietal and, especially, right hemisphere cortical processes.[57] Therefore, the validity of generalizing from rats solving a maze to humans is open to question. Ideally, the same spatial functions could be examined comparatively across species. Since we cannot train a rat to do mental rotation, a reasonable alternative is to train humans to learn mazes instead. In a recent study from my laboratory, we devised a "virtual" maze task that can be administered to humans by computer.[58] Traversing the maze involves learning a route by trial and error over a series of learning trials through a complicated set of alleyways. Only one route through the alleyways leads out of the maze. An exciting outcome of our work is the extremely large size of the male advantage elicited by this task. In terms of either time to completion ($d = 1.6$) or spatial memory errors ($d = 1.4$), the sex differences we found are among the largest ever reported on a spatial task in human beings. We also found a respectably high correlation with scores on the Vandenberg Mental Rotation test, indicating that the two types of tasks, while superficially quite different, do share a significant proportion of common variance.

IV. Evolutionary Significance of the Modulatory Effects

Animal studies suggest that androgen and estrogen levels are the proximate mechanisms responsible in large part for sex differences in spatial behavior. In humans, too, the development and expression of spatial abilities appear to be susceptible to regulation by gonadal steroids. What evolutionary function is served by these modulatory effects?

A number of theories have sought to explain the male advantage in spatial ability in terms of ranging. One of the most influential theories to arise in recent years was proposed by Gaulin and Fitzgerald,[59,60] who hypothesized that the sex difference in spatial abilities is not a universal characteristic among mammalian species, but rather evolved in proportion to navigational demands. A male advantage in spatial ability was hypothesized to evolve only in species where range expansion contributes differentially to the reproductive success of males and females, i.e., where range expansion is a male reproductive tactic. In support of this, Gaulin showed in field studies using radiotelemetry to monitor ranging that a polygynous (*Microtus pennsylvanicus*) but not a monogamous species of vole (*M. ochrogaster* or *M. pinetorum*) showed a sex difference in range size, evident during the breeding season only. In polygynous voles, adult males but not females expanded their home ranges in the breeding season to overlap the home territories of several reproductive females. Because surplus ranging behavior entails energetic and risk costs that must be avoided in the absence of compensatory benefits (e.g., increased mating opportunities), it is not surprising that range expansion occurred in the breeding season only. Correspondingly, only the polygynous species of vole and not the monogamous ones exhibited male superiority on a set of laboratory-administered maze-learning tasks.

In humans, too, certain spatial abilities, notably the ability to form and manipulate mental representations of large-scale three-dimensional space, may have evolved as navigational adaptations. Anthropological data, patterns of wear in leg bones from fossil specimens as far back as the mid-Paleolithic, and other sources of evidence, support the view that we are a species with sexually dimorphic ranging patterns. In a recent review, Sherry and I[61] concluded that of the various evolutionary theories put forth to explain the male advantage in human spatial abilities, sexual selection theories are most viable. These theories postulate greater ranging in males, in search of resources, to achieve status, to compete for mates, or to enhance mating opportunities. One can easily imagine how the organizational effects of steroids, possibly with activation by later hormones, could adapt the male brain for greater ranging. However, an interesting oversight that became apparent in our literature review is that none of the currently proposed evolutionary models predicts the effects of ovarian hormones on spatial abilities that have now been documented in both human and nonhuman species.

Sherry and I proposed a new hypothesis, which we called *Fertility and Parental Care Theory*. We suggested that ranging over long distances, with its associated costs in terms of energy expenditure and heightened risks of predation, may be especially disadvantageous for reproductive females. One important reason for this is that a threshold level of body fat is essential in females for maintaining menstrual cyclicity and optimum fertility and for supporting the caloric demands of lactation. Modern-day empirical data suggest that interference with optimum fertility occurs at surprisingly low levels of physical demands in women. For example, Ellison[25] has shown that ovarian insufficiency begins at moderate levels of exercise in ordinary women. Female athletes, especially those in highly aerobic sports, are prone to menstrual irregularities and oligo- or amenorrhea due to their reduced percentage of body fat. Oligo- and amenorrheic women were clustered among cross-country skiers, long- and middle-distance runners, and rowers in the Stokes and Kimura[28] study cited earlier. Panter-Brick et al.[62] found a relatively high incidence of luteal-phase insufficiency and reduced fertility with seasonal changes in workload among Nepali women. If female foraging or navigation over long distances is disadvantageous, and certain spatial abilities evolved specifically as adaptations to the demands of navigation, then spatial ability might be reduced in females under high estrogen conditions indicative of fertility or parental care investment (e.g., pregnancy) as an adaptation to promote reproductive success. In other words, spatial ability might be reduced in reproductively viable females compared with males because its metabolic costs do not justify its maintenance. If high estrogen levels during the reproductive years serve as a trigger for reduced spatial ability and mobility, then menstrual cycle fluctuations become comprehensible as a by-product of this effect. A similar argument applies to pregnancy where energy balance is also critical, both for optimum development of the fetus and for storing energy for subsequent lactation. At present, almost nothing is known in either humans or nonhuman species about whether spatial ability is in fact suppressed during the high estrogen period of pregnancy.

Sherry and I concluded that the Fertility and Parental Care hypothesis was necessary as an adjunct to traditional views based on sexual selection to account for all the endocrine data. In particular, the inhibitory effects of high estrogen on specific spatial abilities that show a male superiority are not otherwise accounted for by traditional evolutionary theories. We are therefore left with a more complex picture, in which spatial ability may have evolved as an adaptation to ranging but in which endocrine factors in both sexes modulate this function in order to maximize reproductive success.

In closing, I note that most attention in the research literature has been devoted to studying sexually differentiated spatial functions that favor males. Silverman and Eals[63] recently argued that other spatial abilities might have evolved which favor *females* because of greater female involvement in local foraging as part of the sexual division of labor. In fact, Silverman and Eals[63] discovered that women are more accurate than men at remembering the relative locations of static objects within a complex visual scene, a sex difference they suggest may reflect an evolved advantage in the ability to remember the relative positions of food sources or significant landmarks within local arrays of vegetation. This "female" spatial advantage reinforces the view that spatial ability consists of dissociable components that may have quite different evolutionary histories as well as different neurological substrates. Further study of spatial abilities that favor females, and of the possibility that endocrine factors are the proximate causes of these differences as well, is an important direction for future research.

References

1. Harshman, R.A. and Paivio, A., Paradoxical sex differences in self-reported imagery, *Can. J. Psychol.*, 41, 287, 1987.

2. Harshman, R.A., Hampson, E., and Berenbaum, S.A., Individual differences in cognitive abilities and brain organization. Part I: sex and handedness differences in ability, *Can. J. Psychol.*, 37, 144, 1983.

3. Petersen, A.C., Physical androgyny and cognitive functioning in adolescence, *Dev. Psychol.*, 12, 524, 1976.

4. Dawson, J.L.M., Cheung, Y.M., and Lau, R.T.S., Developmental effects of neonatal sex hormones on spatial and activity skills in the white rat, *Biol. Psychol.*, 3, 213, 1975.

5. Zucker, K.J., Bradley, S.J., Oliver, G., Blake, J., Fleming, S., and Hood, J., Psychosexual development of women with congenital adrenal hyperplasia, *Horm. Behav.*, 30, 300, 1996.

6. Hampson, E., Rovet, J.F., and Altmann, D., Spatial reasoning in children with congenital adrenal hyperplasia due to 21-hydroxylase deficiency, *Dev. Neuropsychol.*, 14, 299, 1998.

7. Resnick, S.M., Berenbaum, S.A., Gottesman, I.I., and Bouchard, T.J., Early hormonal influences on cognitive functioning in congenital adrenal hyperplasia, *Dev. Psychol.*, 22, 191, 1986.

8. Grimshaw, G.M., Sitarenios, G., and Finegan, J.K., Mental rotation at 7 years: relations with prenatal testosterone levels and spatial play experiences, *Brain Cogn.*, 29, 85, 1995.

9. Meyer-Bahlburg, H.F.L., Ehrhardt, A.A., Rosen, L.R., Veridiano, N.P., Vann, F.H., and Neuwalder, H.F., Prenatal estrogens and the development of homosexual orientation, *Dev. Psychol.*, 31, 12, 1995.

10. Hines, M. and Shipley, C., Prenatal exposure to diethylstilbestrol (DES) and the development of sexually dimorphic cognitive abilities and cerebral lateralization, *Dev. Psychol.*, 20, 81, 1984.

11. Hines, M. and Sandberg, E.C., Sexual differentiation of cognitive abilities in women exposed to diethylstilbestrol (DES) prenatally, *Horm. Behav.*, 30, 354, 1996.

12. Collaer, M.L. and Hines, M., Human behavioral sex differences: a role for gonadal hormones during early development? *Psychol. Bull.*, 118, 55, 1995.

13. Hampson, E., Spatial cognition in humans: possible modulation by androgens and estrogens, *J. Psychiatr. Neurosci.*, 20, 397, 1995.

14. Hampson, E. and Kimura, D., Reciprocal effects of hormonal fluctuations on human motor and perceptual-spatial skills, *Behav. Neurosci.*, 102, 456, 1988.

15. Hampson, E., Variations in sex-related cognitive abilities across the menstrual cycle, *Brain Cogn.*, 14, 26, 1990.

16. Komnenich, P., Lane, D.M., Dickey, R.P., and Stone, S.C., Gonadal hormones and cognitive performance, *Dev. Psychol.*, 6, 115, 1978.

17. Hampson, E., Estrogen-related variations in human spatial and articulatory-motor skills, *Psychoneuroendocrinology*, 15, 97, 1990.

18. Moody, M.S., Changes in scores on the Mental Rotations Test during the menstrual cycle, *Percept. Mot. Skills*, 84, 955, 1997.

19. Phillips, K. and Silverman, I., Differences in the relationship of menstrual cycle phase to spatial performance on two- and three-dimensional tasks, *Horm. Behav.*, 32, 167, 1997.

20. Saucier, D.M., Kimura, D., Intrapersonal motor but not extrapersonal targeting skill is enhanced during the midluteal phase of the menstrual cycle, *Dev. Neuropsychol.*, 14, 385, 1998.

21. Silverman, I. and Phillips, K., Effects of estrogen changes during the mentrual cycle on spatial performance, *Ethol. Sociobiol.*, 14, 257, 1993.
22. Gordon, H.W. and Lee, P.A., No difference in cognitive performance between phases of the menstrual cycle, *Psychoneuroendocrinology*, 18, 521, 1993.
23. Peters, M., Laeng, B., Latham, K., Jackson, M., Zaiyouna, R., and Richardson, C., A redrawn Vandenberg and Kuse mental rotations test: different versions and factors that affect performance, *Brain Cogn.*, 28, 39, 1995.
24. Metcalf, M.G. and Mackenzie, J.A., Incidence of ovulation in young women, *J. Biosoc. Sci.*, 12, 345, 1980.
25. Ellison, P.T., Measurements of salivary progesterone, *Ann. N.Y. Acad. Sci.*, 694, 161, 1993.
26. Szekely, C., Hampson, E., Carey, D.P., and Goodale, M.A., Oral contraceptive use affects manual praxis but not simple visually guided movements, *Dev. Neuropsychol.*, 14, 399, 1998.
27. Phelps, M.T., Braaksma, D., Masini, L., Wilkie, D.M., and Galea, L.A.M., Fluctuations in cognitive performance in pregnant females, *Brain Behav. Cogn. Sci. Abstr.*, 1998.
28. Stokes, K.A. and Kimura, D., Menstrual Cyclicity and Cognitive Ability Patterns in Athletes, Department of Psychology Research Bulletin 705, University of Western Ontario, London, 1992.
29. Van Goozen, S.H.M., Cohen-Kettenis, P.T., Gooren, L.J.G., Frijda, N.H., and Van De Poll, N.E., Gender differences in behaviour: activating effects of cross-sex hormones, *Psychoneuroendocrinology*, 20, 343, 1995.
30. Gouchie, C. and Kimura, D., The relationship between testosterone levels and cognitive ability patterns, *Psychoneuroendocrinology*, 16, 323, 1991.
31. Shute, V.J., Pellegrino, J.W., Hubert, L., and Reynolds, R.W., The relationship between androgen levels and human spatial abilities, *Bull. Psychon. Soc.*, 21, 465, 1983.
32. Moffat, S.D. and Hampson, E., A curvilinear relationship between testosterone and spatial cognition in humans: possible influence of hand preference, *Psychoneuroendocrinology*, 21, 323, 1996.
33. Christiansen, K., Sex hormone-related variations of cognitive performance in !Kung San hunter-gatherers of Namibia, *Neuropsychobiology*, 27, 97, 1993.
34. Janowsky, J.S., Oviatt, S.K., and Orwoll, E.S., Testosterone influences spatial cognition in older men, *Behav. Neurosci.*, 108, 325, 1994.
35. Van Goozen, S.H.M., Cohen-Kettenis, P.T., Gooren, L.J.G., Frijda, N.H., and Van De Poll, N.E., Activating effects of androgens on cognitive performance: causal evidence in a group of female-to-male transsexuals, *Neuropsychologia*, 32, 1153, 1994.
36. Kimura, D. and Hampson, E., Cognitive pattern in men and women is influenced by fluctuations in sex hormones, *Curr. Dir. Psychol. Sci.*, 3, 57, 1994.
37. Joseph, R., Hess, S., and Birecree, E., Effects of hormone manipulations and exploration on sex differences in maze learning, *Behav. Biol.*, 24, 364, 1978.
38. Stewart, J., Skvarenina, A., and Pottier, J., Effects of neonatal androgens on open-field behavior and maze learning in the prepubescent and adult rat, *Physiol. Behav.*, 14, 291, 1975.
39. Williams, C.L., Barnett, A.M., and Meck, W.H., Organizational effects of early gonadal secretions on sexual differentiation in spatial memory, *Behav. Neurosci.*, 104, 84, 1990.

40. Roof, R.L., Neonatal exogenous testosterone modifies sex difference in radial arm and Morris water maze performance in prepubescent and adult rats, *Behav. Brain Res.*, 53, 1, 1993.

41. Roof, R.L. and Havens, M.D., Testosterone improves maze performance and induces development of a male hippocampus in females, *Brain Res.*, 572, 310, 1992.

42. Isgor, C. and Sengelaub, D.R., Prenatal gonadal steroids affect adult spatial behavior, CA1 and CA3 pyramidal cell morphology in rats, *Horm. Behav.*, 34, 183, 1998.

43. Berry, B., McMahan, R., and Gallagher, M., Spatial learning and memory at defined points of the estrous cycle: effects on performance of a hippocampal-dependent task, *Behav. Neurosc.*, 111, 267, 1997.

44. Luine, V. and Rodriguez, M., Effects of estradiol on radial arm maze performance of young and aged rats, *Behav. Neural Biol.*, 62, 230, 1994.

45. Goudsmit, E., Van De Poll, N.E., and Swaab, D.F., Testosterone fails to reverse spatial memory decline in aged rats and impairs retention in young and middle-aged animals, *Behav. Neural Biol.*, 53, 6, 1990.

46. Woolley, C.S. and McEwen, B.S., Estradiol mediates fluctuation in hippocampal synapse density during the estrous cycle of the adult rat, *J. Neurosci.*, 12, 2549, 1992.

47. Stewart, J. and Kolb, B., Dendritic branching in cortical pyramidal cells in response to ovariectomy in adult female rats: suppression by neonatal exposure to testosterone, *Brain Res.*, 654, 149, 1994.

48. Frye, C.A., Estrus-associated decrements in a water maze task are limited to acquisition, *Physiol. Behav.*, 57, 5, 1995.

49. Warren, S.G. and Juraska, J.M., Spatial and nonspatial learning across the rat estrous cycle, *Behav. Neurosci.*, 111, 259, 1997.

50. Sauve, D., Mazmanian, D., and Woodside, B., Sex differences and the influence of the estrous cycle on activity and spatial memory, *Can. Psychol.*, 31, 360, 1990.

51. Galea, L.A.M., Kavaliers, M., Ossenkopp, K.-P., and Hampson, E., Gonadal hormone levels and spatial learning performance in the Morris water maze in male and female meadow voles, *Microtus Pennsylvanicus, Horm. Behav.*, 29, 106, 1995.

52. Galea, L.A.M., Kavaliers, M., Ossenkopp, K.-P., Innes, D., and Hargreaves, E.L., Sexually dimorphic spatial learning varies seasonally in two populations of deer mice, *Brain Res.*, 635, 18, 1994.

53. Daniel, J.M., Fader, A.J., Spencer, A.L., and Dohanich, G.P., Estrogen enhances performance of female rats during acquisition of a radial arm maze, *Horm. Behav.*, 32, 217, 1997.

54. Korol, D.L., Unik, K., Goosens, K., Crane, C., Gold, P.E., and Foster, T.C., Estrogen effects on spatial performance and hippocampal physiology in female rats, *Soc. Neurosci. Abstr.*, 20, 1436, 1994.

55. Viau, V. and Meaney, M.J., Variations in the hypothalamic-pituitary-adrenal response to stress during the estrous cycle in the rat, *Endocrinology*, 129, 2503, 1991.

56. Fugger, H.N., Cunningham, S.G., Rissman, E.F., and Foster, T.C., Sex differences in the activational effect of ERα on spatial learning, *Horm. Behav.*, 34, 163, 1998.

57. Ditunno, P.L. and Mann, V.A., Right hemisphere specialization for mental rotation in normals and brain damaged subjects, *Cortex*, 26, 177, 1990.

58. Moffat, S.D., Hampson, E., and Hatzipantelis, M., Navigation in a "virtual" maze: sex differences and correlation with psychometric measures of spatial ability in humans, *Evolution Hum. Behav.*, 19, 73, 1998.
59. Gaulin, S.J.C. and FitzGerald, R.W., Sex differences in spatial ability: an evolutionary hypothesis and test, *Am. Nat.*, 127, 74, 1986.
60. Gaulin, S.J.C. and FitzGerald, R.W., Sexual selection for spatial-learning ability, *Anim. Behav.*, 37, 322, 1989.
61. Sherry, D.F. and Hampson, E., Evolution and the hormonal control of sexually-dimorphic spatial abilities in humans, *Trends Cogni. Sci.*, 1, 50, 1997.
62. Panter-Brick, C., Lotstein, D.S., and Ellison, P.T., Seasonality of reproductive function and weight loss in rural Nepali women, *Hum. Reprod.*, 8, 684, 1993.
63. Silverman, I. and Eals, M., Sex differences in spatial abilities: evolutionary theory and data, in *The Adapted Mind*, Barkow, J.H., Cosmides, L., and Tooby, J., Eds., Oxford, New York, 1992, 533.

16

The Luteinizing Hormone–Releasing Hormone System in the Developing Monkey Brain

Ei Terasawa, Laurie A. Abler, and Nancy M. Sherwood

CONTENTS

I. Introduction

When Dr. Matsumoto contacted me regarding publication of a book entitled *Sexual Differentiation of the Brain* in honor of the retirement of Prof. Yasumasa Arai, I responded to him saying that although I am not currently engaged in research on the sexual differentiation of the brain, I would write a chapter

0-8493-1165-9/00/$0.00+$.50
© 2000 by CRC Press LLC

that would be related to Prof. Arai's recent work on the migration of luteinizing hormone–releasing hormone (LHRH) neurons. Perhaps I should explain why we started to study the origin and migration of LHRH neurons in nonhuman primates.

In early 1990, based on articles of Schwanzel-Fukuda and Pfaff,[1] Wray et al.,[2] and Ronnekleiv and Resko,[3] we started to establish a culture system for LHRH neurons. We thought that it would be advantageous to harvest LHRH neurons from the olfactory placode before migration into the brain and culture them.[4] However, because of our unfamiliarity with embryonic materials and the lack of detailed descriptions on the origin and migratory pathway of LHRH neurons in primates, we needed to conduct minimal developmental studies on LHRH neurons. Developmental studies in nonhuman primates, however, require time-mated monkey fetuses by cesarean section, which are very valuable. Thus, we felt that in addition to investigating the origin and the migratory pathway of LHRH neurons, we should study whether other forms of LHRH neurons are present in the primate brain during the early stage of development.

Since the LHRH molecule was first isolated from the pig brain and sequenced,[5] more than a dozen LHRH isoforms have been found in the brain throughout the animal kingdom, and multiple forms of LHRH in the brain of a single species have been reported. Further, based on the amino acid sequence, several scientists have attempted to establish evolutionary relationships for LHRH molecules (see References 6 through 13 for a review).

At this time, we have found that in the developing monkey brain (1) there are two different types of mammalian LHRH neurons that originate at two different developmental stages, originate from probably two different location, and migrate into different areas of the brain; and (2) in addition to mammalian LHRH (mLHRH) neurons, the chicken LHRH-II (cLHRH-II) form is present in the nonhuman primate brain.

II. Two Types of LHRH Neurons in the Forebrain of Monkey Embryos

Using the antibodies GF-6 (supplied by N.M. Sherwood), which is immunoreactive with many forms of LHRH including mammalian and salmon forms, and LR-1 (a gift from R. Benoit, University of Montreal, Canada), which is mainly immunoreactive with the mammalian form, we are able to discern two LHRH cell populations in the brain of monkeys at embryonic Day 36 (E36) and older by comparing adjacent stained sections.[14] One population of cells is immunopositive with both GF-6 and LR-1, GF-6(+)/LR-1(+), whereas the other population of cells is immunopositive with GF-6 but not LR-1, GF-6(+)/LR-1(–); i.e., GF-6 stains both populations, whereas LR-1 stains only one of the populations.[15]

GF-6(+)/LR-1(–) cells first appear in the basal forebrain of embryos at E30, but GF-6(+)/LR-1(+) cells begin to arise in the basal epithelial layers of the medial olfactory pit of embryos at E32. The latter cells migrate along the terminal nerve and appear in the forebrain in embryos at E38. Since the appearance of GF-6(+)/LR-1(–) cells precedes that of GF-6(+)/LR-1(+) cells in embryonic tissues, we have decided to call GF-6(+)/LR-1(–) cells the "early" LHRH population and GF-6(+)/LR-1(+) cells the "late" LHRH cell population.[15]

There are morphological differences between "early" cells and "late" cells. Early LHRH cells are smaller in size (approximately 10×7 μm) and either oval or round in shape, whereas late LHRH cells are larger in size (approximately 15×7 μm) and have a fusiform shape. In addition, the majority of early LHRH cells have a single short neurite, whereas late LHRH cells are bipolar neurons with extensive fiber projections[15] (Figure 16.1).

Late cells are also immunopositive with antisera 1076 and B-6, antibodies targeted to mLHRH, whereas early cells are immunonegative with 1076 and B-6. Similarly, late LHRH cells are weakly immunopositive to the hydroxyproline LHRH antiserum, 939, whereas early LHRH cells are immunonegative with the same antibody. Preabsorption of GF-6 with several different LHRH peptides (except lamprey LHRH-I and -III) blocks immunoreactivity with both LHRH cell populations, whereas preabsorption of LR-1 with mLHRH peptide blocks immunoreactivity with the late cell population.[15]

Late LHRH populations are *bona fide* mLHRH neurons previously described in embryos of the rhesus macaque, mouse, rat, and newt.[1-3,16,17] This conclusion is based on the facts that late LHRH cells (1) are immunopositive to the antimammalian antisera LR-1[1], as well as 1076 and B-6; (2) originate from epithelial cells of the medial olfactory pit and migrate into the brain via the extracranial terminal nerve during early embryonic stages; and (3) complete migration in the septum, preoptic area, and hypothalamus during fetal development (see Section III), similar to the LHRH neuronal migration described by others.[3]

In contrast, early LHRH cells are not exactly mLHRH neurons. They (1) are not immunopositive with the antimammalian antisera LR-1, 1076, or B-6, but are immunopositive to antibody GF-6; (2) are different in shape from late LHRH cells; (3) originate from the olfactory placode prior to the formation of the olfactory pit; (4) migrate into the brain along the olfactory nerve, rather than the terminal nerve; and (5) eventually settle in the striatal and limbic structures of the fetal brain (see Section IV).

III. Distribution Pattern of the Late LHRH Neurons

At E32 to E36, late LHRH neurons differentiate from the olfactory pit and migrate toward the brain. At E34 to 36, late LHRH neurons extend their fibers along the base of the forebrain, reaching into anterior diencephalic areas, and a larger number of 'late' LHRH neurons are seen along the entire extent of the

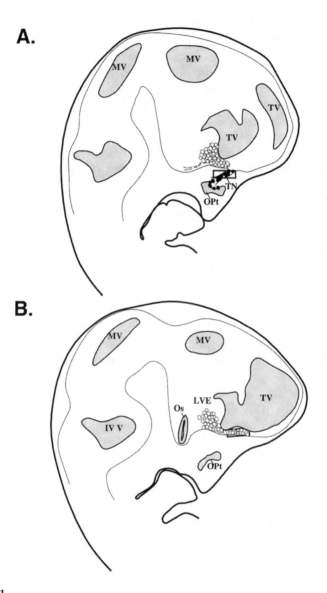

FIGURE 16.1
Schematic drawing of parasagittal sections from an embryo at E36 (A and B). Section A is medial to section B, separated by approximately 100 μm. Medially, a massive number of early LHRH cells (open circles) are found in the base of the forebrain at this age (A). Laterally, early LHRH cells are also found widely from preolfactory areas of the forebrain to more caudal regions, namely, the ventral telencephalon and the lateral ventricular eminence (B). Late LHRH cells (closed circles) migrate along the extracranial terminal nerve, and their fibers (dashed lines) advance caudally within the base of the brain, passing over early LHRH cells (open circles, A). (Modified from Quanbeck et al., *J. Comp. Neurol.*, 380, 293, 1997.)

extracranial terminal nerve. At E38, some late LHRH neurons move into the anterior forebrain, although the majority are still on the extracranial terminal nerve. Between E42 and E52 late LHRH neurons migrate towards their final destinations. In the brain, late LHRH neurons follow one of two (ventral or dorsal periventricular) migratory pathways. The majority of late LHRH neurons take a ventral migratory pathway: Shortly after entering the brain, late LHRH cells disperse the aggregate that they had maintained when migrating on the extracranial terminal nerve and appear to start extending their axons as they migrate individually toward their final destinations. However, most cells seem to maintain an indirect contact with LHRH fibers oriented toward the diencephalon. The minority of late LHRH neurons take the dorsal periventricular pathway: after entering the brain they also lose cell-to-cell contact, but keep migrating dorsally along the wall of the telencephalic vesicles and subsequently toward the third ventricle.

Migration of late LHRH neurons continues into late fetal development, although the basic distribution pattern is established as early as E52. Late LHRH neurons are found in the olfactory bulb, olfactory tubercle, pericommissural areas, the diagonal band of Broca, medial preoptic area, medial septal nucleus, regions around the lamina, suprachiasmatic area, lateral hypothalamus, medial basal hypothalamus, periventricular region of the third ventricle, the region ventrolateral to the arcuate nucleus, and median eminence. At the late fetal stage, densely stained late LHRH nerve fibers are present in the median eminence, and less dense but heavily stained fibers are seen in the organum vasculosum of the lamina terminalis. A detailed description of the migratory pattern and distribution pattern of late LHRH neurons has been reported previously.[14] The final distribution of LHRH neurons in the late fetal age is similar to that described for fetal, juvenile, and adult monkeys.[18-20]

IV. Origin and Distribution of the Early LHRH Neurons

The origin of early LHRH cells is unclear at this time. Although at E30 several neuroblastic cells, weakly stained with GF-6, are present along olfactory nerves, a large number of darkly stained early LHRH neurons with a mature appearance have already appeared in the area of the rostrolateral forebrain. At E32 to E36, increasing numbers of early LHRH cells are distributed throughout wide regions of the ventral forebrain below the telencephalic vesicle.[15] It is possible that a small number of undifferentiated early LHRH cells originate from the olfactory pit and migrate into the forebrain, after which they proliferate for several days, eventually generating thousands of early LHRH cells. Undifferentiated neurons dividing after the onset of migration are seen during neural crest ontogeny.[21] However, it is more likely that the site of early LHRH neurogenesis is the ventricular zone of the telencephalon, since the number of neuroblastic cells observed in the olfactory areas is far fewer than in the ventral forebrain. A recent report by Daikoku and Koide[22] suggests that olfactory placode ablation in rats does not change the number of LHRH cells in the septal region, although it is unclear whether the population of LHRH cells described is equivalent to the early cells in the monkeys. Cell lineage tracer studies with placode ablation will answer the question regarding the origin of early LHRH cells.

At E38 and E42, thousands of early LHRH neurons are distributed medially across a wide area of the basal telencephalon, from septal areas, rostrally, to the preoptic sulcus, caudally. During this developmental stage, unlike the shape and distribution of late LHRH cells, early LHRH cells are small and oval in shape with short, unipolar processes and do not exhibit a morphology indicative of cell migration. By E51, early LHRH cells are scattered widely throughout areas of the telencephalon containing the basal nuclei, such as the lateral septum, stria terminalis, amygdala, internal capsule, putamen, globus pallidus, and claustrum of the fetal brain. Distribution of early LHRH cells in the E77 fetus is similar to that seen in the E51 to E62 fetuses.

If early cells are not moving themselves, how do early LHRH cells reach these locations? It is plausible, at least in part, that early LHRH cells move into these positions along with the development of the forebrain. For example, at E36, the medial and lateral ventricular eminences are the structures comprising the ventral forebrain where early LHRH cells are located. As development proceeds, the two eminences fuse to form one ventricular eminence, which then gives rise to the striatum and amygdala.[23,24] Most of these events take place between E42 and E51 in the rhesus macaque embryo.[25,26]

Additional morphological differentiation of early LHRH neurons occurs along with the formation of the mature telencephalic structures. The appearance of early LHRH cells seen in limbic and striatal structures in the brain of E50 and older fetuses differs from that of early LHRH cells seen at E42 and younger: Cells in striatal structures are large, unipolar, and round in shape with an extensive nucleus surrounded by a thin rim of weakly stained cytoplasm, whereas cells in the limbic system are small, bipolar, and fusiform in shape. Morphological differentiation during late development is well established in the cerebral cortex: e.g., final morphological appearance of neurons migrating from the ventricular zone in an undifferentiated state depends on intercellular interactions with neighboring cells and the surrounding extracellular matrix.[27] Similarly, in the peripheral nervous system after completion of migration, the neural crest cells undergo further morphological differentiation under the influence of growth factors, such as nerve growth factor (NGF).[28]

V. The mLHRH Gene in Early and Late LHRH Neurons

What is the form of the LHRH molecule in early cells? First, it is not likely to be the mLHRH form with its full ten amino acids, since early LHRH cells are not stained with the antibodies LR-1, 1076, or B-6. Further, it is not a post-transcriptionally modified LHRH form, such as hydroxyproline[9]. Hydroxyproline[9] LHRH is found in many species of animals, especially in the fetal brain in rats,[29] and the presence of a relatively higher amount of hydroxyproline[9] LHRH is reported in the hippocampus than in the hypothalamus of adult rats.[30] Although we find early LHRH cells in extrahypothalamic regions in our studies, the observation that early LHRH cells, but not late LHRH cells, are immunonegative with the hydroxyproline[9] LHRH antiserum, 939, suggests that early cells do not contain hydroxyproline[9] LHRH[15]. Second, it does not appear to be chicken-I, chicken-II, salmon, lamprey-I, or lamprey-III LHRH forms, since specific antibodies to those molecules are immunonegative with early LHRH cells. Third, it may not be a form of sea bream LHRH, in which only Arginine[8] is replaced by Serine[8].

In situ hybridization histochemistry with monkey and rat mLHRH cRNA riboprobes suggests that the distribution of late cells expressing mLHRH mRNA (Figure 16.2a) is essentially identical to those stained with GF-6

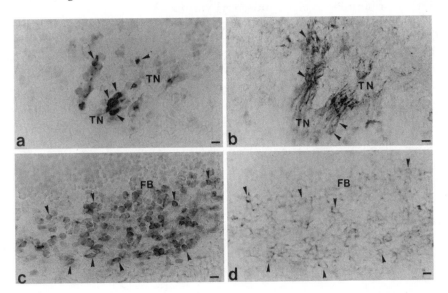

FIGURE 16.2

In situ hybridization for mammalian LHRH mRNA and immunocytochemistry of early and late LHRH neurons using GF-6 in the fetal monkey at E35 (a and b, sagittal sections). Immunopositive cells are visualized using diaminobenzidine, whereas *in situ* hybridization signals are visualized using digoxigenin-11-UTP and alkaline phosphatase. In the fetal head at E35 (a and b), late LHRH neurons (arrowheads) are seen on the terminal nerve (TN), with *in situ* hybridization (a) and immunocytochemistry (b) in adjacent tissue sections. Similarly, early LHRH neurons (arrowheads) are also seen in the ventral forebrain (FB), with both *in situ* hybridization (c) and immunocytochemistry (d) in adjacent sections. Cell size with *in situ* hybridization appears to be smaller, since *in situ* hybridization shrinks tissue sections more than those stained for immunocytochemistry. Approximate locations of a and b are indicated by a square box in Figure 16.1A, and approximate locations of c and d are indicated by a square box in Figure 16.1B. Scale bars: for all = 20 μm.

(Figure 16.2b). The results clearly indicate that late LHRH cells originating from the olfactory pit express mLHRH mRNA as expected. This observation is in agreement with those described by Wray et al.[31] and Ronnekleiv and Resko.[3] Early LHRH cells in the basal forebrain of young fetuses (E35 to E39) and in the striatum and amygdala in older fetuses (E50 to E78) also express mLHRH mRNA. The distribution of mLHRH mRNA positive early cells (Figure 16.2c) in the forebrain is very similar to that seen with GF-6 (Figure 16.2d). However, the expression of mLHRH mRNA in early cells is consistently less than that seen in late cells, suggesting that early cells contain low transcription levels of the mLHRH gene. A double-labeling study with GF-6 and antisense digoxigenin-labeled riboprobes suggest that both late and early LHRH neurons express mLHRH mRNA. In contrast, *in situ* hybridization using sense riboprobe does not hybridize with mLHRH mRNA. In addition, we have confirmed the specificity of the antisense riboprobe hybridization with early and late cells using a "random" digoxigenin-labeled riboprobe, as well as prehybridizing the sections with an excess of unlabeled antisense riboprobe to block antisense hybridization.

The presence of a second population of cells containing the mLHRH gene transcripts in the human brain has been also described by Rance et al.[32] The distribution pattern of these LHRH cells in the adult human brain[32] is strikingly similar to that seen in the rhesus monkey brain after E50, as shown in this and previous studies.[15] Interestingly, the presence of a second population of cells containing the mLHRH gene does not appear to be limited to primates. Two different groups report a second population of neurons transcribing the mLHRH gene in the basal forebrain in fetuses and the septal region in adults using transgenic mice with an mLHRH promoter-driven β-galactosidase reporter gene.[33,34]

VI. LHRH Fragments and the Cleavage Enzyme E.C.3.4.24.15 in Early LHRH Neurons

If early LHRH neurons contain mLHRH mRNA, why are they not stained with LR-1? Previously, we speculated that an N-terminal peptide might be present in early LHRH cells, since GF-6 is N-terminally directed, based on its cross-reactivities with many forms of the LHRH molecule,[15] and on the absence of immunoreactivity with other antisera in early cells. In fact, this could be a product derived from proteolytic cleavage of LHRH, such as LHRH$^{1-5}$ and LHRH$^{4-10}$. To determine the presence of LHRH fragments in the early and late LHRH neurons, the distribution pattern of early and late LHRH cells stained with GF-6 is compared with the distribution pattern stained with antisera raised against LHRH$^{1-5}$ or LHRH$^{4-10}$ (both are gifts of L. Jennes).

The appearance, as well as the distribution, of late LHRH$^{1-5}$ positive neurons at E38 is very similar to late LHRH neurons stained with GF-6. However, late LHRH$^{1-5}$ positive neurons decrease their staining intensity in older fetuses (E50 to E78); i.e., they are immunostained very faintly, or often not at all. In contrast, early LHRH neurons are invariably immunopositive with LHRH$^{1-5}$ in both young and older fetuses. LHRH$^{4-10}$ positive neurons, similar to the GF-6 positive late LHRH neurons, are consistently observed in all fetuses examined; i.e., late LHRH cells arising from the olfactory pit/placode and on the terminal nerve of young (E35 to E39) fetuses are immunopositive with the antibody to LHRH$^{4-10}$. In contrast, staining of early LHRH$^{4-10}$ positive neurons in the base of the telencephalon of younger (E35 and E36) fetuses is either absent or very faint. In older (E50 and older) fetuses, early LHRH$^{4-10}$ positive neurons are distributed in the same regions where GF-6 positive early neurons are found.[35]

Further, preabsorption tests for GF-6 with either the LHRH$^{1-5}$ peptide or LHRH$^{1-10}$ peptide indicate that both early and late LHRH cells are immunonegative (Table 16.1). In contrast, preabsorption of GF-6 with LHRH$^{4-10}$ does

TABLE 16.1

Immunocytochemistry of Antibodies Preabsorbed with Various Peptides

Antibodies	Peptide for Absorption	Late Cells	Early Cells
GF-6	Control	Positive	Positive
	LHRH$^{1-5}$	Negative	Negative
	LHRH$^{1-10}$	Negative	Negative
	LHRH$^{1-3}$	Positive	Positive
	LHRH$^{4-10}$	Positive	Positive
	LHRH$^{5-10}$	Positive	Positive
	Guinea pig LHRH	Negative	Negative
LHRH$^{1-5}$	Control	Weakly positive	Positive
	LHRH$^{1-5}$	Negative	Negative
	LHRH$^{4-10}$	Weakly positive	positive
	LHRH$^{1-10}$	Weakly positive	Positive
	LHRH$^{5-10}$	Weakly positive	Positive
LHRH$^{4-10}$	Control	Positive	Weakly positive
	LHRH$^{4-10}$	Negative	Negative
	LHRH$^{1-10}$	Negative	Weakly positive
	LHRH$^{1-5}$	Positive	Weakly positive
	LHRH$^{5-10}$	Weakly positive	Weakly positive
LR-1	Control	Positive	Negative

not abolish the immunopositive staining of both early and late LHRH neurons. Similarly, preabsorption of GF-6 with either LHRH$^{5-10}$ or LHRH$^{1-3}$ does not abolish the immunopositive staining of both early and late LHRH neurons (Table 16.1). In addition, LHRH$^{1-5}$ immunopositive early cells are blocked by preabsorption of the antibody with the LHRH$^{1-5}$ peptide, but not with any of the other peptides examined, i.e., LHRH$^{1-10}$, LHRH$^{4-10}$, or LHRH$^{5-10}$ (Table 16.1), indicating that early LHRH neurons contain LHRH$^{1-5}$ peptide or a molecule similar to LHRH$^{1-5}$.[35]

If early LHRH neurons contain LHRH$^{1-5}$, they may also contain metalloendopeptidase E.C.3.4.24.15 (EP24.15) peptide, an enzyme that cleaves LHRH at the Try^5-Gly^6 position. To examine this possibility, adjacent sections that immunostained for LHRH$^{1-5}$ and LHRH$^{4-10}$ are exposed to an affinity-purified rabbit polyclonal antibody against EP24.15.[36] Specificity is also tested by exposing the tissues to the EP24.15 antibody after preabsorption with the EP24.15 peptide. By using single staining, late EP24.15 positive cells in fetuses at E35 and E38 are seen on the terminal nerve. However, late LHRH-type neurons, single-stained with the EP24.15 antibody in the older fetus brains (E50 to E78), are difficult to detect, since they are faintly stained, and the cells are mostly scattered. Early EP24.15 positive cells are observed in the basal forebrain of fetuses at E35 and E36 and are consistently seen in the brain of older fetuses at E50 to E78. Immunofluorescein double-labeling indicates that early GF-6 positive neurons are also LHRH$^{1-5}$ positive and EP24.15 positive,[35] and late GF-6 positive cells were faintly immunopositive with EP24.15. Moreover, both early and late LHRH$^{1-5}$ positive neurons are also EP24.15 positive. Therefore, immunopositive staining of early cells with the endopeptidase EP24.15 antibody suggests that early cells also contain the cleavage enzyme for the bond between Tyr^5-Gly^6 position of LHRH$^{1-10}$,[37] and therefore it is likely that early cells contain the LHRH$^{1-5}$ and LHRH$^{6-10}$-like peptides.

VII. Possible Functions of LHRH Fragments and EP24.15

What is the role of neuropeptide breakdown products in modulating neural function? The fact that early LHRH cells express much higher levels of LHRH$^{1-5}$ than late cells, shown by immunocytochemistry and confocal imaging, is indicative of a possible function of early LHRH independent from late LHRH neurons. The existence of LHRH$^{1-5}$ in early LHRH cells in the extrahypothalamic region further suggests that they are not involved in control of LH/FSH release, since the medial basal hypothalamus is the only part of the brain responsible for the maintenance of pulsatile LHRH release and ovulatory LHRH release in primates.[38] A recent report suggests that the LHRH fragment, LHRH$^{1-5}$, alters LHRH release, since LHRH$^{1-5}$, but not LHRH$^{2-10}$, suppresses pulsatile release of LHRH and N-methyl-d-aspartate (NMDA)-induced LHRH release.[39] These authors propose a hypothesis that LHRH$^{1-5}$ is an endogenous antagonist of NMDA receptors. This hypothesis is partly supported by a preliminary report by Moorjani et al.[40] showing that LHRH$^{1-5}$ competes with NMDA binding in membrane preparations from the rat hypothalamus and cortex using a ^{3}H-L-glutamate displacement assay. Therefore, it is possible that LHRH[1-5] is a neurotransmitter in early LHRH neurons. Additional functions of LHRH[1-5] in the fetal brain remain to be investigated.

It has been reported that mLHRH neurons in the adult rat hypothalamus express EP 24.15; i.e., EP24.15 is found in the perivascular space of the median eminence and is secreted into the portal circulation.[41] These authors further show that a specific inhibitor of this enzyme increases the amplitude of the steroid-induced LH surge in ovariectomized rats. In sheep, however, EP24.15 activity does not fluctuate throughout the estrous cycle, and intraventricular infusion of an EP24.15 inhibitor does not block the LHRH release pattern.[42] Nonetheless, the results of Wu et al.[41] indicate a possible neuroendocrine role for EP24.15 in late LHRH neurons in the adult hypothalamus. In addition, it is possible that EP24.15 in late LHRH neurons plays a role in the degradation of other peptidergic inputs on LHRH neurons, since this enzyme also cleaves neuropeptides, such as angiotensin, neurotensin, dynorphin A, and bradykinin.[43] A functional role of EP24.15 in late LHRH neurons in the embryonic brain remains unknown.

VIII. The cLHRH-II Form in the Monkey Brain

cLHRH-II is the most ubiquitous second form of LHRH in the animal kingdom and is not believed to be crucial for the control of gonadotropin secretion.[6,9,13,44,45] The presence of cLHRH-II has been reported in many species such as marsupials,[46] musk shrews,[47] rodents,[48] and humans.[49] The complementary (c) DNA or gene that encodes cLHRH-II has been reported in several fish (see Reference 50), tree shrews,[51] and humans.[49] In contrast to the distribution pattern of mLHRH in the anterior portion of the brain, cLHRH-II cells are consistently found in the posterior portion of the brain, namely, midbrain, of most species.

Early in this series of studies, we started to collaborate with Nancy M. Sherwood. The initial aim was to determine the molecular structure of the early LHRH cells in the monkey fetal forebrain. However, instead, she and her graduate student, David Lescheid, isolated the cLHRH-II peptide by HPLC from the adult and fetal monkey brain.[52] This was a big surprise, because when we were focusing on the forebrain area of monkey fetuses, we did not detect cLHRH-II positive cells with three polyclonal antibodies. Subsequently, we went back to reexamine immunocytochemically stained tissue sections from various developmental stages and found that there were cLHRH-II positive cells in the posterior hypothalamic region of an E34 embryo and in the periaqueductal region of older fetuses.

cLHRH-II positive neurons are round in shape with fine neurites and generally smaller than mLHRH neurons, which are fusiform in shape with thick neurites. They are most commonly distributed in the periventricular region of the posterior portion of the third ventricle to the periaqueductal region of the midbrain, but a small number of cLHRH-II perikarya and fibers are present in the pituitary stalk.[52]

The origin of cLHRH-II cells is currently unknown. However, studies of olfactory ablation suggest that these cells do not originate from the olfactory placode.[23,53] It has been suggested that cLHRH-II neurons originate from the ventricular ependymal cells.[54,55] In primates as well, cLHRH-II neurons appear to originate from precursor cells in the periventricular zone of the third ventricle and aqueduct, because immunopositive cLHRH-II cells in the area most proximal to the ventricle are more round than those away from the ventricle, and the shape becomes more irregular as the cells move distally from the ventricle (E. Terasawa, unpublished observation).

In order to determine the cDNA of cLHRH-II, we carried out RT-PCR on adult and fetal (E85) rhesus monkey brainstem mRNA[56] with primers based on a portion of the upstream signal peptide and the downstream processing region of human and tree shrew cLHRH-II cDNA.[49,51] The resulting PCR product was cloned and sequenced.

The cDNA sequence encoding the cLHRH-II decapeptide in the monkey midbrain is identical to that reported in humans and only differs by one nucleotide substitution from that reported in the tree shrew. However, the cDNA region for the signal peptide in the rhesus monkey midbrain shows 90% sequence identity to that in humans, but only 81% sequence identity to that in tree shrews. The corresponding amino acids, deduced from the signal peptide cDNA of the rhesus monkey, show 88% sequence identity to that in humans and 75% sequence identity to that in tree shrews (Table 16.2). Further, the cDNA region for gonadotropin-releasing hormone associated peptide (GAP) in the monkey brain shows 81% sequence identity to that in humans and 77% sequence identity to that in tree shrews. The corresponding amino acid sequence identity of the GAP region is 70 and 64%, respectively (Table 16.2). From the perspective of cLHRH-II in humans, chicken preproLHRH-II in the rhesus monkey is more closely related to the human sequence than the tree shrew sequence. These data suggest that cDNA and amino acid sequences of the cLHRH-II decapeptide are well conserved, but the sequences of the signal peptide and the GAP region have been modified during evolution. Nonetheless, cLHRH-II is present in the midbrain in the rhesus monkey and this form of LHRH is well conserved throughout vertebrate evolution.

At this point there is little information on the function of cLHRH-II. In a previous study, we have shown that cLHRH-II stimulated LH release *in vivo* in rhesus monkeys.[52] Although there are cLHRH-II fibers with a small number of perikarya present in the basal hypothalamus and the pituitary stalk, the fiber density and staining intensity of mLHRH neurons far exceed those of cLHRH-II. Further, the amount of cLHRH-II in the monkey hypothalamus detected by HPLC is much smaller than that of mLHRH.[52] Therefore, cLHRH-II may not play a major role in stimulating pituitary gonadotropins.

TABLE 16.2

Comparison of the Chicken LHRH-II Amino Acid Sequence Among Human, Rhesus Monkey, and Tree Shrew

```
                        -20          -10           +1           10           20
          Human: MASSRRG--LLLLLLLTAHLGPSEAQHWSHGWYPGGKRALSSAQD

  Rhesus Monkey: MASSRRG-LLLLLMLLTAHPGPSEAQHWSHGWYPGGKRALSSAQD

     Tree Shrew: MASSMLGFLLLLLLLMAAHPGPSEAQHWSHGWYPGGKRASNSPQD

                         30           40           50           60
          Human: PQNALRPPGRALDTAAGSPVQTAHGLPSHALAPLDDSMPWEGRTT

  Rhesus Monkey: PQNALRPPAGSPA-------QATYGLPSDALAHLEDSMPWEGRTT

     Tree Shrew: PQSALRPPAPSAA-------QTAHSFRSAALASPEDSVPWEGRTT

                         70           80           90
          Human: AQWSLHRKRHLARTLLTAAREPRPA

  Rhesus Monkey: AWWSLRRKRYLAQTILTAAREPRPA

     Tree Shrew: AGWSLRRKQHLMRTLLSAAGAPRPA
```

IX. Examination of the Guinea Pig Form of LHRH in the Monkey Brain

Recently, two transcripts, one which encodes a unique form of the decapeptide and another which encodes mLHRH, have been described in the guinea pig brain.[57] Since we have speculated in a previous study that the primary binding site of GF-6 is the N-terminus amino acids, especially LHRH$^{1-5}$, we have tested the cross-reactivity of GF-6 with guinea pig LHRH peptide and examined the immunoreactivity of GF-6 after preabsorption with the guinea pig LHRH peptide. Although guinea pig LHRH differs at the Tyr^2 and Val^7 amino acid positions from His^2 and Leu^7 in mLHRH$^{1-10}$, respectively, cross-reactivity of GF-6 with guinea pig LHRH peptide was equal to that of mLHRH$^{1-10}$ (N.M. Sherwood, unpublished observation). Further, guinea pig LHRH peptide canceled GF-6 immunopositive staining in both late and early cells (see Table 16.1). The His^2 to Tyr^2 change is especially interesting. In all other forms of LHRH discovered to date, His is at the second position. Apparently, this amino acid change suggests that histidine is not critical for binding of GF-6. Nonetheless, preliminary data suggest that the monkey brain does not have guinea pig LHRH; i.e., the cross-reactivity of GF-6 with guinea pig LHRH may not be significant, as brain extracts from an adult or fetal (E85) rhesus monkey did not have GF-6 immunoreactive peaks in the

HPLC fractions where guinea pig LHRH elutes (N.M. Sherwood, unpublished observation).

X. Conclusion

In this chapter we have described the observations in the fetal monkey brain that (1) there are two different types of mLHRH neurons which have the same gene, but originate at two different developmental stages and probably at two different locations and migrate into different areas of the brain, and (2) in addition, the cLHRH-II form is present in the non-human primate brain. The distribution pattern of early LHRH and cLHRH neurons in the extrahypothalamic region clearly suggests that they are not involved in control of gonadotropin release per se, and rather, that they may play a role as peptide neurotransmitters.

Acknowledgments

The authors thank Drs. Robert Benoit (University of Montreal, Montreal, Canada), Mark J. Glucksman and James L. Roberts (Mount Sinai Medical School, New York), and Lother Jennes (University of Kentucky, Lexington, KT) for their generous supplies of antibodies listed in Table 16.1; and Drs. Robert P. Millar (University of Cape Town, Cape Town, South Africa) and Thomas E. Adams (University of California, Davis) for their generous supplies of chicken LHRH-II and hydroproline[9] LHRH antibodies. The authors also thank Laurelee Luchansky for technical assistance as well as her comments on this manuscript and Carol Warby for the cross-reactivity data on GF-6 with guinea pig LHRH peptide.

This study (Publication Number 39-027 from the Wisconsin Regional Primate Research Center) was supported by NIH Grants HD15433, HD11355, and RR00167 (to ET) and Canadian Medical Research Council (to N.M.S.).

References

1. Schwanzel-Fukuda, M. and Pfaff, D.W., Origin of luteinizing hormone-releasing hormone neurons, *Nature*, 338, 161, 1989.

2. Wray, S., Grant, P., and Gainer, H., Evidence that cells expressing luteinizing hormone-releasing hormone mRNA in the mouse are derived from progenitor cells in the olfactory placode, *Proc. Natl. Acad. Sci. U.S.A.*, 86, 8132, 1989.

3. Ronnekleiv, O.K. and Resko, J.A., Ontogeny of gonadotropin-releasing hormone-containing neurons in early fetal development of rhesus macaques, *Endocrinology*, 126, 498, 1990.

4. Terasawa, E., Quanbeck, C.D., Schultz, C.A., Burich, A.J., Luchansky L.L., and Claude, P., A primary cell culture system of luteinizing hormone releasing hormone neurons derived from embryonic olfactory placode in the rhesus monkey, *Endocrinology*, 133, 2379, 1993.

5. Schally, A.V., Arimura, A., Kastin, A.J., Matsuo, H., Baba, Y., Redding, T.W., Nair, R.M., Debeljuk, L., and White, W.F., Gonadotropin-releasing hormone: one polypeptide regulates secretion of luteinizing and follicle-stimulating hormones, *Science*, 173, 1036, 1971.

6. Sherwood, N.M., The GnRH family of peptide, *Trends. Neurosci.*, 10, 129, 1987.

7. King, J.A. and Millar, R.P., Genealogy of the GnRH family, *Prog. Clin. Biol. Res.*, 342, 54, 1990.

8. Lovejoy, D.A., Fischer, W.H., Ngamvongchon, S., Craig, A.G., Nahorniak, C.S., Peter, R.E., Rivier, J.E., and Sherwood, N.M., Distinct sequence of gonadotropin-releasing hormone (GnRH) in dogfish brain provides insight into GnRH evolution, *Proc. Natl. Acad. Sci. U.S.A.*, 89, 6373, 1992.

9. Sherwood, N.M., Lovejoy, D.A., and Coe, I.R., Origin of mammalian gonadotropin-releasing hormones, *Endocrine Rev.*, 14, 241, 1993.

10. Powell, J.F., Zohar, Y., Elizur, A., Park, M., Fischer, W.H., Craig, A.G., Rivier, J.E., Lovejoy, D.A., and Sherwood, N.M., Three forms of gonadotropin-releasing hormone characterized from brains of one species, *Proc. Natl. Acad. Sci. U.S.A.*, 91, 12081, 1994.

11. White, S.A., Bond, C.T., Francis, R.C., Kasten, T.L., Fernald, R.D., and Adelman, J.P., A second gene for gonadotropin-releasing hormone: cDNA and expression pattern in the brain, *Proc. Natl. Acad. Sci. U.S.A.*, 91, 1423, 1994.

12. King, J.A. and Millar, R.P., Evolutionary aspects of gonadotropin-releasing hormone and its receptor, *Cell. Mol. Neurobiol.*, 15, 5, 1995.

13. Muske, L., Evolution of gonadotropin-releasing hormone (GnRH) neuronal systems, *Brain Behav. Evol.*, 42, 215, 1993.

14. Terasawa, E. and Quanbeck, C., Two types of LHRH neurons in the forebrain of the rhesus monkey during embryonic development, in *GnRH Neurons: Gene to Behavior*, Parhar, I.S., and Sakuma, Y., Eds., Brain Shuppan, Tokyo, 1997, 197.

15. Quanbeck, C., Sherwood, N.M., Millar, R.P., and Terasawa, E., Two populations of luteinizing hormone-releasing hormone neurons in the forebrain of the rhesus macaque during embryonic development, *J. Comp. Neurol.*, 380, 293, 1997.

16. Jennes, L., Prenatal development of the gonadotropin-releasing hormone-containing systems in rat brain, *Brain Res.*, 482, 97, 1989.

17. Murakami, S., Kikuyama, S., and Arai, Y., The origin of the luteinizing hormone-releasing hormone (LHRH) neurons in newts (*Cynops pyrrhogaster*): the effect of olfactory placode ablation, *Cell Tissue Res.*, 269, 21, 1992.

18. Witkin, J.W., Nervus terminalis, olfactory nerve, and optic nerve representation of luteinizing hormone-releasing hormone in primates, *Ann. N.Y. Acad. Sci.*, 519, 174, 1987.

19. Goldsmith, P.C., Lamberts, R., and Brezina, L.R., Gonadotropin-releasing hormone neurons and pathways in the primate hypothalamus and forebrain, in *Neuroendocrine Aspects of Reproduction*, Norman, R. L., Ed., Academic Press, New York, 1983, 7.

20. Silverman, A.-J., The gonadotropin-releasing hormone (GnRH) neuronal systems: immunocytochemistry, in *The Physiology of Reproduction*, Knobil, E., and Neill, J., Eds., Raven Press, New York, 1988, 1283.

21. Le Douarin, N.M. and Dupin, E., Cell lineage analysis in neural crest ontogeny, *J. Neurobiol.*, 24, 146, 1993.

22. Daikoku, S. and Koide, I., Spatiotemporal appearance of developing LHRH neurons in the rat brain, *J. Comp. Neurol.*, 393, 34, 1998.

23. Hewitt, W., The development of the human caudate and amygdaloid nuclei, *J. Anat.*, 92, 377, 1959.

24. Humphrey, T., The development of the human amygdala during early embryonic life, *J. Comp. Neurol.*, 132, 135, 1963.

25. Gribnau, A.A.M. and Geijsberts, L.G.M., *Morphology of the Brain in Staged Rhesus Monkey Embryos*, Springer-Verlag, New York, 1984.

26. O'Rahilly, R. and Müller, F., Ventricular system and choroid plexuses of the human brain during the embryonic period proper, *Am. J. Anat.*, 189, 285. 1990.

27. Alvarez-Buylla, A., Commitment and migration of young neurons in the vertebrate brain, multi-author review, *Experientia*, 46, 879, 1990.

28. Bronner-Fraser, M., Cell lineage segregation in the vertebrate neural crest, in *Determinants of Neuronal Identity*, Shankland, M. and Macagno, E.R., Eds., Academic Press, New York, 1992, 359.

29. Gautron, J.-P., Pattou, E., Bauer, K., and Kordon, C., (Hydroxyproline⁹) luteinizing hormone-releasing hormone: a novel peptide in mammalian and frog hypothalamus, *Neurochem. Int.*, 18, 221, 1991.

30. Gautron, J.-P., Pattou, E., Leblanc, P., L'Heritier, A., and Kordon, C., Preferential distribution of c-terminal fragments of [Hydroxyproline⁹] LHRH in the rat hippocampus and olfactory bulb, *Neuroendocrinology*, 58, 240, 1993.

31. Wray, S., Nieburgs, A., and Elkabes, S., Spatiotemporal cell expression of luteinizing hormone-releasing hormone in the prenatal mouse: evidence for an embryonic origin in the olfactory placode, *Dev. Brain Res.*, 46, 309, 1989.

32. Rance, N.E., Young, W.S., and McMullen, N.T., Topography of neurons expressing luteinizing hormone-releasing hormone gene transcripts in the human hypothalamus and basal forebrain, *J. Comp. Neurol.*, 339, 573, 1994.

33. Skynner, M., Sim, J.A., Dyer, R.G., Allen, N., and Herbison, A.E., GnRH transgenics: temporal and spatial expression of GnRH promotor-driven transgenes, *Neurosci. Abstr.*, 23, 1506, No. 591.16, 1997.

34. Spergel, D.J., Krueth, D.F., Hanley, R., Sprengel, R., and Seeburg, P.H., GnRH/LacZ mice exhibit two major populations of β-galactosidase-stained neurons, *Neurosci. Abstr.*, 23, 589, No. 237.9, 1997.

35. Terasawa, E., Sherwood, N.M., Millar, R.P., Jennes, L., and Glucksman, M.J., The presence of a luteinizing hormone releasing hormone (LHRH) fragment in the fetal monkey forebrain, in *Proceedings of the 79th Annual Meeting of the Endocrine Society*, held June 11-14, Minneapolis, MN, No. 383, 1979.

36. Glucksman, M.J., and Roberts, J.L., Strategies for characterizing clonal and expressing soluble endopeptidases, in *Peptidases and Neuropeptide Processing*, Smith, A.I., Ed., Academic Press, San Diego, 1995, 296.

37. Lew, R.A., Tetaz, T.J., Glucksman, M.J., Roberts, J.L., and Smith, A.I., Evidence for a two-step mechanism of gonadotropin-releasing hormone metabolism by prolyl endopeptidase and metalloendopeptidase EC 3.4.24.15 in ovine hypothalamic extracts, *J. Biol. Chem.*, 269, 12626, 1994.

38. Knobil, E., The neuroendocrine control of the menstrual cycle, *Recent Prog. Horm. Res.*, 36, 53, 1980.

39. Bourguignon, J.P., Alvarez-Gonzalez, M.L., Gerard, A., and Franchimont, P., Gonadotropin-releasing hormone inhibitory autofeedback by subproducts antagonist at N-methyl-d-aspartate receptors: a model of autocrine regulation of peptide secretion, *Endocrinology*, 134, 1589, 1994.

40. Moorjani, B., Glucksman, M.J., and Roberts, J.L., The GnRH cleavage product GnRH$^{1-5}$ binding to the hypothalamic NMDA receptor: a pubertal analysis, *Neurosci. Abstr.*, 23, 417, No. 166.12, 1997.

41. Wu, T.J., Pierotti, A.R., Jakubowski, M., Sheward, W.J., Glucksman, M.J., Smith, A.I., King, J.C., Fink, G., and Roberts, J.L., Endopeptidase EC 3.4.24.15 presence in the rat median eminence and hypophysial portal blood and its modulation of the luteinizing hormone surge, *J. Neuroendocrinol.*, 9, 813, 1997.

42. Lew, R.A., Cowley, M., Clarke, I.J., and Smith, A.I., Peptidases that degrade gonadotropin-releasing hormone: influence on LH secretion in the ewe, *J. Neuroendocrinol.*, 9, 707, 1997.

43. Orlowski, M., Michaud, C., and Chu, T.G., A soluble metalloendopeptidase from rat brain: purification of the enzyme and determination of specificity with synthetic natural peptides, *Eur. J. Biochem.*, 135, 81, 1983.

44. King, J.A., Hinds, L.A., Mehl, A.E., Saunders, N.R., and Millar, R.P., Chicken GnRH II occurs together with mammalian GnRH in a South American species of marsupial (*Monodelphis domestica*), *Peptides*, 11, 521, 1990.

45. Millar, R.P. and King, J.A., Plasticity and conservation in gonadotropin-releasing hormone structure, in *Perspectives in Comparative Endocrinology*, Davey, K.G., Peter, R.E., and Tobe, S.S., Eds., National Research Council of Canada, Ottawa, 1994, 129.

46. King, J.A., Mehl, A.E.I., Tyndale-Biscoe, C.H., Hinds, L., and Millar, R.P., A second form of gonadotropin-releasing hormone (GnRH), with chicken GnRH II-like properties, occurs together with mGnRH in marsupial brains, *Endocrinology*, 125, 2244, 1989.

47. Rissman, E.F., Alones, V.E., Craig-Veit, C.B., and Millam, J.R., Distribution of chicken-II gonadotropin-releasing hormone in mammalian brain, *J. Comp. Neurol.*, 357, 524, 1995.

48. Chen, A., Yahalom, D., Ben-Aroya, N., Kaganovsky, E., Okon, E., and Koch, Y., A second isoform of gonadotropin-releasing hormone is present in the brain of human and rodents, *FEBS Lett.*, 435, 199, 1998.

49. White, R.B., Eisen, J.A., Kasten, T.L., and Fernald, R.D., Second gene for gonadotropin-releasing hormone in humans, *Proc. Natl. Acad. Sci. U.S.A.*, 95, 305, 1998.

50. Sherwood, N.M., von Schalburg, C., and Lescheid, D.W., Origin and evolution of GnRH in vertebrates and invertebrates, in *GnRH Neurons: Gene to Behavior*, Parhar I.S. and Sakuma, Y., Eds., Brain Shuppan, Tokyo, 1997, 3.

51. Kasten, T.L., White, S.A., Norton, T.T., Bond, C.T., Adelman, J.P., and Fernald, R.D., Characterization of two new preproGnRH mRNAs in the tree shrew: first direct evidence for mesencephalic GnRH gene expression in a placental mammal, *Gen. Comp. Endocrinol.*, 104, 7, 1996.

52. Lescheid, D.W., Terasawa, E., Abler, L.A., Urbanski, H.F., Warby, C.M., Millar, R.P., and Sherwood, N.M., A second form of gonadotropin-releasing hormone (GnRH) with characteristics of chicken GnRH-II is present in the primate brain, *Endocrinology*, 138, 5618, 1997.
53. Northcutt, R.G., and Muske, L.E., Multiple embryonic origins of gonadotropin-releasing hormone (GnRH) immunoreactive neurons, *Dev. Brain Res.*, 78, 279, 1994.
54. Parhar, I.S., GnRH in tilapia: three genes, three origins and their roles, in *GnRH Neurons: Gene to Behavior*, Parhar, I.S. and Sakuma, Y., Eds., Brain Shuppan, Tokyo, 1997, 197.
55. Parhar, I.S., Soga, T., Ishikawa, Y., Nagahama, Y., and Sakuma, Y., Neurons synthesizing gonadotropin-releasing hormone mRNA subtypes have multiple developmental origins in the medaka, *J. Comp Neurol.*, 401, 217, 1998.
56. Abler, L.A., Sherwood, N.M., Grendell, R.L., Golos, T.G., and Terasawa, E., cDNA of a second form of luteinizing hormone-releasing hormone, chicken LHRH-II, isolated from the non-human primate brain, *Neurosci. Abstr.*, 24, 1607, No. (632.8), 1998.
57. Jimenez-Linan, M., Rubin, B.S., and King, J.C., Examination of guinea pig luteinizing hormone-releasing hormone gene reveals a unique decapeptide and existence of two transcripts in the brain, *Endocrinology*, 138, 4123, 1997.

Index

A

Aggressive behavior, 241–242
Alcohol, 243
5-Alpha Androstanedione, 73
3Alpha-hydroxysteroid dehydrogenase
  deficiency, 265
  function, 43
3Alpha-hydroxysteroid dehydrogenase
    complex, 41
5Alpha-reductase, 41
  DHT induction, 42
  flutamide inhibition, 42
  genetic deficiency, 48
  isoforms, 41
  testosterone induction, 42
γ-Aminobutyric acid, 207
Androgen(s)
  dendritic arborization and, 162
  gap junction regulation, 163
  mechanism of action, 164, 165
  metabolizing enzymes, 37–44
  motoneuron morphology and, 161, 164
  N-methyl-d-aspartate receptor blockade
    and, 169
  perinatal administration, 5, 205, 243
  spatial abilities and, 288
  synapse elimination and, 154–155
Androgen insensitivity syndrome, 264–265
  partial, 259
Androgen receptor(s), 106
  gene expression, 165
  neuroanatomical distribution, 118, 120–121
  prenatal expression, 151
  synapse elimination, 156
Androstanedione, 73
Anovulatory persistent vaginal estrus
    syndrome, 2–3
Anteroventral periventricular nucleus,
    178–179
  BST projections, 179–186
  copulatory regulation, 234
  development, 193
  dopamine neuron, 186–193

function, 204
  gene expression, 186
Antiandrogens, 45
Apoptosis, 205, 238, 239
  estrogen-induced, 245
  hormonal mediated, mechanism,
    245–246
Arcuate nucleus, 207
  neuronal development, 209
  neuropil matrix, 210
  synaptic density, 210
Arimidex, 73, 74, 75
Aromatase, 37, 38–41, 47–49
  brain development and, 60
  castration and, 66
  in fetal brain, 70–72
  gene expression, 65
  induced activity, 66
  inhibition, 73–75, 243
  sexual differentiation in, 69
  steroid regulation, 72–73
Aromatization, 37
Astocytes, 83
Astroglia
  development, 84
  estrogen activation, 87–89
  function, 84, 85–86
  in vitro growth and differentiation, 84

B

Barbiturates, 243
Bed nuclei of stria terminalis, 179–186
Biogenetic law, 113
Brain
  androgenization, 233
  bipotential, 233
  development, 60
  estrogen formation, 65–74
    hormones in, 61–65
  fetal, 70–71
  genetic vs. hormonal organization, 126–127
  as globally male or female, 230
  human, 47–48